Experimental Brain Research Supplementum 4

The Renin Angiotensin System in the Brain

A Model for the Synthesis of Peptides in the Brain

Edited by D. Ganten M. Printz
M.I. Phillips B.A. Schölkens

With 108 Figures and 46 Tables

Springer-Verlag
Berlin Heidelberg New York 1982

Professor Dr. Detlev Ganten, MD, PhD
Deutsches Institut zur Bekämpfung des hohen
Blutdruckes, Pharmakologisches Institut,
Im Neuenheimer Feld 366
D-6900 Heidelberg, Germany

Professor Morton P. Printz, MD
Division of Pharmacology, Department of Medicine M-013,
University of California San Diego
La Jolla, CA 92093, USA

Professor M. Ian Phillips, PhD
College of Medicine,
Department of Physiology, University of Florida
Gainesville, FL 32610, USA

Dr. Bernward A. Schölkens
Hoechst AG, Abteilung für Pharmakologie
D-6230 Frankfurt/Main 80, Germany

The cover of the book shows the lateral view of the brain.
This figure has been published in R. Nieuwenhuys, J. Voogd, Chr. van Huijzen:
The Human Central Nervous System. Second Revised Edition.
Springer-Verlag Berlin Heidelberg New York 1981.
The figure is reproduced with kind permission of R. Nieuwenhuys and Chr. van Huijzen.

ISBN-13:978-3-642-46458-4 e-ISBN-13:978-3-642-46456-0
DOI:10.1007/978-3-642-46456-0

Library of Congress Cataloging in Publication Data. Main entry under title: The Renin angiotensin system. (Experimental brain research supplementum ; 4) Bibliography: p. Includes index. 1. Renin. 2. angiotensin. 3. Brain chemistry. I. Ganten, D. (Detlev), 1941-. II. Series: Experimental brain research Supplement ; 4. QP609.R38R46 612'.814 81-23252 AACR2

Offsetprinting: Beltz Offsetdruck, Hemsbach/Bergstr.
Binding: J. Schäffer OHG, Grünstadt.
2125/3140-543210

Preface

The first publications on the presence of renin, angiotensinogen and angiotensin-like material in brain tissue appeared about ten years ago. This coincided with the discovery of new biological actions of angiotensin in the brain and with the reinvestigation of already known central angiotensin effects. The possibility that angiotensin could be generated within the brain itself had therefore exciting implications. With the recent popularity of brain neuropeptides, interest in this area is even increasing.

The discovery of renin-like activity as well as angiotensinogen and angiotensin in the brain has led, admittedly by inductive inference, to the hypothesis of the existence of a complete endogenous renin-angiotensin system in the brain.[1] For several years there was ardent discussion whether such a renin angiotensin system existed in the brain or not. Inductive inference always contains "an irrational element or a creative intuition".[2] In view of the biological relevance of a brain angiotensin system, deductive testing of the hypothesis was a worthwhile scientific project taken up by a number of investigators As a matter of principle it is not given to science, to reach either truth or falsity, scientific statements can only attain continous degrees of probability whose unattainable upper and lower limits are truth or falsity. However, several of the early singular statements and results turned out to be reproducible and acceptable. The theory has thus so far passed its test and we have found no reason to discard it.

The methodology and characterization of the components of the renin angiotensin system has made major advances in the

---

1 Detlev Ganten, Studies on the existence of an independent Brain Renin-Angiotensin system: A model for Extrarenal Tissue Renin. PhD thesis, McGill University, Montreal, Canada, 1972

2 Karl R. Popper, Logik der Forschung, Julius Springer, Vienna, 1934

last decade and research in the brain renin-angiotensin system has profited from it. Partly due to the active research on the kidney and plasma renin-angiotensin system, our knowledge concerning the brain renin-angiotensin system appears to be more advanced in several aspects than what is known for many other neuropeptides where in general little information is available either on their biosynthetic pathways or on methods of pharmacological intervention. The enzymes, high molecular weight precursor proteins and the effector peptides as well as their receptors are relatively well characterized for the endogenous brain renin-angiotensin system and pharmacological agents are known which interfere with the system in vivo. In this respect the renin-angiotensin system may be considered a model for enzyme-peptide systems in the brain.

The Symposium in Königstein in September 1981 was planned to bring together a small group of scientists who are actively involved in research on the brain renin-angiotensin system so as both to provide a synopsis of the present state of the art and to create an opportunity for open discussion of critical questions. All components of the brain angiotensin system were discussed and are reflected in this volume of 26 original contributions. Apart from papers mainly concerned with the characterization, measurement of components of this enzyme-peptide system, reports are also included which deal with the central effects of angiotensin on various biological functions and with the pathophysiological relevance of brain angiotensin. The volume is preceded by an authoritative brief review which puts the original contributions into context and which makes this book easier to read and also more valuable for those who are not directly involved in this field of research.

The editors wish to thank Hoechst AG for financial support of the meeting, L. Kamphauwe and I. Büchler for their dedicated secretarial assistance.

Heidelberg, January 1982

Detlev Ganten
Morton Printz
M. Ian Phillips
Bernward Schölkens

Contents

List of Contributors

Agnati, L. F.
Department of Human Physiology, University of Modena, I-41100 Modena, Italy

Andersson, K.
Department of Histology, Karolinska Institutet, Box 60400, S-10401 Stockholm, Sweden

Aumann, S. K.
Veterans Administration Medical Center, Oregon Health Sciences University, Portland, OR 97201, USA

Badikov, V. I.
Institute of Normal Physiology, AMS, Moscow, USSR

Barrett, J. D.
Veterans Administration Medical Center, Department of Medicine, UCLA-San Fernando Valley Medical Program, Sepulveda, CA 91343, USA

Bayer, C.
Deutsches Institut zur Bekämpfung des hohen Blutdruckes, Pharmakologisches Institut, Universität Heidelberg, Im Neuenheimer Feld 366, D-6900 Heidelberg, Germany

Bernardi, P.
Department of Human Physiology, University of Modena, I-41100 Modena, Italy

Branisteanu, D.
Department of Physiology, Institute of Medicine and Pharmacy, Iasi, Romania

Bray, J. J.
Department of Physiology, School of Medicine, University of Otago, P.O. Box 913, Dunedin, New Zealand

Brownfield, M. S.
Department of Physiology, University of California, San Francisco, CA, 94143, USA

Buckley, J. P.
Institute for Cardiovascular Studies and Department of Pharmacology, College of Pharmacy, University of Houston, Houston, TX 77004, USA

Calza, L.
Department of Human Physiology, University of Modena, I-41100 Modena, Italy

Celio, M. R.
Abteilung für Anatomie, Universität Zürich, Gloriastraße 19, CH-8006 Zürich, Switzerland

Chen, F. M.
Department of Medicine M-013, University of California San Diego, La Jolla, CA 92093, USA

Clemens, D. L.
Hypertension Center, Department of Biochemistry, Vanderbilt University School of Medicine, Nashville, TN 37232, USA

Csicsmann, J.
Department of Medicine, Austin Hospital, University of Melbourne, Heidelberg, Victoria 3084, Australia

Deleva, J. I.
Department of Pathophysiology, Medical Academy, Sofia 1431, Romania

McDonald, W. J.
Veterans Administration Medical Center, Oregon Health Sciences University, Portland, OR 97201, USA

Doyle, A. E.
Department of Medicine, Austin Hospital, University of Melbourne, Heidelberg, Victoria 3084, Australia

Dzau, V. J.
Hypertension Unit Brigham and Women's Hospital, Harvard Medical School, 75 Francis Street, Boston, MA 02115, USA

Eggena, P.
Veterans Administration Medical Center, Department of Medicine, UCLA-San Fernando Valley Medical Program, Sepulveda, CA 91343, USA

Felix, D.
Brain Research Institute, Universität Zürich, August-Forel-Strasse 1, CH-8029 Zürich, Switzerland

Fuxe, K.
Department of Histology, Karolinska Institutet, Box 60400, S-10401 Stockholm, Sweden

Ganten, D.
Deutsches Institut zur Bekämpfung des hohen Blutdruckes, Pharmakologisches Institut, Universität Heidelberg, Im Neuenheimer Feld 366, D-6900 Heidelberg, Germany

Haas, H. L.
Neurochirurgische Universitätsklinik, CH-8091 Zürich, Switzerland

Haulica, I.
Department of Physiology, Institute of Medicine and Pharmacy, Iasi, Romania

Hawkins, R.
Department of Medicine M-013, University of California San Diego, La Jolla, CA 92093, USA

Hermann, K.
Deutsches Institut zur Bekämpfung des hohen Blutdruckes, Pharmakologisches Institut, Universität Heidelberg, Im Neuenheimer Feld 366, D-6900 Heidelberg, Germany

Hirose, S.
Institute of Applied Biochemistry, University of Tsukuba, Ibaraki 305, Japan

Hökfelt, T.
Department of Histology, Karolinska Institutet, Box 60400, S-10401 Stockholm, Sweden

Hubbard, J. I.
Department of Physiology, University of Otago, School of Medicine, P.O. Box 913, Dunedin, New Zealand

Husain, A.
Research Division, Cleveland Clinic Foundation, 9500 Euclid Avenue, Cleveland, OH 44106, USA

Hutchinson, J. S.
Department of Medicine, Austin Hospital, University of Melbourne, Heidelberg, Victoria 3084, Australia

Inagami, T.
Hypertension Center and Department of Biochemistry, Vanderbilt University School of Medicine, Nashville, TN 37232, USA

Ito, T.
Veterans Administration Medical Center, Department of Medicine, UCLA-San Fernando Valley Medical Program, Sepulveda, CA 91343, USA

Jandhyala, B. S.
Institute for Cardiovascular Studies and Department of Pharmacology, College of Pharmacy, University of Houston, Houston, TX 77004, USA

Johnson, A. K.
Department of Psychology, Cardiovascular Center, University of Iowa, Iowa City, IA 52242, USA

De Jong, W.
Rudolf Magnus Institute for Pharmacology, State University of Utrecht, Medical Faculty, Vondellaan 6, NL-3521 GD Utrecht, The Netherlands

Jung, W.
Hoechst AG, Abteilung für Pharmakologie, D-6230 Frankfurt/ Main 80, Germany

Keil, L. C.
Department of Pharmacology, The Milton S. Hershey Medical Center of the Pennsylvania, State University, Hershey, PA 17033 and (LCK), NASA, Biomedical Research Division, Moffett Field, CA 94035, USA

Lang, R. E.
Deutsches Institut zur Bekämpfung des hohen Blutdruckes, Pharmakologisches Institut, Universität Heidelberg, Im Neuenheimer Feld 366, D-6900 Heidelberg, Germany

Liard, J. F.
Institut de Recherche Cardio-Angéiologique, University of Fribourg, Chemin de Musée, CH-1700 Fribourg, Switzerland

Lind, R. W.
Department of Psychology, Cardiovascular Center, University of Iowa, Iowa City, IA 52242, USA

Lokhandwala, M. F.
Institute for Cardiovascular Studies, Department of Pharmacology, College of Pharmacy, University of Houston, Houston, TX 77004, USA

Mann, J. F. E.
Abteilung für Nephrologie, Abteilung für Innere Medizin, Universität Heidelberg, Bergheimer Strasse 56a, D-6900 Heidelberg, Germany

Mendelsohn, F. A. O.
Department of Medicine, Austin Hospital, University of Melbourne Heidelberg, Victoria, 3084, Australia

Murakami, K.
Institute of Applied Biochemistry, The University of Tsukuba, Ibaraki-Ken 305, Japan

Naruse, K.
Hypertension Center, Department of Biochemistry, Vanderbilt University School of Medicine, Nashville, TN 37232, USA

Neamtu, C.
Department of Physiology, Institute of Medicine and Pharmacy, Iasi, Romania

DiNicolantonio, R.
Department of Medicine, Austin Hospital, University of Melbourne, Heidelberg, Victoria 3084, Australia

Nicolov, N. A.
Department of Pathophysiology, Medical Academy, G. Sofiiski Strasse 1, Sofia 1431, Bulgaria

Ohsawa, T.
Institute of Applied Biochemistry, The University of Tsukuba, Ibaraki-Ken 305, Japan

Okamura, T.
Hypertension Center, Department of Biochemistry, Vanderbilt University School of Medicine, Nashville, TN 37232, USA

Petrescu, G.
Department of Physiology, Institute of Medicine and Pharmacy, Iasi, Romania

Phillips, M. I.
Department of Physiology, University of Florida, J. Hillis Miller Health Center, Box J-274, Gainesville, FL 32610, USA

Poulsen, K.
The Institute of Biochemistry, The Royal Dental College, Copenhagen, Denmark

Printz, M. P.
Division of Pharmacology, Department of Medicine M-013, University of California San Diego, La Jolla, CA 92093, USA

Rascher, W.
Deutsches Institut zur Bekämpfung des hohen Blutdruckes, Pharmakologisches Institut, Universität Heidelberg, Im Neuenheimer Feld 366, D-6900 Heidelberg, Germany

Reid, I. A.
Department of Physiology, University of California, San Francisco, CA 94143, USA

Ritz, E.
Abteilung für Nephrologie, Abteilung für Innere Medizin, Universität Heidelberg, Bergheimer Strasse 56a, D-6900 Heidelberg, Germany

Rix, E.
Medizinische Poliklinik, Universität Heidelberg, Hospitalstrasse 3, D-6900 Heidelberg, Germany

Rosca, V.
Department of Physiology, Institute of Medicine and Pharmacy, Iasi, Romania

Sambhi, M. P.
Veterans Administration Medical Center, Department of Medicine, UCLA-San Fernando Valley Medical Program, Sepulveda, CA 91343, USA

Schelling, P.
Institut de Recherche Cardio-Angéiologique, University of Fribourg, Chemin du Musée, CH-1700 Fribourg, Switzerland

Schölkens, B. A.
Hoechst AG, Abteilung für Pharmakologie, D-6230 Frankfurt/ Main 80, Germany

Schüll, B.
Deutsches Institut zur Bekämpfung des hohen Blutdruckes, Pharmakologisches Institut, Universität Heidelberg, Im Neuenheimer Feld 366, D-6900 Heidelberg, Germany

Severs, W. B.
Department of Pharmacology, The Milton S. Hershey Medical Center of the Pennsylvania State University, Hershey, PA 17033 and (LCK), NASA, Biomedical Research Division, Moffett Field, CA 94035, USA

Sherstnev, V. V.
Institute of Normal Physiology, AMS, Moscow, USSR

Sirett, N. E.
Department of Physiology, University of Otago, School of Medicine, P.O. Box 913, Dunedin, New Zealand

Slatineanu, S.
Department of Physiology, Institute of Medicine and Pharmacy, Iasi, Romania

Smeby, R. R.
Research Division, Cleveland Clinic Foundation, 9500 Euclid Avenue, Cleveland, OH 44106, USA

Speck, G. A.
Deutsches Institut zur Bekämpfung des hohen Blutdruckes, Pharmakologisches Institut, Universität Heidelberg, Im Neuenheimer Feld 366, D-6900 Heidelberg, Germany

Speth, R. C.
Research Division, Cleveland Clinic Foundation, 9500 Euclid Avenue, Cleveland, OH 44106, USA

Stock, G.
Abteilung für Physiologie, Universität Heidelberg, Im Neuenheimer Feld 326, D-6900 Heidelberg, Germany

Stratone, A.
Department of Physiology, Institute of Medicine and Pharmacy, Isi, Romania

Sudakov, K. V.
Institute of Normal Physiology, AMS, Moscow, USSR

Summy-Long, J. Y.
Department of Pharmacology, The Milton S. Hershey Medical Center of the Pennsylvania State University, Hershey, PA 17033 and NASA, Biomedical Research Division, Moffett Field, CA 94035, USA

Takata, Y.
Department of Medicine, Austin Hospital, University of Melbourne, Heidelberg, Victoria 3084, Australia

Takii, Y.
Hypertension Center, Department of Biochemistry, Vanderbilt University School of Medicine, Nashville, TN 37232, USA

Taugner, R.
Abteilung für Physiologie, Universität Heidelberg, Im Neuenheimer Feld 326, D-6900 Heidelberg, Germany

Tonnaer, J. A. D. M.
Rudolf Magnus Institute for Pharmacology, State University of Utrecht, Medical Faculty, Vondellaan 6, NL-3521 GD Utrecht, The Netherlands

Unger, T.
Deutsches Institut zur Bekämpfung des hohen Blutdruckes, Pharmakologisches Institut, Universität Heidelberg, Im Neuenheimer Feld 366, D-6900 Heidelberg, Germany

Villarreal, H.
Department of Nephrology, National Institute of Cardiology, Mexico City, D. F., Mexico

Wiegant, V. M.
Rudolf Magnus Institute for Pharmacology, State University of Utrecht, Medical Faculty, Vondellaan 6, NL-3521 GD Utrecht, The Netherlands

Yokosawa, H.
Hypertension Center Department of Biochemistry, Vanderbilt University School of Medicine, Nashville, TN 37232, USA

Yumatov, E. A.
Institute of Normal Physiology, AMS, Moscow, USSR

Renin and Converting Enzyme

Minireview: The Brain Renin Angiotensin System 1982

M. P. Printz, D. Ganten, T. Unger, and M. I. Phillips

C O N T E N T S

Experimental Brain Research, Suppl. 4

4

PHYSIOLOGY OF THE BRAIN ANGIOTENSIN SYSTEM

INTRODUCTION

The renin-angiotensin system (RAS) has long been considered an exception to general endocrine systems in that the enzyme responsible for prohormone-to-hormone conversion is secreted from the source organ kidney into the blood and the effector peptide angiotensin II (ANG II) is produced extracellularly, thus facilitating its rapid transport to the classical target organs, the adrenal cortex and the arterial smooth muscle.

This concept has now to be revised for the several reasons. Renin has been shown to be present in a number of extrarenal tissues and renin concentrations in these tissues may exceed those in the kidney. The concept of local, presumably intracellular, angiotensin synthesis has been placed on a firmer basis and is supported by the finding of other components of the RAS in extrarenal tissue. Further, pharmacologic and therapeutic interferences with the RAS not only affect the plasma and kidney RAS but also interfere with the local tissue RAS e.g. in the arterial wall and in the brain. This is important first, because inhibitors of the RAS are being used clinically as therapeutic agents and second, because it implies an interrelationship between the plasma and tissue systems.

Thus, the RAS is now recognized as an unusually well defined peptide hormone generating system with a more widespread distribution and probably broader function than previously anticipated.

Different names have been used to designate the extrarenal enzymes. International consensus has been reached, however, that renin (E.C. 3.4.23) is defined as a specific enzyme, that cleaves angiotensin I (ANG I) from its natural homologous renin substrate (angiotensinogen) and which has no angiotensinase activity. The major component of the kidney enzyme serves as renin reference. If the enzyme has not been clearly identified, the ending renin-"like" should be used in such cases.

The first reports of renin-like activity, angiotensinogen and angiotensin in the brain (33,38,39) were subsequently confirmed in the following years by a variety of investigators in different laboratories (52,69). Based on physiological and biochemical experimentation, it has been proposed that the brain RAS participates in the control of blood pressure, water intake, ADH secretion and a variety of other functions. The Königstein Symposium has been held to update current research activities devoted to an elucidation of the brain RAS and its components. This review attempts to place past and present research activities into perspective but is not intended to summarize all the available literature. Since many of the contributions of the Symposium provide updates of very recent research results, they will be referred to often and the reader is advised to go to these chapters for further references, background or different views and interpretations. This review will focus on the biochemistry of the components of the brain RAS leading to angiotensin II (ANG II) and the physiology of this neuropeptidergic system.

THE HIGH MOLECULAR WEIGHT PRECURSOR

In a review of the state of our knowledge on a putative brain precursor of angiotensin in the central nervous system, several questions become immediately apparent: First, may we at this time refer to the precursor as angiotensinogen? Historically, plasma renin substrate was the term applied to the high molecular weight plasma protein precursor of angiotensin I. Over the years the many investigations conducted on the biochemistry and pharmacology of this substance leave little doubt that it is an angiotensinogen or prohormone of angiotensin. Nevertheless, many workers still refer to the plasma prohormone as renin substrate partly because of custom but also because it is an operationally more exact term since the protein is assayed by incubation with exogenous kidney renin and quantitation of the angiotensin I released. Studies of the CNS indicate (as summarized below) that a renin substrate is present in both brain tissue and cerebrospinal fluid. To call this substance "angiotensinogen" is probably correct, although technically it has not yet been established that this protein leads in vivo to the generation of the angiotensin peptides in the CNS. Certainly, the purpose of this conference is to update and summarize our current state of knowledge of the field and to attempt to come closer to an answer of the question whether or not an intrinsic brain renin-angiotensin system exists. For this reason, it may be best to refer to the substance in question as Brain Renin Substrate (BRS).

A second issue that must be considered is why it is important to investigate the existence of a BRS. The demonstration of physiological and pharmacological responses to exogenous angiotensin II (ANG II) which are selectively interfered with by specific antagonists strongly indicates that a brain angiotensin system exists in mammals.The initial observations of centrally mediated cardiovascular effects by intracarotid or intraventricular ANG II subsequently led to findings that exogenous angiotensin provokes a complex but apparently interelated set of physiological responses (12,14). However, if ANG II functions in vivo under normal physiological conditions the peptide must achieve access to these central sites of action. Such access is best moderated if synthesis and/or storage of the peptide is close to its site of action. This argument in no way implies a neurotransmitter function for ANG II since a central neuromodulator would also be expected to exhibit local synthesis, storage and/or release. A complete brain angiotensin system (involving precursor and enzymes) intrinsic to the central nervous system would indicate a local rather than a hormonal pathway for the peptide. Although many studies indicate that blood borne ANG II exerts central effects in some species, these studies may define a set of receptors (or sites of action) which lie in regions of the CNS devoid of a significant blood brain barrier. Further, a local action of ANG II on vascular permeability has been indicated recently in several studies and must be considered as a partial explanation of these observations.

Angiotensin is formed in the periphery by the actions of renin (or renin-like enzymes) on a high molecular weight precursor, angiotensinogen or renin substrate. The peptide therefore derives via a ribosomal system and follows a similar high molecular weight precursor - peptide relationship known for other peptide hormones - both peripheral and central. It is most reasonable to suspect that a central ANG II biosynthetic pathway mirrors that known for the periphery. To identify such a system requires quantitation, isolation, and subsequent characterization of the high molecular weight precursor. This is also essential to establish the origins of this renin substrate. While arguments may be raised against penetration of plasma ANG II to all sites in the CNS, these arguments have stronger merit for the penetration of plasma renin substrate to any site unless a specific transport pathway exists for the protein. A comparison of central and peripheral angiotensinogen would be an essential step in the proof of a central origin of BRS. Alternative approaches to this question have been made by several laboratories and will be summarized below.

Problems in Defining a Brain Renin Substrate

Assay: The primary method currently in use to identify and quantify Plasma Renin Substrate (PRS) is to incubate the preparation with sufficient homologous kidney renin to release all the angiotensin I and then to quantitate the latter by specific radioimmunoassay. Angiotensinases and converting enzyme activity are generally inhibited by the addition of a mixture of protease inhibitors including EDTA (for metal-ion activated proteases, e.g. carboxypeptidases and converting enzyme), PMSF or DFP (for serine proteases of the trypsin/kallikrein class) and a sulfhydryl reagent, (BAL or phenylmercuric acetate for sulfhydryl-dependent proteases). However, the problem of degradation of ANG I, normally focused upon by most laboratories, also extends to renin substrate.Consideration must be given to minimizing non-specific proteolysis of the prohormone. The kinetics for ANG I release by renin from tetradecapeptide are considerably different than from plasma renin substrate; therefore, any protease activity which degrades, even partially, renin substrate could alter the kinetics and thereby lower the estimated RS concentration.

These problems are greatly exacerbated when working with tissues rather than plasma. Homogenization of brain tissue releases a variety of peptidases and proteases many of which are acid proteases. Addition of inhibitors of these enzymes would also result in inhibition of renin activity and interference in the assay. In our hands, significant angiotensinase and renin substrate proteases are found in brain homogenates (75,76,97). The use of inhibitors are essential both during the initial homogenization and subsequent generation of ANG I in the renin incubation. An alternative approach is the separation of the protein from the bulk of proteases. This has been accomplished through the simple addition of ammonium sulfate (AMS) to a brain homogenate to yield a final concentration of 2.3 to 2.5 M AMS (37,75). This precipitated the renin substrate but left many of the degradative enzymes in the supernatant; in general, these enzymes require more than a 2.5 M AMS concentration to precipitate. Centrifugation and use of the resolubilized pellet for renin assay greatly diminished the requirements for exogenous enzyme inhibitors (although we and others routinely employ an inhibitor mixture in incubations) and yielded linear ANG I generation kinetics and stable levels of the released peptide (37,75,76,97). It should be noted that kidney renin generally co-precipitates with other proteins at 2.3 to 2.5 M AMS; however, this does not pose any problem in the assay of RS.

Tissue Source

The tissue used frequently influences the complexity of the assay procedure. In addition to the number of degradative enzymes in brain tissue, recognition must also be given to the fact that the brain is not a homogenous organ but exhibits marked heterogenity both in distribution of functions and in the distribution of constitutive enzymes, proteins and lipids. The use of whole brain, while providing simplicity in the preparation procedures, does obscure regional variations in RS concentration. Such an analysis can provide information only indicating that RS is present; it does not yield a valuable estimate of brain RS concentration. Further, species variations may exist and recognition of this potential variable must be made when comparing results from different laboratories.

Contaminating Plasma Renin Substrate

If a substance that is being quantified in brain tissue also exists in plasma, the contamination with plasma can introduce artifactiously high results. This is a very severe problem with BRS since the concentration of RS in plasma, although quite low in terms of specific concentration (i.e. per mg protein) is still appreciable. Depending upon the endocrine state of the animal and prior surgical procedures, e.g. nephrectomy or adrenalectomy, the PRS can vary from 50 to 300 ng ANG I/ml (in terms of actual protein mass that equates to approximately 2.5 to 150 μg angiotensinogen protein). Higher plasma renin substrate concentrations are found in different conditions and in different species.

There are several approaches to this problem. Traditionally, most laboratories working on such plasma/brain located substances either ignore plasma contamination or perform a transcardiac perfusion of the cerebral vascular bed. The latter procedure involves insertion of a cannula into the aorta via an incision in the left myocardial ventricle, opening the right atria and immediately clamping the descending aorta. Perfusion with 100 ml or more of saline-EDTA or saline-heparin generally yields a relatively whitish brain mass. This perfusion, while esthetically satisfying by the greatly diminished color of the tissue, still leaves much of the contaminating plasma present. In fact, in a recent study, Gregory et al. (47) and Wallis et al. (134) found that such perfusions, even when done carefully and with maximal attention to all details, still left much of the contaminating plasma in the brain tissue as monitored either by radioiodinated albumin or tritiated inulin. An alternative approach is to quantify the actual amount of plasma present in brain tissue and with the knowledge of the plasma RS concentration, substract the latter to obtain net brain RS (47,101). Clearly, such an approach is essential to regional brain RS analysis, but may also be useful for whole brain analyses.

Assay of CSF - Renin Substrate

Renin substrate constitutes 1.5 to 3% of the total cerebrospinal fluid (CSF) protein in several mammalian species examined to date (reviewed recently by Schelling et al.) (113). This selective enrichment (in specific concentration) over plasma is of primary interest. However, from an assay standpoint there is still much less RS in CSF than in plasma. While CSF is a good source of RS it is very limiting especially in smaller mammalian species. Thus, CSF - RS is useful as a potential general monitor of brain RS but probably not the best source either for isolation and purification of the prohormone or to determine minute to minute changes in brain renin substrate levels. The latter would have been a valuable approach to a study of the brain angiotensin system; however, there is a possibility that CSF - RS contains some plasma RS which may enter either via the choroid plexus or directly from brain parenchyma at sites devoid of a blood brain barrier. Studies of the latter phenomena are continuing but caution must be exercised on the use of CSF as a source of brain RS until it is clearly established, in that particular mammalian species, that it consists solely of prohormone from the brain.

Identification and Levels of Renin Substrate in Brain Tissue and CSF

The first report of a renin substrate in dog brain tissue was in 1971 by Ganten et al. (38). They anticoagulated an anesthetized dog, exsanguinated and then perfused the brain with 7 liters of cold saline via the carotid arteries. After homogenization and ammonium sulfate precipitation (2.2 M at pH 5.3) an aliquot of the resolubilized protein yielded (by the rat pressor assay) 30 to 60 ng ANG II/10 g brain tissue. These levels in brain tissue appeared quite low; however, confirmation of these initial results were provided by Reid and Ramsay (102) and by Printz and Lewicki (97). Reid and Ramsay demonstrated that intraventricular injection of renin into the dog led to a sustained elevation of arterial pressure similar to that found with direct ANG II intraventricular injections. They also partially isolated renin substrate from canine CSF and found 90 to 450 ng ANG I activity/ml (104). Printz and Lewicki (97) isolated and partially purified a renin substrate from perfused rabbit brain tissue. They reported 8-9 ng ANG I/g whole brain tissue. Brain renin substrate levels have now been measured in sheep (99), man (27,66,98) primates (103) and rat (47,108,112,113).

In brain tissue, renin substrate appears to reside largely in the soluble or cytosolic fraction (76,86,123). Recently, Sernia and Reid (116) reported on a slow release of renin substrate from brain sections obtained from perfused rats. The tissue was homogenized and the amount of renin substrate which accumulated in the supernatant (following centrifugation of the tissue) was quantitated hourly for 3 hours. They found that supernatant RS concentration increased in a relatively linear fashion during the 3 hours of incubation at 37^0, the increase being almost threefold. The release was negligible at 4^0C compared with 37^0 and was independent of protein synthesis inhibitors. Since previous studies had indicated that brain renin substrate was primarily in soluble fractions, they interpreted their release experiments as indicative of a possible precursor-product time dependent conversion. However, their results could also reflect either (1) dissociation of the renin substrate from a higher molecular weight complex which prevents access of the renin to the leucine-leucine sensitive bond or (2) leakage from plasma trapped in microvascular networks. Further studies must be conducted on the mechanism and significance of this release process.

Many of the studies involving isolation of brain tissue renin substrate have utilized saline perfusion to remove contaminating blood and plasma. As outlined above, concern on whether full removal of plasma was achieved prompted an assessment of the contamination of rat brain tissue by plasma (47,101,134). It was found that transcardiac perfusion is not very effective in the rat and is quite variable. Analysis of BRS on a regional basis indicated that many areas exhibited only minor decreases in their plasma content (10 to 30%) possibly indicative of perfusion shunting around vascular networks. An alternative approach was developed by Timothy Gregory (47, 101) employing radiolabelled inulin to quantitate directly plasma contamination of any given rat brain region. These studies extended earlier studies by Lewicki et al. (75) and permitted a "net" brain renin substrate determination on a regional level. Further, the work of Gregory et al. (47) clearly indicates that BRS is extravascular. It had been proposed (86) that brain tissue renin substrate is extracellular and therefore of CSF origin. However, the regional analysis of Lewicki et al. (75) and that of Gregory et al. (47) do not support this extracellular argument. In fact, calculations indicate that the protein must in fact be intracellular. Taken together, the results from these laboratories would indicate that BRS is most likely in an intracellular but cytosolic compartment.

The first regional distribution analysis of brain renin substrate was reported by Lewicki et al. (75) and indicated both that many areas of rat brain contain the protein and that some areas have an 8 to 15 fold higher concentration than other regions. These studies were extended and replicated by Gregory et al. (47,101), Wallis and Printz (134) and recently Chen et al. (23) following the development of methods to obtain a net brain renin substrate concentration. High concentrations of the protein are found in hypothalamus and thalamus and in central grey regions, while lowest concentrations are in cortex, olfactory bulb and hippocampus. Utilizing the release rate analysis, Sernia and Reid found that highest concentrations were in the pons/midbrain/-diencephalon region and in the medulla and cerebellum while less was released from the cerebral hemispheres and olfactory lobes; lowest concentrations were released from the cerebrospinal cord. This distribution is very similar to that found by Lewicki, Gregory and Chen using direct BRS analysis.

We may conclude from these studies that a brain renin substrate is present in mammalian brain tissue and CSF, that the concentration of the protein exhibits a unique topographical distribution in the rat and rabbit and that release occurs from an extravascular and most likely cytosolic compartment of the brain. The concentrations of the protein in CSF and in some brain areas are appreciable. In CSF it can comprise upwards of 2% of total protein, while in some brain regions the concentration is approximately 0.04 to 0.1% of total soluble protein.

Physico-chemical Comparisons of Brain and Plasma Renin Substrate

Whether brain renin substrate is of central or peripheral origin has been a major issue for some time. The overwhelming evidence supports a central origin for the protein. The first indications were that the renin substrates from both sources are very similar as judged by chromatography using gel filtration, apparent molecular weight and binding to concanavalin A-Sepharose (98,99,102,104). However, differences were noted in the distribution of isoelectric pH of the brain, CSF and plasma renin substrates (97,98,99). This technique provides a measure of isoelectric point microheterogeneity. Since both plasma and brain renin substrates exhibit properties of a glycoprotein (concanavalin A-binding and sensitivity to neuraminidase digestion) it was anticipated that they would show heterogeneity in isoelectric point distributions. Surprisingly, in the human, rabbit, sheep and hog the plasma and brain renin substrates are quite different by isoelectric focusing analysis (97,98,99); rat brain and plasma substrates are much more similar, however differences still are evident (Gregory TJ, Printz MP, in preparation). Since much of the isoelectric point microheterogeneity of a glycoprotein may be attributable to variations in ionizable sialic acid residues, an important test is the convertibility to a similar focusing profile following neuraminidase digestion. As reported by Printz et al. (99), sheep RS from plasma, brain and CSF become somewhat closer in isoelectric focusing pattern following neuraminidase digestion; however, significant differences remain as well as evidence of microheterogeneity. A similar result was reported for human CSF and plasma angiotensinogen by Printz and Gregory (101) and Printz et al. (98). Reid et al. (106) reported finding different carbohydrate compositions between partially purified CSF and plasma dog angiotensinogen. Along these same lines, there is evidence that human CSF renin substrate does not cross-react with antibody directed against human plasma renin substrate (27,66). This finding is supported by an independent study using antisera directed against human plasma angiotensinogen and employing partially purified human CSF angiotensinogen (Printz, MP,and Tewksbury, DA, unpublished results).

At this time, we must conclude that brain and plasma renin substrates exhibit many similarities but important differences. The differences in antigenic cross-reactivity and isoelectric point microheterogeneity are what might be expected for similar proteins which arise from different cellular origins. These findings support a brain origin for brain renin substrate.

Evidence for Differential Regulation of Brain and Plasma Renin Substrate

Studies of the endogenous regulation of brain renin substrate are important for several reasons, one being that such investigations may provide a differentiation between plasma and brain renin substrate. It has been established that plasma renin substrate levels increase following bilateral nephrectomy and treatment with estrogenic steroid hormones while decreases follow adrenalectomy. Glucocorticoids elevate plasma renin substrate and their absence following adrenalectomy probably explains the reason for a decrease in PRS.

Studies utilizing these techniques have been undertaken by several laboratories. Wallis and Printz (133) and Printz et al. (100) reported that selected rat brain areas exhibit marked decreases in apparent BRS following adrenalectomy. Replacement of glucocorticoids restored the levels and in some regions led to elevated BRS concentrations over control values. In these experiments PRS levels decreased by 50%. Since many brain regions exhibited no change following adrenalectomy this provides an additional argument that BRS does not simply reflect PRS contamination of brain tissue; nor is there direct conversion of one to the other up to 14 days following adrenalectomy. Thus, BRS appears to be of central origin. Schelling et al. (114) recently replicated these studies and quantified rat CSF-RS levels following adrenalectomy. They found that while PRS decreased by approximately 50%, CSF-RS decreased by only 20 to 25%; prednisolone increased CSF-RS in a dose related fashion. Schelling demonstrated that the combined effects of adrenalectomy and nephrectomy dissociated the PRS and BRS changes: nephrectomy alone increased both PRS and CSF-RS; however, the combined effects of nephrectomy and adrenalectomy resulted in a decrease in CSF-RS while PRS was still increased over control values. These results parallel similar studies by Gregory and Printz (47,101) who showed that nephrectomy in the rat resulted in a quick elevation of PRS (within 8 hours following surgery) while BRS was unchanged for at least 24 hours. Of importance was their finding that only selected brain regions are significantly elevated following nephrectomy while other regions show no change in "net" BRS.

In contrast to the results of Schelling et al. (114) and Printz et al. (100) Reid and Day (104) found that neither nephrectomy nor dexamethasone resulted in changes in the concentration of CSF-RS while plasma levels were significantly altered. It is not clear why there is a discrepancy in the results from these laboratories, replication of the canine studies would be valuable.

These results do provide support for the use of either brain or CSF renin substrate levels as possible probes of the endogenous regulation of BRS. The marked regional differences following nephrectomy reported by Gregory et al. (47,101) must introduce caution in the use of CSF monitoring only of BRS.

Reported in this conference by Schelling et al. (114) was the finding that administration of a converting enzyme inhibitor, Captopril, for 4 weeks led to a marked decrease in both plasma and CSF renin substrate levels. The changes were similar in magnitude to adrenalectomy and may reflect suppression of circulating glucocorticoids. Clearly, there is strong evidence now from several laboratories that steroid hormones affect brain renin substrate levels differently than plasma renin substrate. The report by Chen et al. (23, this conference) of a probable effect of sex steroids on BRS expands this concept beyond the adrenal glucocorticoids and requires further studies. Chen et al. (23) reported variations in regional BRS levels in female rats during the estrous cycle. Further, they also report that water deprivation leads to a pattern of regional changes in BRS similar to that seen with perturbation by adrenal steroids. Both of these reports may reflect a differential "stress" response of the brain angiotensin system.

Origin of Brain Renin Substrate

The studies summarized above lend strong support to the argument that brain renin substrate is of central origin. Two laboratories have now attempted to determine the extent of penetration of plasma angiotensinogen into the central nervous system. Morris and Reid (20) injected radiolabelled homologous plasma angiotensinogen into the peripheral circulation of the dog and failed to find uptake into CSF. The preparation of plasma RS employed in these studies was only partially purified. However, Lewicki and Printz (summarized in 100) used highly purified (over 90% purity) rabbit plasma RS in a similar experimented protocol. They showed that the clearance kinetics of the PRS were complex with 50% of the material exhibiting a half-time escape from the circulation of 8 to 9 hours. Repetitive injections of radioiodinated PRS into the awake conscious rabbit failed to show significant penetration of label into the CSF. A small amount of radioactivity did appear in CSF; however, the magnitude was far too small to attribute CSF-RS or brain RS to a peripheral origin.

It is clear now that a pool of renin substrate is present in the central nervous system of man, sheep, dog, rats and rabbits. All evidence to date indicates that the protein is of central origin although caution must be exercised until direct intracellular biosynthesis is established. Brain renin substrate probably fluxes into the CSF and constitutes a significant percentage of CSF protein. What is of special interest is the evidence indicating that regional brain CSF is regulated in a highly specific manner by steroids. CSF-RS may reflect short time variations in local synthesis and/or release of brain RS.

It is a decade now since the original finding of a renin substrate in canine brain tissue (38) and we are much closer to a definitive statement as to its origin, localization and regulation. Over the next few years it can be anticipated that these issues will be resolved. A more critical and essential question is whether brain renin substrate is a brain angiotensinogen. When proof of this is established in a rigorous fashion, the presence of an intrinsic brain renin-angiotensin system will also be a fact.

THE ENZYMES: BRAIN RENIN

Methodological Aspects

The direct measurement of brain renin is complicated due to a non-specific renin-like activity of the protease cathepsin D which exists in high concentrations in all tissues and especially in the CNS. The initial observations of a brain renin-like activity were surrounded by controversy as to whether the so-called brain "isorenin" was true renin or cathepsin D (25,49,105,121). It is now clear that many of the initial observations reflected the activities both of a ubiquitous cathepsin and true brain renin. Depending on assay conditions the relative contributions of these two enzymic pathways will be dominant. This non-specific ANG I generating activity may be a laboratory artifact resulting from release of lysosomal enzymes during tissue homogenization. As a result, a specific assay method for true renin and the elimination of the non-specific activity has become essential for the identification of brain renin and the future evaluation of its pathophysiological role.

As a result of recent research results (see Hirose et al., Inagami et al., this volume) a close structural similarity of renin exists with other acid proteases. This similarity is clearly a major obstacle to the distinction of the two enzymes; submandibular renin has a catalytic site almost identical with that of pepsin (63) and exhibits approximately 30% sequence homology with porcine pepsin (55). Hence, it should be closely homologous to cathepsin D.

The resolution of these enzyme activities is a result largely of the application of affinity chromatographic separation advances. Renin in rat and hog brain has been separated from renin-like activity of cathepsin on a column of alpha-casein-Sepharose (55,63). True renin (as reflected by purified kidney or submandibular renin) does not bind to casein-Sepharose and elutes early in the separation. Evidence of this activity was confirmed by direct inhibition with specific antisera to renin. Further, this elution fraction was devoid of proteolytic activity. In contrast, the activity which bound, while exhibiting renin-like activity was not inhibited by the renin antibody and contained high levels of protease activity. Another separation method, Blue-Sepharose, with known affinity for inactive prorenin but not for active renin in the plasma and kidney, was found to bind preferentially active renin of mouse brain but not cathepsin (124). This property also permitted complete separation of renin from cathepsin in mouse brain (41,89, 124). Thus, these affinity techniques provide a reliable methodology for the identification and characterization of brain renin. They also permit a reexamination of the results and conclusions reported for the determination of "isorenin" in the past.

The magnitude of the non-specific renin-like activity of the protease-containing peak is determined largely by the purity and source of the substrate employed. The non-specific action is particularly evident with partially purified sheep or dog angiotensinogen in the pH range of 5.5 to 7 as used in earlier studies. Recent studies (these proceedings) indicate that rat brain (true) renin exhibits a neutral pH activity range (6-8) while the cathepsin activity is acid pH shifted. However, there is a severe overlap of the pH-activity curves of the two enzymes in generating ANG I making it difficult to distinguish their respective contributions merely by the use of neutral or a slightly alkaline pH for angiotensin generation. The extent of overlap reflects the relative concentrations of the two enzymes but when the non-specific activity is very high, the tail of the cathepsin-pH profile can extend to pHs over 7.

The use of the chromatographic separation methods (55,89) is cumbersome and requires large amounts of tissues. Although it has been used to determine regional renin distribution in the hog brain (57), direct identification of renin without tissue fractionation is essential for a large number of samples of a small size. To this end a specific renin activity has been estimated by the extent of inhibition of angiotensin I-generating activity through the use of antibodies directed against hog renin (56). Essential to the success of this method is to minimize the contribution of the non-specific renin-like activity of cathepsin. This can be achieved by using unfractionated plasma from nephrectomized rats as a source of renin substrate and an incubation pH over 7. A cathepsin inhibitor in unfractionated plasma reduces the non-specific renin-like activity considerably. Even at pH 7.5 true renin activity is very difficult to detect when superimposed on the non-specific activity. At a pH below 7, particularly at 5.5 where many studies have been done, renin antibody-sensitive activity (true renin) will be less than a few per cent of the total activity, and its detection is impossible.

Contribution of blood-borne renin in tissue samples cannot be readily dismissed even after bilateral nephrectomy and brain perfusion if one is dealing with a very low level of tissue renin.(See methodological section above of High Molecular Weight Precursor). An alternative method for the identification of specific renin may be accomplished by employing a direct radioimmunoassay method (56,81) if the renin concentration in the brain tissue is sufficient for this method.

Direct visualization of renin-like material by immunohistochemical methods has been used by many investigators in the brain (22,36,64,122), pituitary gland (22,88), placenta (96) and mouse submaxillary gland (48,84). This method is very sensitive but is not quantitative.

In summary, the distinction of true renin from the renin-like activity of cathepsin is much more complex than was thought in the past. The use of affinity chromatographic procedures is a major advance but is not applicable, at present, to mg quantities of tissue. For unfractionated tissue extracts minimization of cathepsin action by the use of unfractionated rat plasma at a pH over 7.3 is essential. This minimal criterion must be satisfied in the identification of renin even in cultured cells as significant catheptic activity is observed in cells in vitro.

Resolution of Brain Renin from Cathepsin

The controversy about the existence of true brain renin different from non-specific acid proteases (49,105) has greatly diminished with the reports first of the convincing separation of these two enzyme activities in brain of man (41), rat (63), dog (89), hog (57), mouse (124) and beef (59). Hirose et al. (59) and Inagami et al. (63) obtained clear evidence for renin in the brain. Using affinity chromatography (summarized above) the ANG I generating activity of brain extracts from nephrectomized rats was separated into two major fractions. Renin was completely neutralized by an antiserum developed against pig kidney renin while cathepsin D-like activity was not. The antibody sensitive renin from rat and hog brain showed neutral pH optimum of enzyme action with rat angiotensinogen. Renin from the brain of nephrectomized mice was separated from cathepsin D-like acid protease on a Matrex blue A column and purified 74,000 fold (124). Antibodies raised against pure mouse submandibular gland renin completely inhibited brain renin activity. Using rat plasma angiotensinogen as substrate, mouse brain renin showed a pH optimum at pH 7.0, while kidney renin had its maximal enzyme activity at pH 8.5. Mouse brain renin was detected as a single sharp peak following isoelectric focusing at pH 5.6, whereas kidney renin was represented by two peaks of activity at pH 5.6 and 6.7. The in vivo activity of mouse brain renin was studied by injection of the purified enzyme into the lateral cerebral ventricle of rats. Formation of ANG I from endogenous brain angiotensinogen was observed, resulting in increased water uptake and a rise in arterial blood pressure (124).

Recently Hirose et al. (59) achieved complete purification of bovine pituitary renin and an 850,000 fold enrichment of the enzyme. The pure renin had a specific activity of 146 μMoles ANG I per mg per hour and a Km of 1 μMol with rat angiotensinogen. The molecular weight was 36,000, the pI 5.25, the pH optimum was between 6.5 and 7.5. The amino acid composition was similar but not identical to kidney renin with a high content of Leu, Gly, Glu, and relatively low content of Cys, Met and His; but a lower Val and Ser content was found in brain renin than in kidney renin. The pituitary enzyme did not bind to agarose bound lectins; antibodies raised against pituitary renin cross-reacted with bovine and hog renal renin.

The concentration of renin in the brain is relatively low. There are marked species differences, however, with respect to brain renin content, e.g. in bovine and hog anterior pituitary being high (45.3, 42.7 ng ANG I/mg protein/h) and rabbit, sheep, rat pituitary exhibiting low renin activity (5.2, 1.3, 0.7, respectively) (59). A contamination with plasma renin is unlikely, because nephrectomy has repeatedly been shown to cause no change or even an increase in brain renin concentration (41). The amino acid composition, pH optimum, the pI values and other physico-chemical characteristics of purified brain renin are different from kidney renin; in addition brain renin has been localized intracellularly and the existence of prorenin in the brain is indicative of its local synthesis (57,124). Haulica et al. (52) suggested the de novo endogenous formation of brain renin in monolayer cultures of pineal glands which produced increasing amounts of the enzyme when kept over 30 days in culture. Recent reports on the presence of renin and other components of the RAS in neuroblastoma cells also support the local synthesis of renin in nervous tissue.

Distribution of Renin in Brain

The distribution of renin and angiotensin in the central nervous system has been investigated by biochemical analysis of extracts from individual brain regions and by immunohistochemistry. In histochemical studies using antibodies directed against submaxillary gland renin, we have found renin-like immunoreactivity in the nuclei paraventricularis, periventricularis and supraopticus of the hypothalamus in rats and mice (36). Immunoreactive cells were also observed in the cerebellar cortex, i.e., in the Purkinje cells and their dendrites, and in the medulla oblongata in the vicinity of the vestibular nuclei. Renin appeared to be associated with oxytocin containing cells while angiotensin immunofluorescence was found in the vasopressin positive cells (36). Inagami et al. (65) also described immunopositive staining in the perikarya of neurons in hippocampus, cortex, thalamus, septum as well as the hypothalamic and cerebellar nuclei of rat and mouse listed above. In addition many nuclei in brain stem such as nucleus ambiguus, nucleus tractus solitarii were also renin positive. Renin-like immunoreactivity was also found in luteotrophic cells of adenohypophysis of rat and in nerve and glial cells of human brain (see 22, 36, 65, 122).

The presence of true renin in these brain regions has been quantified by biochemical analysis of extracts from hog brain (81). After isolation and characterization of the enzyme, the highest concentrations of renin activity were found in the pineal gland, the adenohypophysis and choroid plexus, followed by hypothalamus, cerebellum and amygdala. Lower levels were obtained in pons, medulla oblongata, thalamus, hippocampus, spinal cord and cerebral cortex.

Brain Angiotensin

Biological activity resembling angiotensin has been reported in brain extracts and its identity with angiotensin was probable, since it could be neutralized by antibodies against ANG I or ANG II. Recently it has become possible to extract ANG I and ANG II from the brain of nephrectomized rats. The peptides have been characterized on high performance liquid chromatography (HPLC) capable of separating all angiotensin fragments including [Val^5]-ANG II from [Ile^5]-ANG II. The peptides extracted from rat brain corresponded to synthetic [Ile^5]-ANG II and [Ile^5]-ANG I, with small amounts of [Ile^5]-ANG(2-8) (ANG III) being present in the brain. Brain angiotensin thus appears to have the same amino acid sequence as plasma angiotensin and the same peptide was cleaved from brain and plasma angiotensinogen upon incubation with renin in vitro (54).

For the cerebrospinal fluid, most investigators report angiotensin values undetectable or little above the sensitivity of the analytical methods. The values range from undetectable to 200 pg/ml. It should be noted, however, that in most studies, the identity of the material measured in CSF was not established beyond the fact it reacted with angiotensin antibodies. Using HPLC in combination with radioimmunoassays for peptide characterization (54) we have demonstrated the presence of [Ile^5]-ANG I and [Ile^5]-ANG II in human CSF.

Distribution of Angiotensin in Brain

Localization of ANG II-like immunoreactivity in brain has been demonstrated by immunohistochemical techniques (36,42,65,94). There is general agreement as to the distribution of angiotensin, although different antibodies and techniques have been employed in the various studies.

High density of ANG II positive nerve terminals exists within the external layer of the median eminence, probably arising from ANG II positive nerve cell bodies in the nucleus paraventricularis. The supraoptic nucleus has also been reported to contain angiotensin-like material. Nucleus supraopticus and paraventricularis contain vasopressin, but when, for comparison, slices were stained for vasopressin, no evidence was found that the same cells contain both vasopressin and angiotensin. This is in contrast to the observation recently reported by Kilcoyne et al. (68) who found ANG II-like immunoreactivity in the same magnocellular neurons where vasopressin was located. It was suggested that ANG II shares antigenic determinants with a vasopressin precursor.

Further ANG II positive brain areas were a) High density: Substantia gelatinosa of the spinal cord, nucleus tractus spinalis nervi trigemini, nucleus amygdaloideus centralis, sympathetic lateral column; b) Low-to-moderate density: Nucleus dorsomedialis hypothalami, locus coeruleus, ventral and caudal part of nucleus caudatus putamen; c) Scattered terminals: Periventricular mesencephalic gray, hypothalamus, preoptic area, subcortical limbic structures (amygdaloid cortex, septal area), limbic cortex, thalamus (midline area), ventral midbrain, substantia nigra, reticular formation including the region of the norepinephrine and epinephrine cell groups A1 and C1, ventrolateral reticular formation of medulla oblongata, raphe region, nucleus tractus solitarii, nucleus dorsalis motorius nervi vagi, periventricular area of pons and medulla oblongata; d) No immunofluorescence: Parts of the neocortex, cortex cerebelli (59).

The fact, that ANG II-like immunoreactive material has been demonstrated in brain cells of rat fetuses kept in monolayer culture (94) and that angiotensin II was chromatographically identified in cloned and cultured neuroblastoma cells (65) indicates that the ANG II-like material described in these studies is an endogenous brain peptide not originating from the plasma.

In summary, evidence indicates the existence of an independently functioning RAS in the brain with local ANG II formation as the main effector peptide and the presence of a true renin enzyme. Endogenous brain ANG may well be important in the maintenance of high blood pressure in experimental hypertension. The widespread distribution of components of the RAS in the brain suggests its participation in additional, to date unknown, biological functions. This remains to be studied in the future.

BRAIN CONVERTING ENZYME (CE) AND RELATED PEPTIDASES

Converting Enzyme

Converting enzyme (EC 3.4.15.1) is a dipeptidyl carboxypeptidase which cleaves the COOH-terminal dipeptides from various substrates. It was first described by Skeggs et al. (120) as the enzyme that splits the carboxy-terminal histidyl-leucine (His-Leu) dipeptide from the decapeptide angiotensin I (ANG I) thus generating the effector peptide of the renin-angiotensin system (RAS) angiotensin II (ANG II). Converting enzyme has been demonstrated to be ubiquitously distributed in the organism of various species including man with particularly high concentrations in lung, plasma, kidney and testis (28). Besides these peripheral tissues, CE is also present in the brain. In the rat high enzyme activities were found in the striatum, posterior lobe of the hypophysis and cerebellum, and lower activities in midbrain and medulla oblongata and in the hypothalamus (109,135).

A particularly rich source of CE in the brain is the chorioid plexus (3,109), but high CE activity has also been reported from cortical microvessels of calf brain (45) and rabbit brain (13). CE was shown to be bound to vascular endothelial cells (e.g. lung), to epithelial cells from the brush borders of proximal tubules (kidney) or of epithelial plexus cells facing the ventricular space (plexus chorioideus), but it also exists in a soluble form in plasma, seminal fluid, cerebrospinal fluid (CSF), and lymph (28,109,112).

Table 1 summarizes converting enzyme activities as measured in our laboratory with a fluorometric assay according to Depierre and Roth (25) in various brain areas as well as lung, kidney, plasma and CSF.

Table 1. Converting enzyme activity in various tissues of the rat

Values are expressed as pmol His-Leu per min incubation per mg protein	
Striatum	794 ± 50
Hypothalamus	154 ± 12
Medulla oblongata	114 ± 16
Neurohypophysis	4 500 ± 110
Adenohypophysis	1 660 ± 120
Chorioid plexus	77 040 ± 625
Lung	6 195 ± 189
Aorta	96 ± 7
Kidney	45 ± 11
Plasma	1 031 ± 16
Cerebrospinal fluid	9 767 ± 2884

Values represent means of three or more experiments

When brain tissue was subjected to subcellular fractionation, the highest CE activities were associated with the microsomal fractions containing the synaptosomes while there was little activity in the supernatant (8,135). Converting enzyme isolated and purified from brain tissue exhibited similar characteristics as CE from different peripheral sources with respect to molecular weight, pH optimum, chloride dependency, and inhibition by chelating agents such as EDTA and o-phenantroline or by peptide inhibitors like SQ 20881 (3,8,135).

Fuxe et al. (35) reported on a 50% decrease of CE activity in the corpus striatum following unilateral lesions of this brain area with ibotenic acid, an agent producing selective degeneration of nerve cells within the injected area without damaging glial elements, extrinsic nerve terminals and axons of passage. The authors concluded that CE has a neuronal localization within the corpus striatum, a finding which is in agreement with the depletion of the enzyme in the corpus striatum of patients with Huntington's chorea (2).

As with peripheral CE, little is known so far on the regulation of CE activity in the brain. Arregui and Barer (1) have demonstrated that chronic hypoxia leads to a reduction of CE activity in the substantia nigra and, to a larger extent, in the striatum, hypothalamus and cerebellum.

In our laboratory we have found increased CE activity in blood pressure controlling brain regions, such as the medulla oblongata, following prolonged oral treatment with the CE inhibitor captopril (131). This suggests a product-dependent regulation of CE with an induction of the enzyme when the local angiotensin II levels are lowered.

Enkephalinase Activity

CE not only cleaves the carboxy-terminal His-Leu dipeptide from ANG I, but has a broad spectrum of endogenous substrates. It is identical with a major kinin degrading enzyme, kininase II (28).

Swerts et al. (128) raised the question whether CE was identical with an enkephalin degrading enzyme from brain tissue demonstrated first by Malfroy et al. (78). This was based on the finding that a dipeptidyl carboxypeptidase from mouse striatum which released [^{3}H]-Tyr-Gly-Gly from [^{3}H]enkephalins had some properties in common with purified CE, e.g. its activity was inhibited by chelating agents like EDTA and restored by the addition of zinc ions; and bradykinin or Hippuryl-His-Leu had similar affinities to both enzymes. On the other hand, "enkephalinase" could not be activated by chloride ions, as is the case with CE, and the CE inhibitors SQ 20881 or SQ 14225 (captopril) were less potent against "enkephalinase" than against CE.

Erdös et al. (29) demonstrated the hydrolysis of enkephalins by a purified peptidyl dipeptidase which they claimed to be identical with CE. Benuck and Marks (9) reported on a dipeptidyl carboxypeptidase from membranes of rat brain that inactivated enkephalins. Based on the anatomical distribution in brain regions and the properties of the dipeptidyl carboxypeptidase after purification by immunoaffinity chromatography, the authors claimed a coidentity of the enkephalin degrading enzyme and CE.

The possibility that ANG II formation and opioid peptide metabolism might be regulated by one single enzyme proved to be a great stimulus for research because of its far-reaching possible consequences concerning the interaction of both peptide systems. Further purification of the enkephalin metabolizing carboxypeptidases from brain tissue was achieved by Gorenstein and Snyder (46), who separated three different enkephalin degrading enzymes from solubilized membrane fractions of rat brain:first an aminopeptidase yielding a free tyrosine from the amino terminal end of enkephalins, second a carboxydipeptidase, generating Tyr-Gly-Gly which they designated "enkephalinase A", and finally an enzyme designated "enkephalinase B" which gave rise to a Tyr-Gly fragment. The "enkephalinase A" activity could be separated without overlap from the CE activity by DEAE column chromatography. Furthermore, the addition of 0.1 M NaCl to the assay reduced "enkephalinase A" activity by 40% but stimulated CE activity twofold, and SQ 20881 (10 μM) almost abolished CE activity while having little effect on "enkephalinase A". Thus the two enzymes could be clearly differentiated.

Apparently the "enkephalinase" isolated by Benuck and Marks (9) was a different enzyme from "enkephalinase A" as described by Gorenstein and Snyder (46). However, more recently, Benuck and Marks (10) isolated a dipeptidyl carboxypeptidase distinct from CE from rabbit brain membrane preparations. This enzyme has many characteristics in common with "enkephalinase A" and, interestingly, not only cleaves the carboxy-terminal dipeptide Phe-Met or Phe-Leu respectively from enkephalins but also the Phe-Arg-dipeptide from bradykinin while being inactive against ANG I or the synthetic CE substrate Hippuryl-His-Leu.

The present data seem to indicate that a number of enzymes associated with membrane structures in the brain are able to cleave the carboxy-terminal dipeptide from enkephalin.

Among those, "enkephalinase A" has a high affinity to enkephalin and is clearly distinct from CE. Converting enzyme, on the other hand, does release the carboxy-terminal dipeptide from enkephalins, however, with a lower affinity (129). Conversely, the CE inhibitors SQ 20881 and captopril can inhibit enkephalinase activity from brain preparations, although their inhibitory potency against CE is 1000 times higher (129).

In vivo evidence for the inhibition of enkephalins by CE inhibitors stems from the results of Stine et al. (126): Captopril injected into the cerebral ventricles (i.c.v.) at a dose of 900 μg inhibited the degradation of i.c.v. infused [^{3}H]-methionine-enkephalin in the rat striatum and potentiated the methionine-enkephalin induced analgesia. Moreover, captopril itself induced a dose-related analgesia following i.c.v. injection, and this effect was prevented by prior administration of the opiate antagonist naltrexone (10 mg/kg i.p.).

Substance P Degrading Enzymes

Prominent among the other peptides that have been reported to interfere with CE or CE inhibitors, is substance P. This undecapeptide with a wide distribution in the brain and peripheral tissue was reported to inhibit CE activity from rat striatal homogenates being about 100 times more potent than Leucine-enkephalin (80). In a study by Lee et al. (73) the CE inhibitor SQ 20881 strongly inhibited soluble substance P degrading activity from rat brain homogenates, but had only a weak action on the particulate enzyme fraction, while the CE inhibitor captopril did not inhibit the soluble enzyme and had only a weak action on the particulate enzyme.

A substance P degrading enzyme from human brain membranes was further purified and characterized by the same authors (74). The enzyme acted on substance P, leaving at the internal peptide bonds Glu^{6}-Phe^{7}, Phe^{7}-Phe^{8} and Phe^{8}-Gly^{9} but had no exopeptidase activity. It was characterized as a neutral metallo-endopeptidase with a molecular weight of 40 000 - 50 000 and could be distinguished from previously described substance P degrading peptidases. CE activity and enkephalinase activity could be separated from substance P degrading enzyme. Captopril had no inhibitory effect at a concentration of 1 mM, while SQ 20881 was a weak inhibitor of the enzyme (IC_{50} : 1.8 mM).

The authors suggested that the membrane-bound and the cytosolic substance P degrading enzyme activities may represent different enzymes. This could explain the differences in potency of several inhibitors including the CE inhibitors SQ 20881 and captopril towards substance P degrading activities from soluble and particulate fractions of the brain.

Interactions between inhibitors and substance P were also found in the periphery. Cannon et al. (2) reported that the addition of the CE inhibitor SQ 20881 to human plasma caused an increase of endogenous immunoreactive substance P and partially prevented the degradation of synthetic substance P. Couture and Regoli (24) observed that in rat plasma the degradation of substance P, and its C-terminal fragments Octa-substance P (4-11) and Hexa-substance P (6-11) was completely blocked by captopril at a concentration of 10 µg/ml plasma.

Despite these findings, substance P is not likely to be degraded by the carboxypeptidase CE, since substance P has a C-terminal amide and not a free carboxyl group. However, its inhibitory actions on brain CE activity and, conversely, the inhibitory effects of CE inhibitors on substance P degrading activities from brain tissue and plasma in vitro, raise the possibility that interactions of substance P with CE and related enzymes may occur in vivo. Indeed, Hanson and Lovenberg (50) have recently reported increased concentrations of substance P in distinct brain areas of the rat following the intracranial application of CE inhibitors.

Table 2 summarizes the data on brain CE and related peptidases discussed in this chapter.

Table 2. Similarities and dissimilarities between converting enzyme and related brain peptidases

Enzyme≠	Source	Distribution	Subcellular Localization	Purification	Km
Converting Enzyme (CE)	Rat brain	Str>Pit>Cer >Mo>Hyp>Cor	Microsomes ≫ Mitochondria Supernatant	no	-
CE	Rat brain	-	Microsomes Synaptosomes ≫ Cytosol	partial; 70-110 fold	2.2 mM (Hip-His-Leu)
"Enkephalinase" high) low) affinity	Mouse striatum	-	particulate (high-low aff.) soluble fraction (low aff.)	no	90 nM (^{3}H-Leu ENK)
Enkephalin degrading peptidyl-dipeptidase (=CE)	Hog kidney Human endothelial cell cultures	-	-	to homogeneity (kidney) no (cell cultures)	-
CE ENK-degrading enzyme	Human brain	-	particulate fraction	partial; separation of CE and ENK-DE with overlap	-
"Enkephalinase"	Mouse striatum	-	-	no	-
CE and "Enkephalinase"	Mouse brain Rat brain	CE: Str>Cer >Cor "ENK" Str>Cer >Cor	-	no	-
"Enkephalinase A" "Enkephalinase B" CE	Rat brain	-	membrane fractions solubilized	partial; separation of CE from ENK'ase A without overlap	-
CE	Bovine Rat brain	microvessels	-	no, but 30 fold enrichment in microvessels vs. cortical homogenate	-

Table 2 (continued)

Chloride activation	Inhibition by EDTA (100%)	Inhibitors (IC_{50})	Reference
yes	10^{-3} M	SQ 20881 10^{-7} M; -	(135)
yes	10^{-4} M	Saralasin 40% inhib. at 10^{-5} M; SQ 20881 $\sim10^{-7}$ M; -	(11)
-	-	Leu-ENK[+] 7×10^{-8} M; Substance P[+] $\geq10^{-6}$ M; -; Bacitracin[+] 2×10^{-5} M	(78)
-	-	-; SQ 20881 100 % inhib. at 10^{-4} M; -	(29)
-	-	<u>CE</u> / <u>ENK-DE</u>: SQ 20881: 10^{-3} M / 10^{-3}; Captopril: 10^{-5} M 100% inhib. / $\sim10^{-5}$	(4)
-	$IC_{50}\sim10^{-4}$ M activity restored by Zn^{2+}	Met-ENK $\sim10^{-6}$ M; SQ 20881 $\sim10^{-4}$ M; Bradykinin $\sim10^{-5}$ M; Captopril 10-5 M	(128)
-	-	<u>CE</u>: Leu-ENK 10^{-3} M; SQ 20881 $\sim10^{-7}$ M; Met-ENK $\sim10^{-3}$ M; Captopril $\sim10^{-8}$ M. <u>ENK'ase</u>: Leu-ENK 10^{-5} M; SQ 20881 10^{-4} M; Met-ENK 10^{-6} M; Captopril $\sim10^{-5}$ M	(129)
CE: yes ENK'ase A: no	CE, ENK'ase A: similar inhib. (90-97%) at 10^{-5} M	<u>CE:</u> SQ 20881 90% inhib. at 10^{-5} M; <u>ENK'ase A:</u> 13% inhib. at 10^{-5} M; Pyromycin, Bacitracin inhib. ENK'ase only	(46)
yes	10^{-3} M	Bradykinin 10^{-6} M; SQ 20881 $\sim10^{-7}$ M; Captopril $\sim10^{-9}$ M	(45)

Table 2 (continued)

Enzyme≠	Source	Distribution	Subcellular Localization	Purification	Km
CE ≙ ENK degrading enzyme (ENK-DE)	Rat brain	CE and ENK-DE Pit>Cer>Str> Mo>Hyp>Cer	membrane fractions solubilized	partial; immunoadsorption to purif. CE antibody	0.5 mM (Hip-His-Leu) 0.1 mM Met-ENK
"Dipeptidyl Carboxydipeptidase A" (CE) ≙ "Dipeptidyl Carboxypeptidase B" (=ENK-DE)	Rabbit brain	-	membrane fractions solubilized	partial; immunoadsorption to CE antibody to separate CE from ENK-DE++	ENK'ase 0.14 mM (Met-ENK)
CE	Canine lung	-	particulate fraction	-	-
"Peptidyldipeptidase B" (≠ CE)	Rabbit kidney lung		membrane fractions solubilized	See: Ref. ≠ 6	80 µM (Leu-ENK)
"Enkephalinase"	Rat brain and periph. organs	Lung⪢Testis> Kidney, Str> Pit	particulate fractions	no	-
Substance P degrading enzyme (SP-DE)	Rat brain	-	particulate fraction soluble fraction	no	-
CE	Rat striatum	-	-	-	-
SP degrading Endopeptidase	Rat plasma	-	-	-	-
SP degrading neutral Metallo-endopeptidase	Human diencephalon	-	particulate membrane fraction solubilized	1000 fold; separation from CE and ENK'ase activities	29 µM

≠ Nomenclature as proposed by the authors in the original articles

\+ High affinity peptidase from particulate fraction

++ Both enzymes cleave Tyr-Gly-Gly from Leu-ENK and Phe-Arg from bradykinin, but ENK-DE hardly cleaves His-Leu from Hip-His-Leu or ANG I

∤ Abbreviations: Str: Striatum; Pit: Pituitary gland; Cer: Cerebellum; MO:Medulla oblongata; Hyp: Hypothalamus; Cor: Cortex; ENK: Enkephalin; Leu-ENK: Leucine-enkephalin Met-ENK: Methionine-enkephalin

Table 2 (continued)

Chloride activation	Inhibition by EDTA (100%)	Inhibitors (IC_{50})	Reference
CE: yes ENK-DE: little	-	ENK-DE: Captopril Ki: 10^{-7} M	(10)
no for ENK degradation both enzymes	CE and ENK-DE both 40-60% at 10^{-3} M	CE: SQ 20881, Captopril: 90-100% inhib. at ~10^{-6} M ENK-DE: 0- 10% inhib. at ~10^{-6} M	(9)
yes	2×10^{-5}	ß-Endorphin Ki:3.20×10^{-5}M Met-ENK Ki:1.18×10^{-4}M Leu-ENK Ki:1.48×10^{-4}M	(110)
no	20 % at 10^{-3}M	Gly-Gly-Phe-Met no inhibition by SQ 20881 or Captopril	(8)
-		Phe-Ala - Thiorphan	(77)
no	EDTA: no PCMBA 5×10^{-5} partic. 5×10^{-6} sol.	SQ 20881 partic.: 30% sol.: 5×10^{-6}M Captopril 25% 10^{-3}M no inhib. Bacitracin inhib.: partic.> sol.	(73)
-	-	Substance P 5×10^{-5} M Leu-ENK ~10^{-3} M - -	(80)
-	-	Captopril 100% at 5×10^{-5} M -	(24)
-	5×10^{-3}	ANG I Bradykinin weak inhibitors SQ 20881 2×10^{-3} Captopril no inhibition	(74)

PHYSIOLOGY OF THE BRAIN ANGIOTENSIN SYSTEM

One of the most compelling biological effects produced by a peptide is the drinking induced by injections of angiotensin II (ANG II) into the brain. This appears to be a direct effect and is an absolutely reliable phenomenon which totally alters an animal's behavior to seek water and drink. For this reason alone, angiotensin in the brain is attractive to study, but in addition it raises blood pressure and releases hormones. The rise in blood pressure appears to be indirect since it is mediated by two processes which are triggered simultaneously by ANG II. These are the release of vasopressin and the increase in sympathetic activation. Another action is the release of ACTH by ANG II but this again may be indirect since vasopressin is known to release ACTH. Recently, Ganten (this volume) has demonstrated that ANG II also releases oxytocin and this appears to be direct. Whether we study the direct drinking effect or the indirect effects, they are robust, reliable phenomena produced every time that small doses are injected into the brain ventricles.

The question of how truly physiological these effects are is still a vexing one. Although very low doses injected into the third ventricle have been shown to produce the ANG II effects (90), the concentrations are still in the upper limits of normal concentrations found in CSF. If, however, the CSF levels merely represent a spillover from brain tissue then the doses used may be considered to be physiological. When one reflects on the concentration of neurotransmitter released into the synaptic cleft of neurons, the levels must be extremely high. The synaptic cleft is 200 Å wide, offering only a minute volume of extracellular fluid into which the neurotransmitter or hormone can be diluted. Therefore, in order to achieve similar concentrations at the receptor sites for ANG II, much higher concentrations have to be applied to the CSF.

ANG II is the end product of the cascade of the renin-angiotensin system and therefore the various components of this system should also produce the same physiological effects. In this review we will consider the biological effects of the system and then the sites of action and some newly discovered actions of ANG II in the brain.

Biological Effects of the Components of the RAS

Renin

Injection of renin into the brain ventricles was shown to produce drinking and blood pressure effects in both rats (34), dogs (102) and monkeys (Schölkens, this volume). Recently, there has been interest in the possibility that nerve growth factor (NGF) also produces these effects. A long-standing dispute concerning the existence of an RAS in the brain was the concern that Cathepsin D may have been the substance isolated and not renin since Cathepsin D can cleave renin substrate to form angiotensinogen (105). The biochemical aspects of this controversy are discussed above. Previous in vivo experiments,however,have been carried out with impure renin and NGF. We have now studied the biological effects of pure renin extracted from mouse submaxillary gland by Dr. T. Inagami, pure NGF purified by Dr. D.M. Young and Cathepsin D.Different doses of substances were tested and compared to the effects of vehicle after injection into the third ventricle of rats prepared for drinking tests and blood pressure tests at the same time. Cathepsin D had no effect on either drinking or blood pressure at any of the doses used. This was also found by Tonnaer (this volume). Purified nerve growth factor also had no effect on drinking or pressor activity. Purified renin not only produced an increase in drinking and blood pressure acutely, but had an extraordinary long-lasting effect on the water intake and a longer-lasting effect than angiotensin II on the blood pressure (Fig. 1). The results allow us to dismiss nerve growth factor and Cathepsin D as dipsogenic or angiotensin II producing hormones. Previous studies with NGF have used impure hormone which contains renin and that has undoubtedly accounted for the effects reported. Cathepsin D has no action on angiotensinogen in the CSF because the pH is high in the CSF whereas the optimal pH for Cathepsin D, which is active in lysosomes, is low. Tonnaer (this volume) has indicated that brain renin is active by using a different approach. He blocked the drinking to tetradecapeptide i.v.t. by pretreatment with N-acetyl pepstatin and concluded that it had inhibited an endopeptidase which could be brain renin.

Renin Substrate

The injection of angiotensinogen (renin substrate) to the brains of rats produced a limited but significant amount of drinking when tested in one laboratory (94). In a study by Reid et al. (107) no effect was seen when natural substrate was injected. Both studies were in agreement that synthetic substrate tetradecapeptide (TDP) produced copious amounts of drinking. The reason for the effect of TDP was also agreed to be due to converting enzyme which could directly cleave the synthetic substrate to form angiotensin II and therefore was not an appropriate test for the existence of renin in the brain. The difference between the two studies in the use of natural substrate however was disconcerting. As we understood more about the distribution of renin and the components of the RAS in the brain a reasonable explanation for the results has become apparent. In the Reid et al. study the substrate was injected into the ventricles. In the Phillips et al. study the substrate was injected into the brain tissue close to the ventricles. Immunocytochemical studies of the distribution of renin have clearly shown that it is within cells (122). Therefore, in order for the substrate to be activated by renin it has to be placed in the vicinity of cells which have been sufficiently disrupted to release the renin.

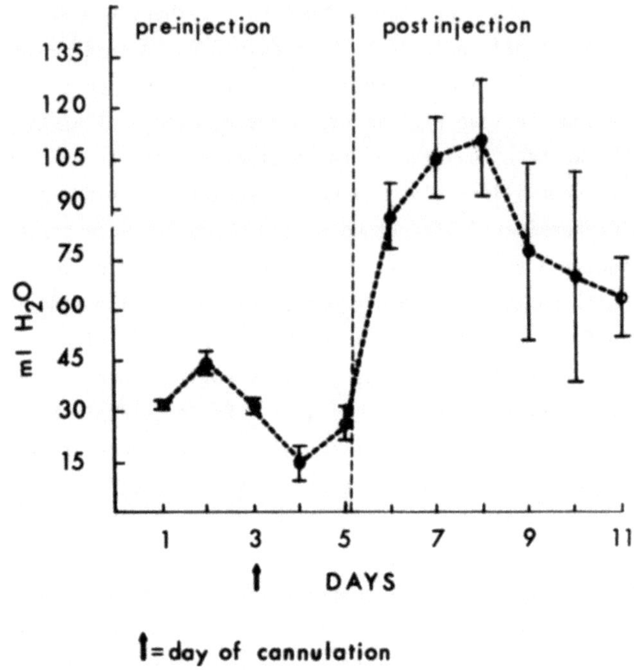

Fig. 1. The effect of purified mouse renin injected into the brain ventricle (1 µg in 1 µl) on water intake over several days. Water intake was measured at 24 hour interval

This could account for the difference in the studies because when the cannula is placed in brain tissue there is some destruction of cell membranes, but when the cannula is placed into the CSF, no cells near the tip of the cannula will be effected. Thus, the evidence for natural substrate being cleaved by renin in vivo is acceptable. Reid (this volume) has argued that the distribution of renin substrate does not watch the distribution of renin and therefore the two could not interact. This is based on a single immunocytochemical study, however, and Printz (this volume) using more direct biochemical assay has shown angiotensinogen in several brain areas where renin has been found.

Converting Enzyme

Numerous experiments have now been carried out on the role of converting enzyme in the brain by the use of converting enzyme inhibitors. Captopril injected into the brain in spontaneously hypertensive rats produced a decrease in blood pressure (125). The decrease in blood pressure was greater than that produced by the same dose of Captopril given i.v. Therefore, the central injection was not leaking into the periphery since otherwise the effect should have been smaller. This led us to hypothesize that Captopril crosses the blood brain barrier and reaches brain sites of high RAS activity as part of its blood pressure lowering effect. Unger (this volume) reports more marked effects of Captopril when given chronically. Recently we have used a new inhibitor, MK 421, and found substantially the same results. MK 421 lowered blood pressure in SHR rats when injected directly into the brain at a dose of 10 micrograms which does not produce a response when injected into the blood (Fig. 2). The argument that injecting these substances into the blood would cause greater dilution than injecting into the ventricles misses the point. If these substances did not act in the brain but acted by leaking into the periphery, then the effects should be much greater by the i.v. route than by the i.v.t.route. Since the opposite is the case, the inhibitors must be acting on brain RAS. In addition to the acute effects of lowering blood pressure, MK 421 continued to keep blood pressure down after its converting enzyme inhibitory effects were past. This may reflect another property of the converting enzyme inhibitor.

Angiotensin II

Finally, we come to the active peptide, angiotensin II. The effects on drinking and blood pressure have by now been so well characterized and so thoroughly studied with inhibitors and analogs that it is undisputedly the activator of drinking behavior and the other effects. Angiotensin III is not quite as potent as angiotensin II and none of the smaller fragments of the octapeptide are as potent (34). Although not all substituted fragments have been used, the C terminal phenylalanine in position 8 seems to be critical to activating the receptors. The minimum number

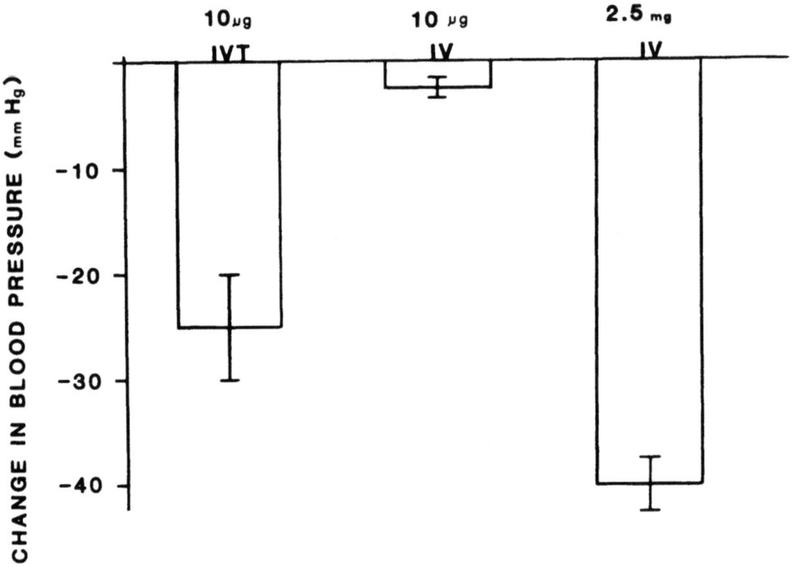

Fig. 2. The blood pressure lowering effect of a converting enzyme inhibitor (MK 421). The effect of 10 µg intraventricularly is much greater than the effect of the same dose intravenously. Therefore, there is an effect in the brain which is not due to leakage from the brain into the peripheral circulation. An equivalent fall in pressure by i.v. MK 421 requires 2000 x more MK 421 than i.c.v.

of amino acids for the effect is 6 in the 3-8 hexapeptide. The N-terminal amino acid appears to determine the duration of action, the hydrophobic side chains Val, Ile, Pro, stabilize the conformation and Tyr, His are necessary for binding. In addition to its dipsogenic and pressor effect, ANG II in the brain also appears to alter the physiology of renin release from the kidneys. Several authors have observed lower plasma renin levels after i.v.t. ANG II (see Schölkens, this volume) and this has led Buckley (this volume) to propose that plasma renin levels are suppressed in hypertensive states whenever brain ANG II is elevated.

Other Peptides

ANG II seems to be a unique peptide in producing drinking behavior, but other peptides can induce a pressor effect when injected centrally into the brain. These include neurotensin, bradykinin, substance P (Unger, this volume), enkephalin (Hutchinson, this volume) and vasoactive intestinal peptide (VIP). An intriguing and illustrative aspect of these responses is that we find VIP, neurotensin and bradykinin all produce depressor effects when injected intravenously (i.v.) but pressor effects when injected i.v.t. This finding confirms the impression gained in the previous angiotensin studies that there are two compartments separate from one another, one on the blood side and one on the brain side accessible only to intracerebral peptides (93). Because the responses are opposite, there is a built-in control for the effects of leakage from brain into the periphery when these peptides are given. Despite this common central peptidergic stimulation (43), there is a quantitative difference between these peptides and angiotensin II. In all cases that we have tested, the other peptides have to be given in much greater doses than for ANG II (Fig. 3). Furthermore, the pressor response to these peptides is not inhibited by the ANG II antagonist, Saralasin. Therefore, there is not a single receptor for these pressor inducing peptides and the effect is not non-specific activation of an ANG II receptor.

The Sites of Action

The sites of action for ANG II in the brain have been discussed and argued over at great length for some years. Currently, two areas in the rat brain and one area in the dog brain have been accepted as receptor sites. In the rat brain, the subfornical organ (SFO) appears to be the site of action for blood-borne angiotensin to exert its manifold effects. The organum vasculosum laminae terminalis (OVLT) appears to have the dual ability to detect both blood-borne angiotensin II and central or ventricular angiotensin II (93). This was proposed over five years ago and the evidence for it has been accumulating since. Plugging the surface of the OVLT but not of the SFO prevents i.v.t. ANG II from having any action (60).

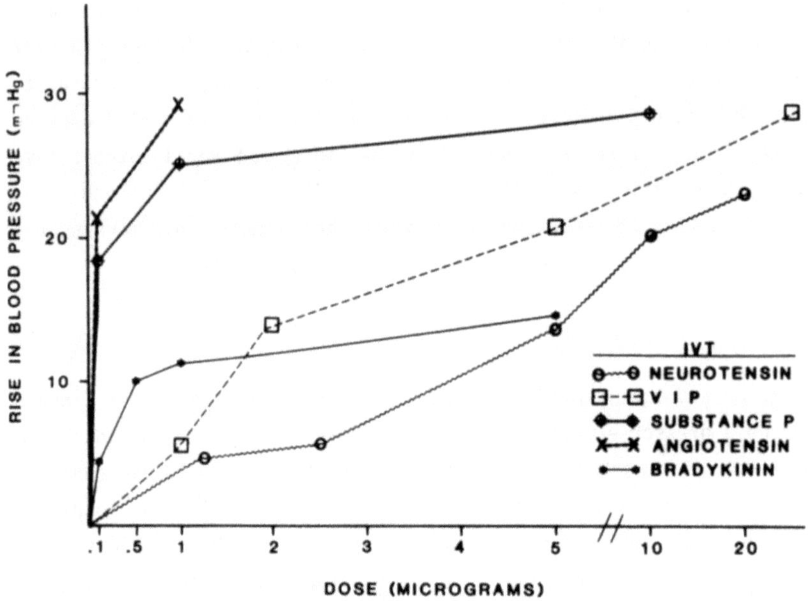

Fig. 3. The comparative dose response curve of neuropeptides which cause a pressor effect when injected in 1 µl volumens into the lateral ventricles of rat brains. Note the high potency of angiotensin II

Labelled angiotensin injected into the blood arrives at the SFO and OVLT but does not spread to other parts of the brain (132). Fluorescent labelled angiotensin injected into the brain ventricles and tested for its physiological action was found to bind to the OVLT exclusively (72). Microiontophoretic application of ANG II to the OVLT directly in anesthetized rats (94) and in unanesthetized brain slices (70) produced predominantly excitatory effects on the neurons. Neurons in the SFO were also excited by ANG II and specifically inhibited by ANG II antagonist, Saralasin (31,91). The AV3V lesions which include the OVLT abolish both i.v. and i.v.t. responses to ANG II (16). In addition, the lesions make rats hypernatremic, adipsic and abolish hypertension, if present (16), and vasopressin release (7). This would suggest that the AV3V lesion destroys the receptors for ANG II or the pathways from the receptors or alternatively the blood supply to the receptor areas. Autoradiographic and horseradish peroxidase studies of the connections from the SFO and OVLT (20,82) demonstrate an intimate connection between the two organs so that a lesion of the OVLT could disrupt the efferents of the SFO as well as eliminating the receptors in the OVLT. The dynamics of brain structure after lesioning are altered and the technique is not desirable for proving the presence or absence of normal receptor sites. A hole in the brain may in some cases open up new access to areas outside the normal reach of intraventricular peptides. This could explain the results presented by Johnson (this volume).

The other ANG II receptor site is the area postrema which is apparently of great importance in the dog but of no known importance in the rat (53). This may represent a difference in the evolution of the brains of these species and ultimately the significance of these structures will have to be clarified in the human brain. The close proximity of the AP to the nucleus tractus solitarii (NTS) situates it in a key position to modify baroreceptor reflexes. However, these reflexes are associated with short-term changes in blood pressure and yet long-term infusions of angiotensin II produce continuous hypertension. Clearly, the mechanism of action after the ANG II reception on these receptor sites has yet to be better understood before any conclusions can be drawn. In addition to these circumventricular organs there are areas with concentrated binding sites for ANG II in the septum superior colliculus and lateral hypothalamus (Sirett, this volume). Further work of the type described here should show if these are receptor sites that lead to a physiological response to ANG II.

Other Effects of the RAS

Memory

Angiotensin II has been shown to have a disruptive effect on memory (71,85) and was tested in a passive avoidance learning situation where the animals would not enter a chamber where 24 hours before they had received a punitive stimulus. If at the same time as the shock, renin was injected i.v.t., the rats appeared to have forgotten and entered the compartment 24 hours later. The impaired response was abolished by captopril showing that ANG II was the effective peptide. Other peptides such as vasopressin have been shown to enhance memory in a similar test (26). Part of this effect may be due to an action by vasopressin on the hippocampus. The hippocampus has been associated with short-term memory based on neurological studies in humans. Since angiotensin II releases vasopressin, one might have predicted that ANG II would have enhanced memory via AVP release. Immunocytochemical studies on AVP distribution point to two different systems (17), one involving the supraoptic nucleus and paraventricular nucleus which release AVP via the median eminence into the neurohypophysis. The second system emanates from a suprachiasmatic nucleus which courses dorsally and caudally through the brain to the NTS. It is probably the second system that innervates the hippocampus and the results on the memory test would imply that ANG II does not activate release of vasopressin via this system but activates only the former hypophyseal system. Within the hippocampus we have noted one or two isolated cells containing ANG II (95). Sirett et al. (119) have also observed binding in the hippocampus. Generally, cells in the hippocampus are not excited by microiontophoretic application of ANG II (31). The projections of the hippocampus via the fimbria to the septum may be modified by ANG II action, since the septum contains both a concentration of immunoreactive ANG II (95), ANG II binding sites (119), and neurons excited by ANG II (62). The functional role of these receptors is far from understood, particularly since lesioning the septum results in polydipsia and not adipsia as one might expect, if the ANG II receptors were involved in thirst.

Serotonin

The work of Haulica et al. (51) on the pineal gland has demonstrated that high renin levels are present in the gland particularly during hours of darkness. This diurnal rhythm follows the same diurnal variation observed with the serotonin content of the pineal gland. The inhibition of serotonin synthesis with para-chlorophenylalanine abolished or reversed the low mid-day values of renin. One study has claimed that the dipsogenic response to ANG II is mediated by serotonin (87). We have not been able to support this claim in our own studies where central serotonin levels were reduced but had no effect on the pressor or the dipsogenic response to ANG II i.v.t. (19). Perhaps the cyclic phenomena discovered by Haulica needs to be taken into consideration before deciding on the influence of serotonin on the angiotensin responses.

Norepinephrine

The synergic interaction between norepinephrine and ANG II in the periphery has long been established. Such an interaction in the brain has been studied by Fuxe et al.(this volume). One of the assumptions in our work has been that ANG II centrally activates the sympathetic nervous system. Therefore, we could expect changes at the brain level and also in the blood levels of norepinephrine. Preliminary studies by Sumners in our laboratory indicate that the levels of NE in blood are raised by intraventricular ANG II injections as has also been shown by Schölkens et al. (this volume). In addition, he has performed a micropunch study on the brain after injecting ANG II i.v.t. and found changes in the NE turnover levels in different structures but not in the dopamine levels in those structures (see Table 3). The pressor response can be separated from the drinking response by alpha adrenergic blockade (19). When this is done, the pressor response is diminished but the drinking response is untouched. Therefore, the pressor response may be mediated by norepinephrine in the brain, leading to an increase in sympathetic activation. The release of vasopressin by norepinephrine is debatable because one group using anesthetized dogs have found that norepinephrine inhibits AVP release (118) whereas we have found that in unanesthetized rats norepinephrine stimulates the release of AVP (18). The technical and species difference between these two sets of data have to be taken into account to arrive at a true picture of the role of norepinephrine in vasopressin release.

Opiate Peptides

Recently, Barnes and Ferrario (5) and we (61) have discovered that naloxone inhibits the actions of angiotensin II. Barnes and Ferrario studied the anesthetized dog preparation and we studied the unanesthetized rat (Fig. 4). In this case, both sets of results are in the same direction. Naloxone significantly reduced pressor and drinking actions of central angiotensin II. This implies that angiotensin expresses its effects via an enkephalin component at some stage in the neural circuitry. Enkephalin alone increases blood pressure when injected into the brain but inhibits vasopressin release (111). Therefore, the pressor effect of the opiate would seem to be via sympathetic activation. Whether this enkephalin component comes before, after, or at the same time as the alpha-adrenergic component in the circuit has not yet been established. Other evidence that the opioid peptides are involved when angiotensin activates brain receptors is the observation of Haulica et al. (this volume) that ANG II induces analgesia. This analgesia was measured by the tailflick method and could be antagonized by naloxone. One of the actions of opioid peptides is to produce immobility. This may account for a reduction in drinking reported when enkephalins were injected i.v.t. together with ANG II (127). A common link between angiotensin and enkephalins is the possibility that converting enzyme has similarities with enkephalinase (9) (see discussion above). Captopril which is a converting enzyme inhibitor and could therefore prevent the hydrolysis of enkephalins by the enkephalinase activity, also produces analgesia (126,128). Consequently kininase inhibition by captopril increases the amount of bradykinin and prolongs its effects (130). Converting enzyme is also an important link between bradykinin and angiotensin II because in its role as kininase, it brings about the degradation of bradykinin.

Table 3. Effect of angiotensin II (100 ng i.v.t.) upon norepinephrine turnover measured by alpha methyl tyrosine (MT) method in brain stem and hypothalamus

Brain Area	NE Turnover	Control	Alpha MT	Alpha MT + ANG II
A1	Increase	15.58 ± 1.29	4.96 ± 0.82	1.68 ± 0.53
Nucleus Tractus Solitarii	-	10.60 ± 1.30	4.78 ± 0.57	3.37 ± 0.70
Raphe Magnus	Increase	11.49 ± 1.50	5.14 ± 0.95	1.73 ± 0.56
A5	-	14.80 ± 1.59	5.03 ± 0.64	6.13 ± 0.51
Locus Coeruleus	Increase	36.30 ± 2.23	10.85 ± 1.08	4.79 ± 1.02
A7	Increase	12.10 ± 0.72	5.73 ± 0.67	4.23 ± 0.63
Hypothalamus	Increase	1.44 ± 0.19	0.92 ± 0.08	0.58 ± 0.06

n = 6; tissue pooled from 12 animals; increases = $p < 0.05 - 0.001$

Hypothalamus is µg NE/gm tissue; all others pg NE/µg protein; saline treatment without ANG II had no effect

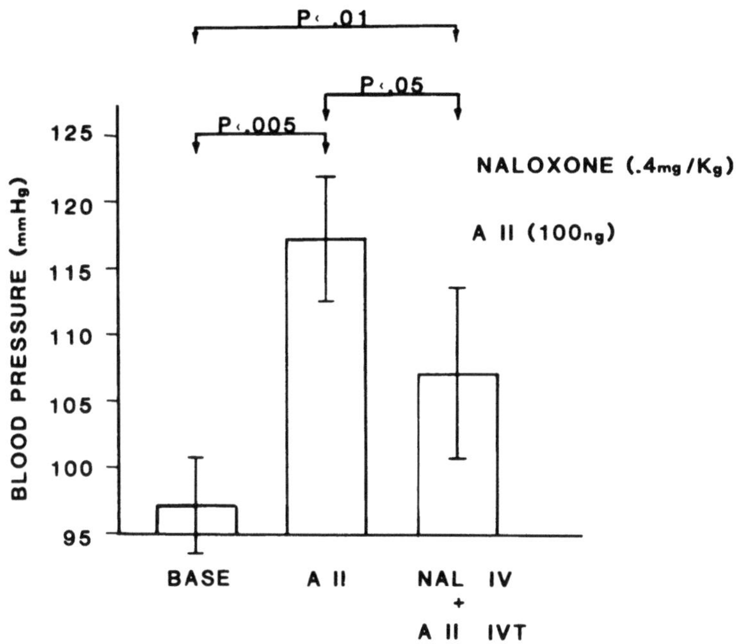

Fig. 4. Naloxone i.v. inhibits the rise in blood pressure due to angiotensin II injected into the brain. Naloxone had no effect on its own

Bradykinin and Prostaglandins

As mentioned earlier, bradykinin induces a pressor response centrally and a depressor response peripherally. Unger (this volume) showed that the pressor effect of 2 μg bradykinin can be enhanced when Captopril is given. This was assumed to be due to prolonged bradykinin action by Captopril's kininase inhibitory effect. However, an alternative explanation is that ANG II inhibits bradykinin even though the response is the same. We tested whether the bradykinin pressor response was due to reduced angiotensin II by applying Saralasin i.v.t., the angiotensin antagonist, at the same time as bradykinin i.v.t. The pressor response was not reduced but increased (Fig. 5). The two peptides, ANG II and bradykinin appeared to interact reciprocally, even though both were producing the same response in the brain.

The possibility of peptide-peptide interactions at the synaptic level opens new ways for neural control. In the kidney, it is accepted that bradykinin is released by prostaglandins and another link between ANG II and bradykinin is that ANG II is also modified by prostaglandins. Prostaglandin E_1, by itself inhibits drinking (67) but causes a pressor response although at much higher doses than those for ANG II. Inhibition of prostaglandins synthesis by meclofenamate prolonged the dipsogenic activity of ANG II (92). Finally, ANG II applied to endothelial tissue releases prostaglandins (44). These data imply that ANG II in the brain releases prostaglandins which in turn inhibit the dipsogenic response to ANG II and thereby act in a regulatory fashion as a feedback on the RAS. The prostaglandins separate the drinking effects of brain ANG II from its pressor effects.

To summarize the state of knowledge on the biological actions of a brain RAS, substantial amount of data has now been accumulated to implicate the brain RAS in a number of different functions; in particular water balance, memory, diurnal rhythm and interactions with other substances in the brain such as peptides, catecholamines, serotonin, bradykinin and prostaglandins. In addition, there are indications that brain ANG II inhibits the peripheral renin-angiotensin system. Although some of the evidence is weak while other data are quite convincing, the general picture is of a central peptide which has evolved as a prime mover of behavior in redressing impaired water balance but at the same time being an important modifier of neurotransmitter and other peptide activity. Finally, whereas renin may be the rate limiting step in the production of angiotensin, it is the converting enzyme that appears to be the rate limiting factor in the functional role of ANG II in these peptide interactions. This is because of its multiple actions on bradykinin and possibly enkephalins, and the reciprocal interactions between these peptides and angiotensin II.

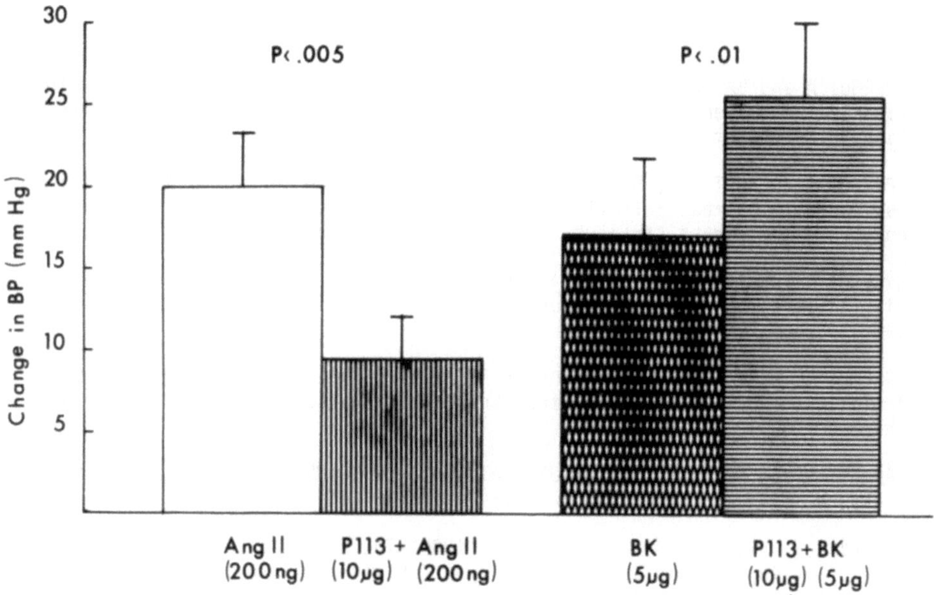

Fig. 5. Change in blood pressure with i.c.v. injections of angiotensin II and bradykinin (BK) in combination with the angiotensin antagonist Saralasin (P 113). Whereas angiotensin II is inhibited by the antagonist, the pressor effect of bradykinin was enhanced. Therefore the two peptides have a reciprocal relationship on the same response (R. Lewis and M.I. Phillips, unpublished data)

REFERENCES

1. ARREGUI A and BARER GR (1980) Chronic Hypoxia in Rats: Alterations in Stria-to-Nigral Angiotensin Converting Enzyme, GABA, and Glutamic Acid Decarboxylase. J Neurochem 34:740-743

2. ARREGUI A, BENNETT JP, BIRD ED, YAMAMURA HI, IVERSEN LL and SNYDER SH (1977) Huntington's chorea: Selective depletion of activity of angiotensin converting enzyme in the corpus striatum. Ann Neurol 2:294-298

3. ARREGUI A and IVERSEN LL (1978) Angiotensin Converting Enzyme: Presence of High Activity in Chorioid Plexus of Mammalian Brain. Europ J Pharmacol 52: 147-150

4. ARREGUI A, LEE C-M, EMSON PC and IVERSEN LL (1979) Separation of Human Brain Angiotensin Converting Enzyme from Enkephalin-Degrading Activity. Europ J Pharmacol 59:141-144

5. BARNES KL and FERRARIO CM (1980) Angiotensin and CNS Regulation of the Blood Pressure. Clin Exp Hypertension 2 (344):465-477

6. BARNEY CC, KATOVICH MJ and FREGLY MJ (1980) The Effect of Acute Administration of an Angiotensin Converting Enzyme Inhibitor, Captopril (SQ 14,225) on Experimentally Induced Thirsts in Rats. J Pharmac Exp Therap 212:53-57

7. BEALER SL, PHILLIPS MI, JOHNSON AK and SCHMID PG (1979) Effect of Anteroventral Third Ventricle Lesions on Antidiuretic Responses to Central Angiotensin II. Am J Physiol 236:E610-E615

8. BENUCK M, BERG MJ and MARKS N (1981) A Distinct Peptidyl Dipeptidase that Degrades Enkephalins: Exceptionally High Activity in Rabbit Kidney. Life Sci 28:2643-2650

9. BENUCK M and MARKS N (1980) Characterization of a Distinct Membrane-Bound Dipeptidylcarboxypeptidase Inactivating Enkephalin in Brain. Biochem and Biophys Res Commun 95:822-828

10. BENUCK M and MARKS N (1979) Co-Identity of Brain Angiotensin Converting Enzyme with a Membrane-Bound Dipeptidylcarboxypeptidase Inactivating Met-Enkephalin. Biochem and Biophys Res Commun 88:215-221

11. BENUCK M and MARKS N (1978) Subcellular Localization and Partial Purification of a Chloride-Dependent Angiotensin I Converting Enzyme from Rat Brain. J Neurochem 30:729-734

12. BICKERTON RK and BUCKLEY JP (1961) Evidence for a central mechanism in angiotensin induced hypertension. Proc Soc Exp Biol Med 106:834-836

13. BRECHER P, TERCYAK A and CHOBANIAN A (1981) Properties of Angiotensin Converting Enyzme in Intact Cerebral Microvessels. Hypertension 3:198-204

14. BUCKLEY JP, SMOOKLER HH, SEVERS WB, DEUBEN RR (1978) A central site of action of angiotensin II and its possible role in the central regulation of the cardiovascular system. In: BUCKLEY JP and FERRARIO CM (eds) Central Actions of Angiotensin and Related Hormones. Pergamon Press, New York, p 149-155

15. BUGGY J, FISHER AE, HOFFMAN WE, JOHNSON AK and PHILLIPS MI (1975) Ventricular Obstruction: Effect on Drinking Induced by Intracranial Injection of Angiotensin. Science 190:72-74

16. BUGGY J and JOHNSON AK (1978) Angiotensin-Induced Thirst: Effects of Third Ventricle Obstruction and Periventricular Ablation. Brain Res 149:117-128

17. BUIJS RM, SWAAB DF, DOGTERON J and VANLEEUWEN FW (1978) Intra- and Extrahypothalamic Vasopressin and Oxytocin Pathways in the Rat. Cell Tiss Res 186: 423-433

18. CAMACHO A and PHILLIPS MI (1978) The Role of Catecholamines in Central Angiotensin II-Induced Responses. The Physiologist 21: 16

19. CAMACHO A and PHILLIPS MI (1981) Separation of Drinking and Pressor Responses to Central Angiotensin by Monoamines. Am J Physiol 240:R106-R113

20. CAMACHO A and PHILLIPS MI (1981) Hypothalamic and Extra-Hypothalamic Connections of the Organum Vasculosum of the Lamina Terminalis (OVLT). Neurosci Lett 25:201-204

21. CANNON DEB, SKRABANEK P and POWELL D (1979) Difference in Behaviour between Synthetic and Endogenous Substance P in Human Plasma. Naunyn Schmiedeberg's Arch Pharmacol 307:251-255

22. CELIO MR, CLEMENS DL, INAGAMI T (1980) Renin in anterior pituitary, pineal and neuronal cells of mouse brain. Biomed Res 1:427-431

23. CHEN FM, HAWKINS R and PRINTZ MP (1982) Evidence for a functional independent brain angiotensin system. Correlation between brain angiotensin receptors, angiotensin and drinking in the estrous cycle of rats. In: Experimental Brain Research. The renin angiotensin system: a model for the synthesis of peptides in the brain. GANTEN D, PRINTZ MP, PHILLIPS MI, SCHÖLKENS B (eds), Springer Verlag Heidelberg, Suppl 4

24. COUTURE R and REGOLI D (1981) Inactivation of Substance P and its C-terminal Fragments in Rat Plasma and its Inhibition by Captopril. Can J Physiol Pharmacol 59:621-625

25. DEPIERRE D and ROTH U (1975) Fluorometric determination of dipeptidylcarboxypeptidase (angiotensin I converting enzyme). Enzyme 19:65-70

26. DEWIED D (1976) Hormonal Influence on Motivation, Learning and Memory Processes. Hosp Pract 11:123-131

27. EGGENA P, ITO T, BARRETT ID, VILLARREAL H, SAMBHI MP (1982) A comparison of human renin substrate in plasma and cerebrospinal fluid. In: Experimental Brain Research. The renin angiotensin system: a model for the synthesis of peptides in the brain. GANTEN D, PRINTZ MP, PHILLIPS MI, SCHÖLKENS B (eds), Springer Verlag Heidelberg, Suppl 4

28. ERDÖS EG (1975) Angiotensin I Converting Enzyme. Circ Res 36,2:247-254

29. ERDÖS EG, JOHNSON AR and BOYDEN NT (1978) Hydrolysis of Enkephalin by Cultured Human Endothelial Cells and by Purified Peptide Dipeptidase. Biochem Pharmacol 27:843-848

30. ERDÖS EG (1979) Kininases. In: Bradykinin, Kallidin and Kallikrein. EG ERDÖS (ed), Springer Verlag, New York, Vol. XXV, p 428-488

31. FELIX D and PHILLIPS MI (1978) Effects of Angiotensin II on Central Neurons. In: Iontophoresis and Transmitter Mechanisms in the Mammalian Central Nervous System. RYALL and KELLY (eds), Elsevier/North-Holland, Biomedical Press, p 104-106

32. FELIX D and PHILLIPS MI (1979) Inhibitory Effects of Luteinizing Hormone Releasing Hormone (LHRH) on Neurons in the Organum Vasculosum Laminae Terminalis (OVLT). Brain Res 169:204-208

33. FISCHER-FERRARO C, NAHMOD VE, GOLDSTEIN DJ, FINKIELMAN S (1971) Angiotensin and Renin in Rat and Dog Brain. J Exp Med 133:353-361

34. FITZSIMONS JT (1980) Angiotensin Stimulation of the Central Nervous System. Rev Physiol Biochem Pharmacol 87:117-167

35. FUXE K, GANTEN D, KÖHLER Ch, SCHÖLL B and SPECK G (1980) Evidence for Differential Localization of Angiotensin I Converting Enzyme and Renin in the Corpus Striatum of Rat. Acta Physiol Scand 110:321-323

36. FUXE K, GANTEN D, HOEKFELT T, LOCATELLI V, POULSEN K, STOCK G, RIX E, TAUGNER R (1980) Renin-like immunocytochemical activity in rat and mouse brain. Neurosci Lett 18:245- 250

37. GANTEN D (1972) Studies on the existence of an independent brain renin-angiotensin system: A model for extrarenal tissue renin. Ph.D. THESIS, McGill University, Montreal, Canada

38. GANTEN D, MARQUEZ-JULIO A, GRANGER P, HAYDUK K, KARSUNSKY KP, BOUCHER R, GENEST J (1971) Renin in dog brain. Am J Physiol 221:1733-1737

39. GANTEN D, MINNICH JL, GRANGER P, HAYDUK K, BRECHT HM, BARBEAU A, BOUCHER R, GENEST J (1971) Angiotensin-Forming Enzyme in Brain Tissue. Science 173: 64-65

40. GANTEN D, SCHELLING P, GANTEN U (1977) Tissue Isorenins. In: Hypertension, Physiopathology and Treatment. GENEST J, KOIW E, KUCHEL O (eds), McGraw-Hill Book Comp New York, p 240-56

41. GANTEN D, SPECK G (1978) The brain renin-angiotensin system: A model for the synthesis of peptides in the brain. Biochem Pharmacol 27:2379-2389

42. GANTEN D, FUXE K, PHILLIPS MI, MANN JFE, GANTEN U (1978) The Brain Isorenin Angiotensin System: Biochemistry, Localization, and Possible Role in Drinking and Blood Pressure Regulation. In: Frontiers in Neuroendocrinology, GANONG WF, MARTINI L (eds) Raven Press, New York, Vol 5, Chapter 3, p 61-99

43. GANTEN D, UNGER TH, ROCKHOLD R, SCHAZ K and SPECK G (1979) Central Peptidergic Stimulation: Its Possible Contribution to Blood Pressure Regulation. In: Radioimmunoassay of Drugs and Hormones in Cardiovascular Medicine. ALBERTINI A, DAPRADA M.and PESKAR BA (eds), Elsevier/Nortn Holland, Amsterdam, p 33-43

44. GIMBRONE MA Jr and ALEXANDER RW (1975) Angiotensin II Stimulation of Prostaglandin Production in Cultured Human Vascular Endothelium. Science 189: 219-220

45. GIMBRONE MA, MAJEAU GR, ATKINSON WJ, SADLER W and CRUISE A (1979) Angiotensin Converting Enzyme Activity in Isolated Brain Microvessels. Life Sci 25:1075-1084

46. GORENSTEIN Ch and SNYDER SH (1979) Two Distinct Enkephalinases: Solubilization, Partial Purification and Separation from Angiotensin Converting Enzyme. Life Sci 25:2065-2070

47. GREGORY TJ, WALLIS CJ and PRINTZ MP (1981) Regional Changes in rat brain angiotensin following bilateral nephrectomy. Hypertension (submitted)

48. GRESIK E, MICHELAKIS A, BARKA T and ROSS T (1978) Immunocytochemical localization of renin in the submandibular gland of the mouse. J.Histochem Cytochem 26:855-861

49. HACKENTHAL E, HACKENTHAL R and HILGENFELD U (1978) Purification and partial characterization of rat brain protease (isorenin). Biochem Biophys Acta 522: 561-573

50. HANSON GR and LOVENBERG W (1980) Elevation of Substance P-like Immunoreactivity in Rat Central Nervous System by Protease Inhibitors. J Neurochem 35: 1370-1374

51. HAULICA I, BRANISTEANU DD, ROSCA V, STRATONE A, BERBELEU V, BALAN G and IONESCU L (1975) Renin-like Activity in Pineal Gland and Hypophysis. Endocrinology 96:530-534

52. HAULICA J, HEFCO E, ROSCA V, BRANISTEANU D, PETRESCU GH, STRATONE A, COCULESCU M, BORDEA J (1977) A Renin-like Activity in the Human Hypophysis. Rev Roum Med Endocrinol 15: 51-54

53. HAYWOOD JR, BUGGY J, PHILLIPS MI and BRODY MJ (1980) The Area Postrema Plays No Role in the Pressor Action of Angiotensin in the Rat. Am J Physiol 239: H108-H113

54. HERMANN K, GANTEN D, BAYER C, UNGER TH, LANG RE, RASCHER W (1982) Definitive Evidence for the Presence of [Ile^5]-Angiotensin I and [Ile^5]-Angiotensin II in the Brain of Rats. In: Experimental Brain Research. The renin angiotensin system: a model for the synthesis of peptides in the brain. GANTEN D, PRINTZ MP, PHILLIPS MI, SCHÖLKENS B (eds), Springer Verlag Heidelberg, Suppl 4

55. HIROSE S, YOKOSAWA H, INAGAMI T (1978) Immunochemical identification of renin in rat brain and distinction from acid proteases. Nature 274:392-393

56. HIROSE S, WORKMAN RJ, INAGAMI T (1979) Specific antibody to hog renal renin and its application to the direct radioimmunoassay of renin in various organs. Circ Res 45:275-279

57. HIROSE S, YOKOSAWA H, INAGAMI T, WORKMAN RJ (1980) Renin and prorenin in hog brain: ubiquitous distribution and high concentration in the pituitary and pineal. Brain Res 191:489-499

58. HIROSE S, NARUSE M, OHTSUKI K, INAGAMI T (1981) Totally inactive renin zymogen and different forms of active renin in hog brain tissues. J Biol Chem 256:5572-5576

59. HIROSE S, OHSAWA T, MURAKAMI K, INAGAMI T (1982) Pituitary Renin: Purification and Characterization. In: Experimental Brain Research. The renin angiotensin system: a model for the synthesis of peptides in the brain. GANTEN D, PRINTZ MP, PHILLIPS MI, SCHÖLKENS B (eds), Springer Verlag Heidelberg, Suppl. 4

60. HOFFMAN WE and PHILLIPS MI (1976) Regional Study of Cerebral Ventricular Sensitive Sites to Angiotensin II. Brain Res 110:313-330

61. HOFFMANN LH and PHILLIPS MI (1981) Inhibition of Angiotensin II Pressor Effect by Naloxone. The Physiologist 24 (4):24

62. HUWYLER T and FELIX D (1980) Angiotensin II-Sensitive Neurons in Septal Areas of the Rat. Brain Res 195:187-195

63. INAGAMI T, YOKOSAWA H, HIROSE S (1978) Definitive evidence for renin in rat brain by affinity chromatographic separation from proteases. Clin Sci Mol Med 55:121s-123s

64. INAGAMI T, CELIO MR, CLEMENS DL, LAN D, TAKII Y, KASSELBERG AG, HIROSE S (1980) Renin in rat and mouse brain: immunohistochemical identification and localization. Clin Sci 59:49s-51s

65. INAGAMI T, CLEMENS DL, CELIO MR, BROWN A, SANDRU L, HERSCHKOWITZ N, HOFFMANN LH, KASSELBERG AG (1980) Immunohistochemical localization of renin in mouse brain. Neurosci Lett 18:91-98

66. ITO T, EGGENA P, BARRETT JP, KATZ D, METTER J, SAMBHI MP (1980) Studies on angiotensinogen of plasma and cerebrospinal fluid in normal and hypertensive human subjects. Hypertension 2:432-436

67. KENNEY NJ and EPSTEIN AN (1978) Antidipsogenic Role of the E-Prostaglandins. J Comp Physiol Psychol 92:204-219

68. KILCOYNE MM, HOFFMAN DL, ZIMMERMAN EA (1980) Immunocytochemical localization of angiotensin II and vasopressin in rat hypothalamus: Evidence for production in the same neuron. Clin Sci 59:57s-60s

69. KIPROV D and DIMITROV T (1976) Renin Activity in the prehypertensive and early hypertensive stage of spontaneously hypertensive Rats. Comptes rendus de L'Académie bulgare des Sciences 29:1543-1546

70. KNOWLES WD and PHILLIPS MI (1980) Angiotensin II Responsive Cells in the Organum Vasculosum Laminae Terminalis (OVLT) Recorded in Hypothalamic Brain Slices. Brain Res 197:256-259

71. KOLLER M, KRAUSE HP, HOFFMEISTER F and GANTEN D (1979) Endogenous Brain Angiotensin II Disrupts Passive Avoidance Behavior in Rats. Neurosci Lett 14: 71-75

72. LANDAS S, PHILLIPS MI, STAMLER JF and RAIZADA MK (1980) Visualization of Specific Angiotensin II Binding Sites in the Brain by Fluorescent Microscopy. Science 210:791-793

73. LEE CM, ARREGUI A and IVERSEN LL (1979) Substance P Degradation by Rat Brain Peptidases: Inhibition by SQ 20881. Biochem Pharmacol 28:553-556

74. LEE CM, SANDBERG BEB, HANLEY MR and IVERSEN LL (1981) Purification and Characterization of a Membrane-Bound Substance P-Degrading Enzyme from Human Brain. Europ J Biochem 114:315-327

75. LEWICKI JA, FALLON JH and PRINTZ MP (1978) Regional distribution of angiotensinogen in rat brain. Brain Res 158:359-371

76. LEWICKI JA (1979) Angiotensinogen in Central Compartments: Characterization, Distribution and Relationship to Plasma Prohormone. Doctoral Dissertation, University of California, San Diego

77. LLORENS C and SCHWARTZ J-C (1981) Enkephalinase Activity in Rat Peripheral Organs. Europ J Pharmacol 69:113-116

78. MALFROY B, SWERTS JP, GUYON A, ROQUES BP and SCHWARTZ J-C (1978) High Affinity Enkephalin-Degrading Peptidase in Brain is Increased after Morphine. Nature 276:523-526

79. MANGIAPANE ML and SIMPSON JB (1980) Subfornical Organ Lesions Reduce the Pressor Effect of Systematic Angiotensin II. Neuroendocrinology 31:380-384

80. McGEER EG and SINGH EA (1979) Inhibition of Angiotensin Converting Enzyme by Substance P. Neurosci Lett 14:105-108

81. MICHELAKIS AM, YOSHIDA H, MENZIE J, MURAKAMI K, INAGAMI T (1974) A radioimmunoassay for the direct measurement of renin in mice and its application to submaxillary gland and kidney studies. Endocrinology 94:1101-1105

82. MISELIS RR, SHAPIRO RB and HAND PJ (1979) Subfornical Organ Efferents to Neural Systems for Control of Body Water. Science 205:1022-1025

83. MISONO KS, INAGAMI T (1980) Characterization of the active site of mouse submaxillary gland renin. Biochemistry 19:2616- 2622

84. MENZIE JW, HOFFMAN LH, MICHELAKIS AM (1978) Immunofluorescent localization of renin in mouse submaxillary gland and kidney. Am J Physiol 234:E480-E483

85. MORGAN JM and ROUTTENBERG A (1977) Angiotensin Injected into the Neostriatum After Learning Disrupts Retention Performance. Science 196:87-89

86. MORRIS BJ and REID IA (1978) The distribution of angiotensinogen in dog brain studied by cell fractionation. Endocrinology 103:492-500

87. NAHMOD VE, FINKIELMAN S, BENARROCH E and PIROLA CJ (1978) Angiotensin Regulates Release and Synthesis of Serotonin in Brain. Science 202:1091-1093

88. NARUSE K, TAKII Y, INAGAMI T (1982) Immunohistochemical localization of renin in luteinizing hormone containing cells of rat pituitary. Proc Nat Acad Sci USA (in press)

89. OSMAN MY, SMEBY RR, SEN S (1979) Separation of dog brain renin-like activity from acid protease activity. Hypertension 1:53-60

90. PHILLIPS MI (1978) Angiotensin in the Brain. Neuroendocrinology 25:354-377

91. PHILLIPS MI and FELIX D (1976) Specific Angiotensin II Receptive Neurons in the Cat Subfornical Organ. Brain Res 109:531-540

92. PHILLIPS MI and HOFFMAN WE (1976) Sensitive Sites in the Brain for the Blood Pressure and Drinking Responses to Angiotensin II. In: International Symposium on the Central Actions of Angiotensin and Related Hormones, BUCKLEY JP and FERRARIO C (eds), Pergamon Press, New York, p 325-356

93. PHILLIPS MI, FELIX D, HOFFMAN WE and GANTEN D (1977) Angiotensin-Sensitive Sites in the Brain Ventricular System: In: Society for Neuroscience Symposia, COWAN WM and FERRENDELLI JA (eds), 2:308-339

94. PHILLIPS MI, WEYHENMEYER J, FELIX D, GANTEN D, HOFFMAN WE (1979) Evidence for an endogenous brain renin-angiotensin system. Fed Proc 38:2260-2266

95. PHILLIPS MI, QUINLAN JT and WEYHENMEYER J (1980) An Angiotensin-Like Peptide in the Brain. Life Sci 27:2589-2594

96. POISNER AM, WOOD GW, POISNER R, INAGAMI T (1982) Localization of renin in trophoblasts in human chorion leave at term pregnancy. Endocrinoloy (in press)

97. PRINTZ MP and LEWICKI JA (1977) Renin substrate in the central nervous system: Potential significance to central regulatory mechanisms. In: BUCKLEY JP and FERRATIO CM (eds), Central Actions of Angiotensin and Related Hormones. Pergamon Press, New York, p 57-64

98. PRINTZ MP, PRINTZ JM and CHANGARIS DA (1977) Identification and isolation of angiotensinogen from human cerebroventricular fluid and evidence of renin activity. Fed Proc 36:1015 (Abstr.)

99. PRINTZ MP, PRINTZ JM and GREGORY TJ (1978) Identification of angiotensinogen in animal brain homogenates. Circ Res Suppl I, 43:21-27

100..PRINTZ MP, LEWICKI JA and WALLIS CJ (1980) Brain angiotensinogen: Origin and evidence for an influence by adrenal corticosteroids. In: GROSS F and VOGEL G (eds), Raven Press, New York, p 193-207

101. PRINTZ MP and GREGORY TJ (1981) Brain angiotensinogen: Evidence for an independent and functional central angiotensin system. In: BUCKLEY JP and FERRARIO CM (eds), Central Nervous System Mechanisms in Hypertension, p 311-326

102. REID IA and RAMSAY DJ (1975) The effects of intracerebroventricular administration of renin on drinking and blood pressure. Endocrinology 97:536-542

103. REID IA (1976) Interaction between angiotensin II and the central nervous system in the regulation of arterial pressure. In: Proc 5th Int Congr Endocr Vol I, p 34-39

104. REID IA and DAY RP (1977) Interactions and properties of some components of the renin-angiotensin system in brain. In: BUCKLEY JP and FERRARIO CM (eds), Central Actions of Angiotensin and Related Hormones, Pergamon Press, New York, p 267-282

105. REID IA (1977) Is there a brain renin-angiotensin system? Circ Res 41:147-153

106. REID IA, MOFFAT B and MORRIS BJ (1978) Two forms of angiotensinogen in dog cerebrospinal fluid. ICRS Med Sci 6:383

107. REID IA (1979) The Brain Renin-Angiotensin System: A Critical Analysis. Fed. Proc 38:2255-2259

108. RETTIG R, SPECK G, SIMON W, SCHELLING P, FAHRER A and GANTEN D (1978) In vivo enzyme activity of purified human brain renin. Klin Wochenschr 56 (Suppl I): 43-45

109. RIX E, GANTEN D, SCHÜLL B, UNGER TH and TAUGNER R (1981) Converting enzyme in the Chorioid Plexus, Brain and Kidney. Immunocytochemical and Biochemical Studies in Rats. Neurosci Lett 22:125-130

110. SANDER GE, LORENZ PE and VERMA PS (1980) Inhibition of Partially Purified Canine Lung Angiotensin I Converting Enzyme by Opioid Peptides. Biochem Pharmacol 29:3115-3118

111. SCHAZ K, SPECK G, SIMON W, SCHLÜR KH, UNGER TH, ROCKHOLD R and GANTEN D (1980) Enkephalin Effects on Blood Pressure, Heart Rate and Baroreceptor Reflex. Hypertension 2:395-407

112. SCHELLING P, GANTEN U, SPONER G, UNGER TH and GANTEN D (1980) Components of the renin-angiotensin system in the cerebrospinal fluid of rats and dogs with special consideration of the origin and the fate of angiotensin II. Neuroendocrinology 31:297-308

113. SCHELLING P, SPECK G, UNGER TH and GANTEN D (1980) The brain angiotensin system: Biochemistry, localization and functional aspects. In: PARVEZ H and PARVEZ S (eds), Advances in Experimental Medicine: A centenary Tribute to Claude Bernard. Elsevier/North Holland, p 243-288

114. SCHELLING P, FELIX D and LIARD JF (1982) Angiotensinogen in cerebrospinal fluid and plasma: Some regulatory aspects. In: Experimental Brain Research. The renin-angiotensin system: a model for the synthesis of peptides in the brain. GANTEN D, PRINTZ MP, PHILLIPS MI, SCHÖLKENS B (eds), Springer Verlag Heidelberg, Suppl 4

115. SEPULVEDA P, MARCINIZYN J jr, LIN D and TANG J (1975) Primary structure of porcine pepsin. J Biol Chem 250:5082-5088

116. SERNIA C and REID IA (1980) Release of angiotensinogen by rat brain in vitro. Brain Res 192:217-225

117. SEVERS WB, SUMMY-LONG J, TAYLOR JS and CONNOR JD (1970) A Central Effect of Angiotensin: Release of Pituitary Pressor Material. J Pharmacol Exp Ther 174: 27-34

118. SMEBY RR, HUSAIN A and SPETH RC (1982) Properties and subcellular localization of brain renin and its substrate. In: Experimental Brain Research. The renin-angiotensin system: a model for the synthesis of peptides in the brain. GANTEN D, PRINTZ MP, PHILLIPS MI, SCHÖLKENS B (eds), Springer Verlag, Heidelberg, Suppl 4

119. SHARE L (1979) Interrelations Between Vasopressin and the Renin Angiotensin System. Fed Proc 38:2267-2271

120. SIRETT NE, McLEAN AS, BRAY JJ and HUBBARD JI (1977) Distribution of Angiotensin II Receptors in Rat Brain. Brain Res 122:299-312

121. SKEGGS LT, KAHN JR and SHUMWAY NP (1956) Preparation and Function of the Hypertensin Converting Enzyme. J Exp Med 103:295-299

122. SKEGGS LT, LENTZ KE, KAHN JR, DORER FE and LEVINE M (1969) Pseudorenin: A new angiotensin-forming enzyme. Circ Res 25:451-462

123. SLATER EE, DEFONDINI R and ZIMMERMANN EA (1980) Wide distribution of immunoreactive renin in nerve cells of human brain. Proc Nat Acad Sci 77:5458-5460

124. SPECK G, POULSEN K, UNGER TH, RETTIG R, BAYER C, SCHÖLKENS B and GANTEN D (1981) In vivo activity of purified mouse brain renin. Brain Res 219:371-384

125. STAMLER JF, BRODY MJ and PHILLIPS MI (1980) The Central and Peripheral Effects of Captopril (SQ 14,225) on the Arterial Pressure of the Spontaneously Hypertensive Rat. Brain Res 186:499-503

126. STINE SM, YANG H-Y and COSTA E (1980) Inhibition of In Situ Metabolism of ^{3}H (met^5)-Enkephalin and Potentiation of (met^5)-Enkephalin Analgesia by Captopril. Brain Res 188:295-299

127. SUMMY-LONG JY, KEIL LC, DEEN K and SEVERS WB (1981) Opiate Regulation of Angiotensin-Induced Drinking and Vasopressin Release. J Pharmacol Exp Ther 217: 630-637

128. SWERTS JP, PERDRISOT R, MALFROY B and SCHWARTZ J-C (1979a) Is "Enkephalinase" Identical with "Angiotensin Converting Enyzme "? Europ J Pharmacol 53:209-210

129. SWERTS JP, PERDRISOT R, PATEY G, de la BAUME S and SCHWARTZ J-C (1979b) "Enkephalinase" is Distinct from Brain "Angiotensin Converting Enyzme". Europ J Pharmacol 57:279-281

130. UNGER TH, ROCKHOLD RW, YUKIMURA T, RETTIG R, RASCHER W and GANTEN D (1980) The Possible Role of Kinins and Substance P in Blood Pressure Regulation of Normotensive and Spontaneously Hypertensive Rats. In: Proceedings of the Symposium: Central Nervous System Mechanisms in Hypertension, Houston, Texas, Raven Press

131. UNGER TH, HÜBNER D, SCHÖLL B, YUKIMURA T, RASCHER W, LANG RE and GANTEN D Dissociation between in vivo and in vitro measurements of converting enzyme activity following chronic oral treatment with captopril in rats. Am J Cardiol in press

132. VanHOUTTEN M, SCHIFFRIN EL, MANN JFE, POSNER BI and BOUCHER R (1980) Radioautographic Localization of Specific Binding Sites for Blood-Borne Angiotensin II in the Rat Brain. Brain Res 186:480-485

133. WALLIS CJ and PRINTZ MP (1980) Adrenal regulation of regional brain angiotensinogen content. Endocrinology 106:337-342

134. WALLIS CJ, GREGORY TJ and PRINTZ MP (1981) submitted

135. YANG H-Y T and NEFF NH (1972) Distribution and Properties of Angiotensin Converting Enzyme of Rat Brain. J Neurochem 19:2243-2250

Pituitary Renin: Purification and Characterization

S. Hirose, T. Ohsawa, K. Murakami, and T. Inagami

INTRODUCTION

Centrally administered angiotensin II (ANG II) elicits multiple physiological responses directed to blood pressure elevation as well as positive fluid and electrolyte balance (35). For example, it increases adrenergic activity with a consequent increase in cardiac output and total peripheral resistance (41), dipsogenesis (10), increased sodium appetite (1) and stimulation of vasopressin release (17,29).

Since these responses are mediated by the central nervous system (35), the observation of renin-like activity in the brain of the dog and rat by Ganten et al. (15) and Fischer-Ferraro et al. (9) suggested the presence of an endogenous angiotensin generating system in the brain (14,36). However, subsequent studies (38,39) raised doubts as to whether the renin-like activity demonstrated in brain extracts was really due to true renin since it displayed a low pH optimum of about 4.5 with little activity above pH 6.0. Indeed, Day and Reid (6) and Hackenthal et al. (19) showed that most of the renin-like activity in the brain is due to the lysosomal acid proteinase, cathepsin D. Thus, the assignment of a physiological role for renin in the brain awaited the demonstration of the existence of brain renin that is distinct from nonspecific cathepsin D. Before 1978, such a demonstration was difficult because of the co-purifying properties of acid proteinases with renin. Hirose, Yokosawa and Inagami (20), Ganten and Speck (14) and Osman et al. (33), using different methods, achieved separation of renin from cathepsin D, providing the first convincing evidence that renin is present in the brain.

IDENTIFICATION OF RENIN IN RAT BRAIN

Failure to obtain a complete separation of renin from acid proteases by various conventional chromatographic methods led us to design a more specific affinity chromatographic separation method. Making use of the fact that hemoglobin and casein are good substrates for cathepsins while renin has limited interaction with them, agarose-bound hemoglobin or casein was prepared as affinity gels. Casein-agarose affinity gel

Experimental Brain Research, Suppl. 4

worked satisfactorily; as described in Fig. 1 of the chapter by Inagami et al. in this volume, true renin and cathepsin D were clearly separated by this affinity chromatography.

With this method renin concentration in the brain was estimated to be approximately 250 pg ANG I/g wet tissue/h. The possibility remained that the renin activity determined by the above method may simply represent renin associated or entrapped in vascular wall (40). However, later studies eliminated such a possibility (2,13,18,27, 28,42).

ALTERNATIVE METHODS FOR MEASURING RENIN ACTIVITY IN SAMPLES CONTAINING BOTH RENIN AND ACID PROTEINASES

A rapid and simplified method was developed (22) for determining renin activity in crude brain extracts without separating renin from cathepsin D. Several aspects of this procedure deserve comment. The assay was carried out at pH 7.5 to reduce the contribution of acid proteinases and anephric rat plasma was used as the substrate without further purification to make use of plasma inhibitors of cathepsin D (7,26). Under these conditions, the contribution of acid proteinases to the total ANG I-generating activity in bovine pituitary extracts was less than 4%.

A direct radioimmunoassay method for the determination of renin was developed by Hirose, Workman and Inagami (21) and applied to studies of the regional distribution of renin in hog brain. Although the sensitivity of detection by this method is limited, this approach specifically determines renin in the presence of cathepsin D (21).

BRAIN RENIN IN HOG AND OTHER MAMMALS

Application of these methods to hog brain revealed that the hog pituitary contains a relatively large amount of renin (21), 0.2 ng renin/mg protein (20 ng ANG I/mg protein/h). This is more than 1,000 times the concentration of renin detected in the rat brain. Table 1 summarizes the regional distribution of renin in the hog brain determined both by the direct radioimmunoassay and by measuring the enzyme activity. A widespread distribution throughout the entire brain was evident with particularly high concentrations being found in the pineal and the pituitary (22). The renin concentration was highest in the pineal; however, the total amount of renin was largest in the anterior pituitary.

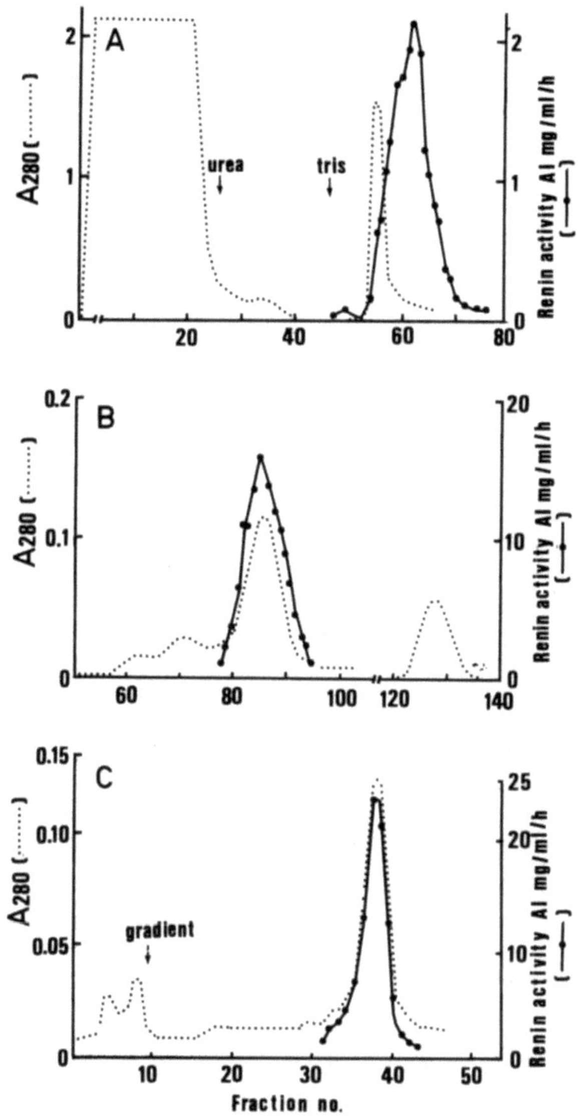

Fig. 1. Chromatography of bovine pituitary renin on pepstatin-aminohexyl-agarose (A), Ultrogel AcA 44 (B), and CM-cellulose (C) columns. Details of the separations are described in the text

Table 1. Renin concentration in various regions in hog brain (22)

Region	Renin activity ng ANG I/mg protein/h	Renin concentration ng/mg protein
Pineal	325	1.96
Adenohypophysis	32.4	0.41
Choroid plexus	21.6	0.07
Hypothalamus	0.482	
Cerebellum	0.312	
Pons	0.174	
Medulla oblongata	0.197	
Spinal cord (intracranial)	0.143	
Cerebral cortex	0.100	

Renin activity was not proportional to the amount of immunoreactive renin as determined by the direct radioimmunoassay. This led to another interesting finding that the brain contains an inactive renin precursor which can be activated by proteolysis (23). We also demonstrated the presence of a renin-binding protein (RBP) or renin inhibitor in the hog brain which complexes renin and renders it inactive (32). In our original report on the RBP, we suggested that RBP might not be a renin inhibitor; however, subsequent careful examination of experimental conditions clearly established that RBP is an endogenous renin inhibitor (43). These results support the idea that renin is synthesized and functions in the brain.

PURIFICATION OF ANTERIOR PITUITARY RENIN

Purification of brain renin is hampered by several factors, including a relatively low concentration in the starting material and co-purification of cathepsin D owing to its remarkable similarity to renin (6,19). Earlier attempts to isolate brain renin using whole brain extracts resulted in only partial purifications: a 70-fold increase in the specific activity of brain renin and a sixfold decrease in that of cathepsin D was obtained from saline-perfused dog brain by Osman et al. (34) using ammonium sulfate fractionation and CM-cellulose chromatography; Ganten et al. (16) achieved a 3,600-fold purification of human brain renin by a combination of gel filtration on Sephadex G-100 and affinity chromatography on pepstatin-Sepharose. Their preparation was virtually free of cathepsin activity and had a specific activity of 560 pmoles ANG I/mg protein. To circumvent those problems mentioned above, we chose anterior pituitary as the starting material which is the richest source of brain renin (22) even if it does not represent neuronal tissues.

Comparison of renin contents in the anterior pituitary of various mammalian species indicated bovine and porcine pituitaries contain higher levels of renin than other species (Table 2). Consideration of supply logistics made us select bovine anterior pituitary as the source organ for the purification.

Table 2. Renin content in anterior pituitaries of various species

Source of pituitary	Renin activity (ng ANG I/mg protein/h)
Bovine (fresh)	45.3
Hog (fresh)	42.7
Rabbit (frozen)	5.2
Sheep (frozen)	1.3
Rat (fresh)	0.7

The following procedure (see below) was developed and found optimal for the purification of pituitary renin after a systematic examination of the conditions of extraction, concentration, affinity chromatography (Fig. 1A), gel filtration (Fig. 1B), and ion exchange chromatography (Fig. 1C). Starting from 50 kg of the bovine anterior pituitary, 0.5 mg of renin was obtained according to the procedure with all steps conducted at $4^{o}C$. The overall purification was 1,670,000-fold and the recovery of activity was 11%. The purified enzyme gave a single protein band on sodium dodecyl sulfate-polyacrylamide gel electrophoresis (SDS-PAGE). The contaminating proteinase activity decreased through the purification, and the final preparation showed no general proteolytic activity.

PROPERTIES

The purified bovine pituitary renin had a specific activity of 146 μmoles ANG I/mg/h at pH 7.5 and $37^{o}C$ against rat angiotensinogen (2.7 μM). The K_M for rat angiotensinogen was approximately 1 μM, a value comparable to those of kidney renins. The molecular weight of enzyme is 36,000, as determined by gel filtration on Ultrogel AcA 44 and SDS-PAGE (Fig. 2B). The enzyme has an isoelectric point of 5.25 (Fig. 2A). The pH optimum for purified renin is between 6.5 and 7.5 with little or no activity displayed below pH 5.0. Amino acid analysis showed a low mole per cent of Cys, Met, and His and a high mole per cent of Leu, Gly, and Glu. The amino acid composition is very similar to those of hog (5,25), rat (31) and dog (8) kidney renins and mouse submaxillary gland renin (4). The ANG I-generating activity of the bovine brain renin was completely neutralized by antibodies against hog renal renin. The pituitary renin failed to bind to five different agarose-bound lectins although all the kidney renins so far examined had an affinity for concanavalin A. These results indicate

Procedure of Purification of Renin from Bovine Anterior Pituitary

Bovine Anterior Pituitary (50 kg)

STEP 1 Freeze-dried and pulverized

Powder (10 kg)

STEP 2 Extracted in 20 mM imidazole/2 mM EDTA/2 mM PMSF/2 mM K-tetrathionate, pH 6.5
Centrifuged at 5,000 x g for 30 min

Crude Extract (140 l)

STEP 3 Absorbed to DE-52, washed with 20 mM acetate (pH 5.5)
Eluted in 0.1 M acetate/0.2 M NaCl, pH 5.0
Precipitated with ammonium sulfate (75%)
Collected by centrifugation
Dialyzed against 20 mM imidazole/0.1 M NaCl, pH 6.5

Extract (15 l)

STEP 4 Affinity Chromatography on a Pepstatin-aminohexyl-agarose Column (4.4 x 11 cm) (Fig. 2A)
Applied to the affinity column, washed with 20 mM imidazole/0.5 M NaCl (pH 6.5) and with 1 M urea
Eluted with 0.5 M tris-HCl, pH 8
Concentrated to 30 ml by ultrafiltration
Dialyzed against 50 mM pyridine, pH 5.5

STEP 5 DEAE-Sephacel Chromatography
Applied to a column (0.5 x 25 cm) of DEAE-Sephacel
Developed by a 0 - 0.2 M NaCl gradient
Concentrated to 1.5 ml using a collodion bag

STEP 6 Gel Filtration (Fig 2B)
Chromatographed on an Ultrogel AcA 44 column (1.5 x 95 cm)
Dialyzed against 20 mM acetate, pH 5.05

STEP 7 CM-Cellulose Chromatography (Fig. 2C)
Absorbed to CM-52 packed in a 0.5 x 25 cm column
Eluted by a linear gradient of NaCl (0 - 0.2 M)

Pure Renin (0.5 mg)

that brain renin is very similar to renal renins in many aspects such as molecular weight, isoelectric point, amino acid composition, chromatographic behavior, and enzyme activity except for the sugar residues.

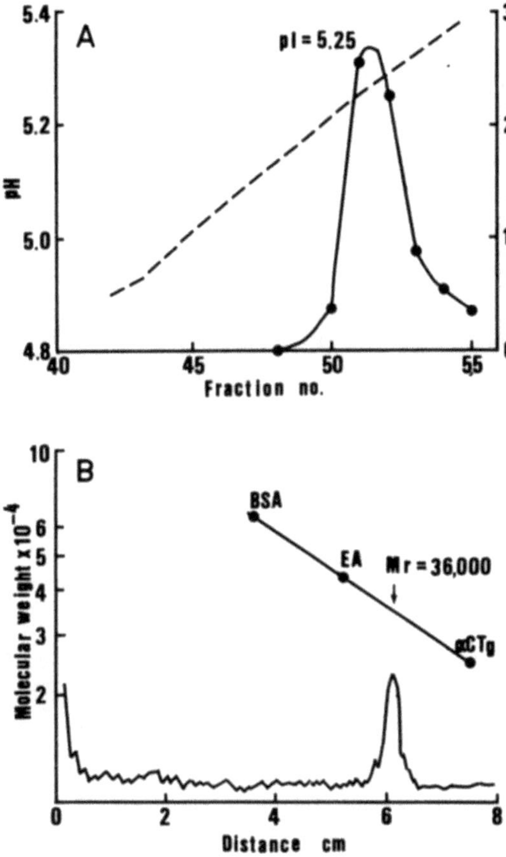

Fig. 2. Electrophoretic and electrofocusing analysis of purified bovine pituitary renin. A, isoelectric focusing profile; B, densitometric scan of a SDS-PAGE gel. BSA, bovine serum albumin (M_r = 68,000); EA, egg albumin (M_r = 45,000); αCTg, α-chymotrypsinogen (M_r = 24,500)

ANTIBODIES TO PITUITARY RENIN

Antibodies against bovine pituitary renin were raised in rabbits by subcutaneous injections of the purified enzyme. Using the immune sera, we compared the antigenic properties of renin from different species and organs. The antibodies cross-reacted with bovine and hog renal renin. Also, a weak reaction was observed with mouse submaxillary gland renin. Antibodies with the ability to neutralize renin activity specifically would be a powerful tool in identifying the renin activity present in the extrarenal tissues such as adrenal medulla, peripheral neurons, and structures in the brain. The adrenal medulla (44) and certain neurons (3,12,36) are the newly discovered sites of action of ANG II. The origin of ANG II acting on these tissues, especially adrenal medulla, remains to be determined.

THE BRAIN RENIN-ANGIOTENSIN SYSTEM: SPECULATIONS ON ITS ROLES

In the foregoing sections, we reviewed evidence from our laboratories that brain contains authentic renin. An inactive renin precursor and a protein inhibitor of renin have also been demonstrated in the brain. These facts together with the ubiquitous distribution of renin throughout the brain indicate that brain renin participates in the regulation of a wide range of physiology through the formation of ANG II. Although a large amount of work remains to be done to establish the physiological roles of brain ANG II, we would like to speculate on its roles and present our current view as a working hypothesis.

PITUITARY, PINEAL, AND CHOROID PLEXUS RENIN

These sites are located outside the blood-brain barrier and renin is found in a relatively high concentration (Table 1). The physiological significance of these renins is not clear. Our speculation is that the renin may participate in the autoregulation of the respective organs acting as a local hormonal system analogous to the intrarenal renin-angiotensin system (30). ANG II constricts the afferent arteriole and glomerular mesangium, and this leads to decreased renal circulation and glomerular filtration (11,37); similar situations might occur in the pituitary, pineal and especially in the choroid plexus by the action of ANG II locally formed. Another possible role of ANG II in the hormone-rich organs is that it might somehow be involved in the control of hormone biosynthesis and secretion, e.g. through nuclear binding. The possibility that renin produces hormones other than ANG II in these organs appears to be unlikely since renin has a highly restricted substrate specificity.

REFERENCES

1. BRYANT RW, FLUHARTY SJ, EPSTEIN AN (1978) Excessive drinking of sodium solutions by rats receiving continuous intracranial infusion of angiotensin. Federation Proc 37:323 (abstract)

2. CELIO MR, CLEMENS DL, INAGAMI T (1980) Renin in anterior pituitary, pineal, and neuronal cells of mouse brain: immunohistochemical localization. Biomedical Res 1:427-431

3. CHANGARIS DG, KEIL LC, SEVERS WB (1978) Angiotensin II immunohistochemistry of the rat brain. Neuroendocrinology 25:257-274

4. COHEN S, TAYLOR JM, MURAKAMI K, MICHELAKIS AM, INAGAMI T (1972) Isolation and characterization of renin-like enzymes from mouse submaxillary glands. Biochemistry 11:4286-4292

5. CORVOL P, DEVAUX C, ITO T, SICARD P, DECLOUX J, MENARD J (1977) Large scale purification of hog renin: physicochemical characterization. Circ Res 41:616-622

6. DAY RP, REID IA (1976) Renin activity in dog brain: enzymological similarity to cathepsin D. Endocrinology 99:93-100

7. DORER F, LENTZ DE, KAHN JR, LEVINE M, SKEGGS LT (1978) A comparison of substrate specificity of cathepsin D and pseudorenin. J Biol Chem 253:3140-3142

8. DZAU VJ, SLATER EE, HABER E (1979) Complete purification of dog renal renin. Biochemistry 18:5224-5228

9. FISCHER-FERRARO C, NAHMOD VE, GOLDSTEIN DJ, FINKIELMAN S (1971) Angiotensin and renin in rat and dog brain. J Exp Med 133:353-361

10. FITZSIMONS JT (1972) Thirst. Physiol Rev 52:408-559

11. FREEMAN RH, DAVIS JO (1979) Physiological actions of angiotensin II on the kidney. Federation Proc 38:2276-2279

12. FUXE K, GANTEN D, HÖKFELT T, BOLME P (1976) Immunohistochemical evidence for the existence of angiotensin II-containing nerve terminals in the brain and spinal cord in the rat. Neurosci Lett 2:229-234

13. FUXE K, GANTEN D, HÖKFELT T, LOCATELLI V, POULSEN K, STOCK G, RIX E, TAUGNER R (1980) Renin-like immunocytochemical activity in the rat and mouse brain. Neurosci Lett 18:245-250

14. GANTEN D, SPECK G (1978) The brain renin-angiotensin system: a model for the synthesis of peptides in the brain. Biochem Pharmacol 17:2379-2389

15. GANTEN D, MINNICH JL, GRANGER P, HAYDUK K, BRECHT HM, BARBEAU A, BOUCHER R, GENEST J (1971) Angiotensin-forming enzyme in brain tissue. Science 173:64-65

16. GANTEN D, SPECK G, MEYER D, LOOS H-E, SCHELLING P, RETTIG R, UNGER Th (1980) Biochemical and functional aspects of brain renin. In: GROSS F, VOGEL G (eds) Enzymatic release of vasoactive peptides. Raven, New York, pp 31-49

17. GREGG CM, MALVIN RL (1978) Localization of central sites of action of angiotensin II on ADH release in vitro. Am J Physiol 234: F 135-140

18. HAAS HL, FELIX D, CELIO MR, INAGAMI T (1980) Angiotensin II in the hippocampus: a histochemical and electrophysiological study. Experentia 36:1394-1395

19. HACKENTHAL E, HACKENTHAL R, HILGENFELDT U (1978) Isorenin, pseudorenin, cathepsin D and renin: a comparative enzymatic study of angiotensin-forming enzymes. Biochim Biophys Acta 522:574-588

20. HIROSE S, YOKOSAWA H, INAGAMI T (1978) Immunochemical identification of renin in rat brain and distinction from acid proteases. Nature 274:392-393

21. HIROSE S, WORKMAN RJ, INAGAMI T (1979) Specific antibody to hog renal renin and its application to the direct radioimmunoassay of renin in various organs. Circ Res 45:275-281

22. HIROSE S, YOKOSAWA H, INAGAMI T, WORKMAN RJ (1980) Renin and prorenin in hog brain: ubiquitous distribution and high concentration in the pituitary and pineal. Brain Res 191:489-499

23. HIROSE S, NARUSE M, OHTSUKI K, INAGAMI T (1981) Totally inactive renin zymogen and different forms of active renin in hog brain tissues. J Biol Chem 256:5572-5576

24. HÖKFELT T, JOHANSSON O, LJUNGDAHL AKE, LUNDBERG JM, SCHULTZBERG M (1980) Peptidergic neurones. Nature 284:515-521

25. INAGAMI T, MURAKAMI K (1977) Pure renin: isolation from hog kidney and characterization. J Biol Chem 252:2978-2983

26. INAGAMI T, YOKOSAWA H, HIROSE S (1978) Definitive evidence for renin in rat brain by affinity chromatographic separation from protease. Clin Sci Mol Med 55: 121s-123s

27. INAGAMI T, CELIO MR, CLEMENS DL, LAU D, TAKII Y, KASSELBERG AG, HIROSE S (1980) Renin in rat and mouse brain: immunohistochemical identification and localization. Clin Sci 59:49s-51s

28. INAGAMI T, CLEMENS DL, CELIO MR, BROWN A, SANDRU L, HERSCHKOWITZ N, HOFFMAN LH, KASSELBERG AG (1980) Immunohistochemical localization of renin in mouse brain. Neurosci Lett 18:91-98

29. KEIL LC, SUMMY-LONG J, SEVERS WB (1975) Release of vasopressin by angiotensin II. Endocrinology 96:1063-1065

30. LEVENS NR, PEACH MJ, CAREY RM (1981) Role of the intrarenal renin-angiotensin system in the control of renal function. Circ Res 48:157-167

31. MATOBA T, MURAKAMI K, INAGAMI T (1978) Rat renin: purification and characterization. Biochim Biophys Acta 526:560-571

32. MURAKAMI K, CHINO S, HIROSE S, HIGAKI J (1980) Specificity and localization of renin-binding proteins. Biomedical Res 1:476-481

33. OSMAN MY, SEN S, SMEBY RR (1979) Separation of renin activity from acid protease activity in brain extracts. Federation Proc 37:354 (abstract)

34. OSMAN MY, SMEBY RR, SEN S (1979) Separation of dog brain renin-like activity from acid protease activity. Hypertension 1:53-60

35. PHILLIPS MI (1978) Angiotensin in the brain. Neuroendocrinology 25:354-377

36. PHILLIPS MI, WEYHENMEYER J, FELIX D, GANTEN D, HOFFMAN WE (1979) Evidence for an endogenous brain renin-angiotensin system. Federation Proc 38:2260-2266

37. PLOTH DW, NAVAR LG (1979) Intrarenal effects of the renin-angiotensin system. Federation Proc 38:2280-2285

38. RAMSAY DJ (1979) The brain renin-angiotensin system: a re-evaluation. Neuroscience 4:313-321

39. REID IA (1977) Is there a brain renin-angiotensin system. Circ Res 41:147-152

40. REID IA (1979) The brain renin-angiotensin system: a critical analysis. Federation Proc 38:2255-2259

41. SCROOP GC, KATIC F, JOY MD, LOWE RD (1971) Importance of central vasomotor effects in angiotensin-induced hypertension. Brit Med J 1:324-326

42. SLATER EE, DEFENDINI R, ZIMMERMAN EA (1980) Wide distribution of immunoreactive renin in nerve cells of human brain. Proc Natl Acad Sci USA 77:5458-5460

43. UENO N, MIYAZAKI H, HIROSE S, MURAKAMI K (1981) A 56,000-dalton renin-binding protein is an endogenous renin inhibitor. J Biol Chem 256: in press

44. ZIMMERMAN BG (1978) Action of angiotensin on adrenergic nerve endings. Federation Proc 37:199-202

Identification, Characterization and Evidence for Intraneuronal Function of Renin in the Brain and Neuroblastoma Cells

T. Inagami, T. Okamura, S. Hirose, D. L. Clemens, M. R. Celio, K. Naruse, Y. Takii, and H. Yokosawa

INTRODUCTION

The presence of all the components of the renin-angiotensin system and angiotensin receptors in the brain and multiple physiological responses elicited by centrally administered angiotensin II are well recognized and point to the presence of a local angiotensin II generating system in the brain. Thus, the observation of renin-like activity ("iso-renin") in the brain by Ganten et al. (8) and Fischer-Ferraro et al. (6) fitted well in this context. However, the acidic pH optimum of its activity and persistent co-purification with acid proteases raised question on the identity of "isorenin" (5,11,27).

Against this background we investigated (i) the identity of the brain renin-like activity (ii), its origin in the brain rather than being blood-borne renin,(iii) its presence and function in neuronal and neuroblastoma cells and (iv) its localization in various brain regions. These studies were directed to the elucidation of the mechanism of angiotensin II formation as a general model of peptide hormone formation in the brain.

NATURE OF RENIN-LIKE ACTIVITY

Against the evidence that renin-like activity of the brain is due to the nonspecific activity of cathepsin D (5,11,27), we have been able to demonstrate the presence of small amounts of true renin by separating it from cathepsin by affinity chromatography on α-casein-Sepharose (14,18). That the angiotensin I generating activity in the first peak (Fig. 1A) was inhibited by anti-renin antibody but lacked protease activity (Fig. 1C) and that its renin activity has a neutral pH optimum were taken as evidence for true renin; whereas the second peak, with a strong acid protease activity but refractory to inhibition by anti-renin antibody was considered as the non-specific action of cathepsin D. Note that the use of partially purified sheep angiotensinogen enormously enhanced this nonspecific activity almost completely overshadowing true renin activity (Fig. 1B) (18). The nonspecific activity is par-

Experimental Brain Research, Suppl. 4

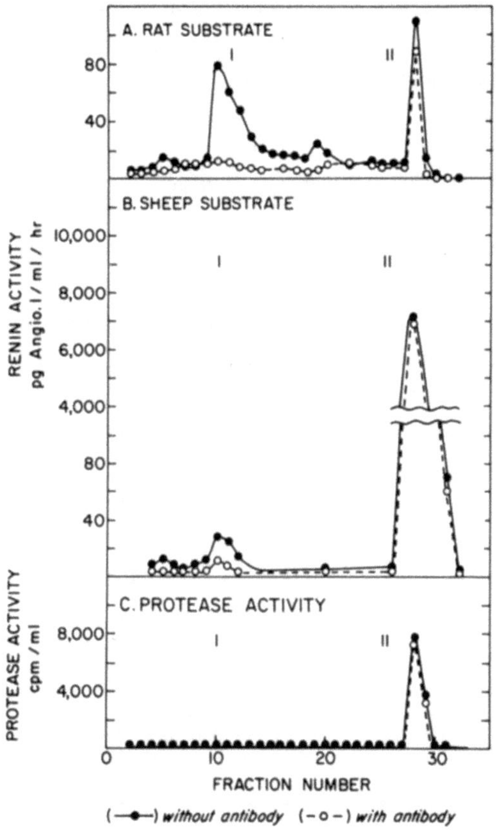

Fig. 1. Separation of renin in rat brain extract by affinity chromatography on α-casein-Sepharose at pH 3.5.Panel (A) Renin-like activity determined using rat plasma as substrate without (●) and with (o) preincubation of the extract with anti-rat renin antibody. Panel (B) The activity was determined with sheep plasma. Panel (C) Protease activity determined using ^{14}C-labeled hemoglobin as substrate. For details of methodology see refs. 14 and 18.(Reproduced with permission of Nature and Clinical Science)

ticularly pronounced at pH 5.5 - 6.5. The separation of brain renin from cathepsin D was also accomplished by pepstatin affinity chromatography by Ganten and Speck (9) and by ion exchange chromatography by Osman et al. (24).

DISTRIBUTION OF RENIN IN VARIOUS REGIONS OF THE BRAIN

Although the affinity chromatography on α-casein-Sepharose indicated the presence of renin in many regions of the brain, in order to overcome the cumbersome chromatographic procedure, a more rapid and simpler method was devised which permitted the specific determination of renin in small amounts of brain tissues. Use of a neutral pH of 7.5 and unfractionated plasma of nephrectomized rats minimized the contribution of cathepsin. Results were checked by a direct radioimmunoassay for renin (13). The activity of true renin was quantified by specific inhibition with anti-renin antibody. Results obtained with the brain of nephrectomized and exhaustively saline-flushed hog (Table 1) indicated wide distribution of renin in practically all regions of the brain including even the olfactory bulb (15).

Table 1. Renin activity, renin concentration and prorenin in various regions of hog brain

	Renin activity ± S.D. ng ANG I/mg protein/h[a] n = 4	Renin conc. ± S.D. ng/mg protein[b] n = 3	Inactive prorenin[c] Active renin
Pineal	325 ± 39	1.96 ± 0.16	0.29
Adenohypophysis	32.4 ± 3.1	0.41 ± 0.04	2.75
Choroid plexus	21.6 ± 2.7	0.07 ± 0.02	0
Hypothalamus	0.482 ± 0.035		0.87
Cerebellum	0.312 ± 0.033		0.03
Amygdaloid nucleus	0.284 ± 0.024		0.16
Pons	0.174 ± 0.040		1.30
Medulla oblongata	0.197 ± 0.038		0.21
Thalamus	0.161 ± 0.035		0.45
Hippocampus	0.164 ± 0.029		0.18
Spinal cord	0.143 ± 0.035		1.53
Cerebral cortex	0.100 ± 0.029		0.37
Caudate nucleus	0.097 ± 0.007		0.35
Mesencephalon	0.048 ± 0.003		1.29
Neurohypophysis	0.017 ± 0.002		0.14
Kidney cortex	18400. ± 1900.	80. ± 12	
Normal plasma	0.714 ± 0.074		
Anephric plasma	0.003 ± 0.0001		

Taken from ref. 15 (with permission by Brain Research).

[a] Renin activity inhibitable by antihog renin antibody which is equivalent to true renin activity. [b] Absolute concentration determined by the direct radioimmunoassay of hog renin according to Hirose, Workman and Inagami, ref. 13. [c] Ratio of inactive prorenin to active renin. Inactive prorenin was determined by the gain of activity upon trypsin treatment

Tissues outside of the blood brain barrier such as the choroid plexus, anterior pituitary and pineal contain renin at markedly elevated concentrations. In contrast, posterior lobe contains very low levels of renin as is in agreement with immunohistochemical studies described below. Various regions of the brain have low but finite levels of renin. Wide distribution at low levels might be interpreted as being due to plasma renin entrapped in tissue homogenates. However, it is definitely higher than that of nephrectomized plasma. A convincing conclusion awaits more definitive evidence as discussed below.

PROPERTIES OF BRAIN RENIN

Overall properties such as molecular weight (36,000) and optimal pH of enzyme activity (pH 6.5 - 7) (Fig. 2) are practically identical with renal or plasma renin. However, significant differences exist in the multicomponent isoelectric focusing patterns (Fig. 3). Compared with generally higher pI values of renal and plasma renin, pituitary and brain renin have lower pI values (15). The structural basis for the multiple peaks and differences in pI values is not clear.

The presence of inactive prorenin was suggested by non-parallel difference in renin activity and renin concentration determined by direct radioimmunoassay (Table 1) (13). The presence of totally inactive prorenin was demonstrated by separating it by pepstatin-Sepharose which has no affinity to inactive prorenin (20). Different tissues seem to have different proportions of prorenin as shown in Table 1. The inactive prorenin of hog pituitary has a molecular weight of 49,000 - 50,000 and pI's of 4.8, 4.9, 5.0. On the other hand active renin in the same tissue has a molecular weight of 40,000 and more acidic pI values of 4.5, 4.7, 4.85 and 5.0 (12). The presence of inactive prorenin and different pI values of brain renin was taken as evidence for the endogenous origin of renin in the brain and neuroendocrine systems.

RENIN IN NEUROBLASTOMA CELLS

A rigorous answer to the question whether renin in the brain is synthesized endogenously or derived from the blood-borne renin of renal origin has been difficult to obtain because we use homogenates of the whole brain tissues which contain blood vessels. Use of brain cells in culture was expected to eliminate most, if not all, of the complicating factors. We have screened a large number of neuroblastoma cell lines with neuronal characteristics for the presence of renin. Among many cloned cell lines of mouse and rat origin tested, many of them listed in Table 2 were shown to contain renin which was inhibited specifically by anti-renin antibody (23).

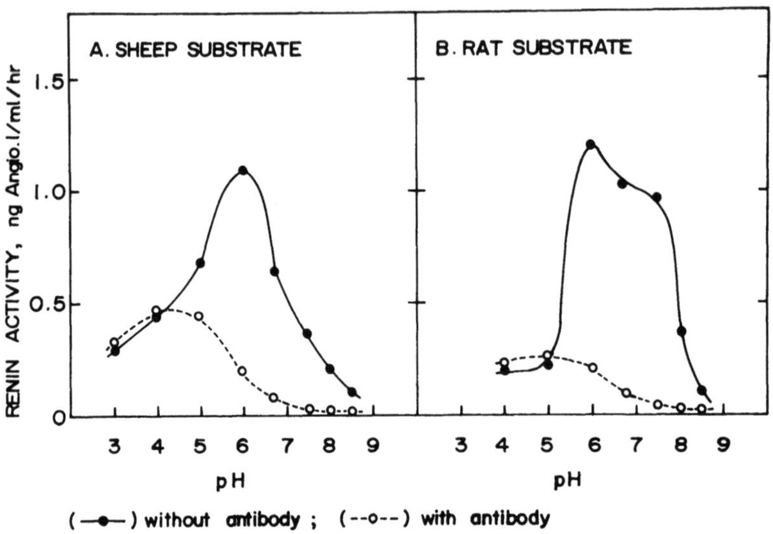

Fig. 2. pH-Dependence of renin activity of the rat brain extract fractionated by affinity chromatography on α-casein-Sepharose (peak I of Fig. 1). The activity was determined before preincubation with anti-rat renin antibody (•) and after preincubation (o). For detailed method see ref. 15 (Reproduced with the permission of Nature)

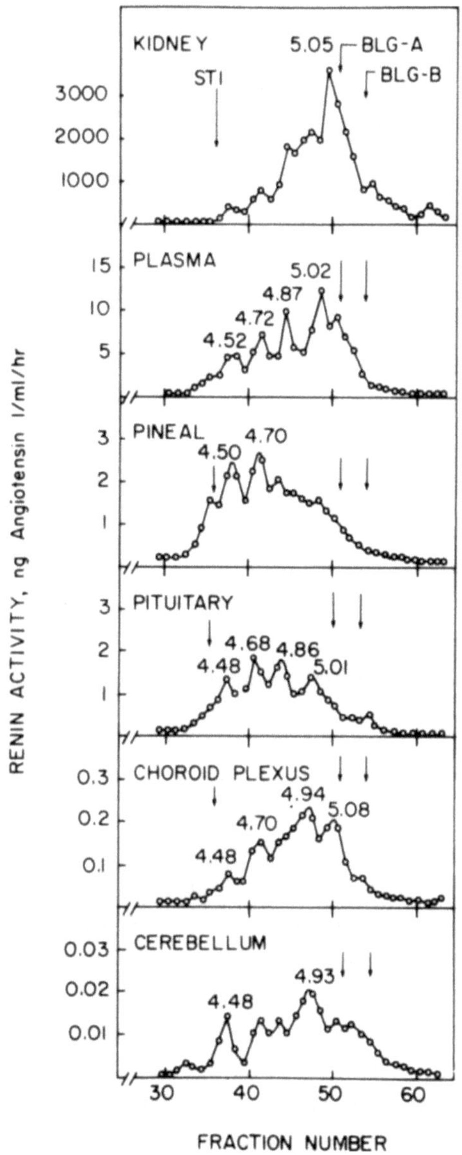

Fig. 3. Isoelectric focusing of renin in various regions of the brain, ref. 15 (Reproduced with permission of Brain Research)

Table 2. Cloned neuroblastoma cells

	Renin pg ANG I/h/ 10^6 cells	Renin % of total Renin-like activity	CE^+ pmol/min/ 10^6 cells	ANG I pg/ 10^6 cells	ANG II/III* pg/ 10^6 cells
Mouse cells					
Neuro-2a	156	93	351	11.8	4.3
N41A3	502	93	762	8.6	4.5
N4TG1	1.3	58	---	4.7	0
Rat cells					
B82	3.6	61	202	9.0	5.2
B50	2.8	43	---	5.3	0
RT4E4	1.9	47	849	29.3	7.2
B103	2.0	0	0	0	0
B65	0	0	---	5.6	1.3

Taken from ref. 23 (with permission of Proc. Nat. Acad. Sci. U.S.A.)
+Angiotensin converting enzyme activity determined by hippuryl-his-leu. The activity was completely inhibited by captopril, specific antibodies and deletion of chloride

*Determined in the absence of captopril

Since small amounts of renin are present in sera used in culture media to support cell growth, it is possible that it may be taken up by cultured cells. However, we were fortunate that one of our anti-mouse renin antibodies had a very high degree of species specificity directed exclusively to mouse renin without crossreactivity with renin from other species (20). Using this antibody we were able to demonstrate almost complete inhibition of renin in mouse neuroblastoma cell lines Neuro-2a and NB41A3. Moreover, renin activity was greatly increased when serum was removed from culture medium. This observation supports the concept of endogenous synthesis of renin. Even if one may argue that neuroblastoma cells are not normal neuronal cells, the finding of renin in these cells can be taken as a strong evidence for the endogenous synthesis of renin in neuronal cells. This discovery represents the first identification of cloned neuronal cell lines which maintain differentiated capability of renin synthesis after repeated cell divisions and passages. These cells will be of great utility for the study of renin synthesis.

Cloned mouse glial cells, C6, and glial cells in 14 day old primary mouse brain cell culture did not contain renin. Therefore, we propose that renin production is a property of neuronal cells and glial cells may possibly perfom a renin disposal function.

SOURCE OF ANGIOTENSINOGEN - A PRELIMINARY OBSERVATION

The source of angiotensinogen used in the formation of angiotensin in the neuroblastoma is not known. Two alternative sources may be considered, namely,intracellular synthesis or endocytosis of extracellular substrate. Neuroblastoma cells, maintained for 5 days in serum-free medium, show markedly increased renin levels. It is also accompanied by a significant level of angiotensins which seem to suggest the intracellular synthesis of angiotensinogen.

IMMUNOHISTOCHEMICAL LOCALIZATION OF RENIN IN THE BRAIN

The brain consists of continuous but functionally differentiated regions without being discretely separated by capsules or membranes. Different types of cells are intermixed in a given region. The specific identification of renin containing cells is possible only by immunohistochemical methods. We have purified renins from mouse submaxillary gland and rat kidney to states which satisfy multiple criteria of purity (4,19) and raised antibodies to these renin preparations. The specificity of these antibodies was demonstrated by the lack of their crossreaction with human renin, cathepsin D from rat and mouse kidneys and by single bands in double diffusion on agar plates.

Rat and mouse brains were perfused with cold saline, fixed with cold Bouin's solution or phosphate-buffered 4% paraformaldehyde, dehydrated in graded alcohols, cleared in xylene and embedded in paraffin. Sections were cut 6 μm thick and stained by the unconjugated peroxidase method of Sternberger (31).

Wide distribution of renin positive cells was observed. Staining was mostly in neuronal cells. An earlier report on the staining of glial cells (17) is being reinvestigated for possible error in classification of cells or staining of degradation product of renin in glial cells. Some renin positive cells showed granular staining in perinuclear regions (3,16). Among the many renin positive cells were cerebellar Purkinje cells, deep cerebellar nucleus, medial cerebellar nucleus, many nuclei in different regions in the brain stem, including nucleus tractus solitarii, raphe nucleus, and locus coeruleus. Nuclei in the septal region, thalamus and hypothalamus, which include supraoptic nucleus, suprachiasmic nucleus and paraventricular nucleus were also renin positive. Pyramidal cells of hippocampus and some cells in the cortex and olfactory bulb were stained. The pineal gland and adenohypophysis were also positively stained. Renin was localized in LH-gonadotroph (unpublished observation). Choroid plexus also showed strong staining. On the other hand the two regions, area postrema and subfornical organ well recognized for their role in mediating pressor effect of blood-borne angiotensin II were not stained by antibody to renin. Intraneuronal localization seems to eliminate the possibility that renin in the brain is the entrapped blood-borne renin. Although it is not possible to determine whether it is synthesized in neuronal cells by immunohistochemistry alone, the granular staining in perinuclear regions strongly suggest intraneuronal synthesis. This conclusion was further supported by the observation of renin in cultured neuroblastoma cells as discussed in the previous section. Intraneuronal staining of renin has been observed independently by Fuxe et al. (7) and Slater et al. (30). However, wide disagreements exist among these authors concerning the exact localization of renin containing cells.

INTRANEURONAL FUNCTION OF RENIN

It has not been clear whether renin exhibits its angiotensin generating activity in the neuron or is released into the interstitium to function in extracellular fluid in a manner similar to renin in plasma.

We have observed that renin containing neuroblastoma cells also contain angiotensins as well as angiotensin converting enzyme (Table 2) (23). Angiotensin converting enzyme activity determined by using hippuryl-His-Leu as substrate was distinguished from enkephalinase, another dipeptidyl carboxypeptidase, by its strong chloride ion dependence, inhibition by captopril and inhibition by specific antiserum to lung converting enzyme. Angiotensins were determined after extraction and chromatography by cellulose thin layer chromatography. When cells were lysed in the presence of captopril, angiotensin I was the predominant form whereas angiotensin II/III became the predominant form when the cells were lysed in the absence of captopril or EDTA.

These neuroblastoma cells possess highly differentiated neuronal characteristics and can be considered as models of neuronal cells (1,28,29,32). The results discussed above indicate that renin containing neuronal cells are equipped with all the enzymes required for the formation of angiotensin II. Thus, it is highly likely that an intracellular mechanism of angiotensin formation exists in certain neuronal cells. This is in contrast to the extracellular mechanism in the plasma. However, we now have evidence that the intracellular mechanism may not be an exceptional one. We have recently reported that an intracellular angiotensin is present in the juxtaglomerular cells of the kidney (2). Both renin and angiotensin II are localized in gonadotrophs of the adenohypophysis, supraoptic nucleus, paraventricular nucleus, and other neurons. It is likely that angiotensin II is formed intracellularly. Thus, angiotensin seems to be produced by dual mechanisms, an intracellular mechanism compatible with hormone production in other endocrine glands and by an extracellular mechanism. The latter may be considered an exception. This example suggests that the various peptide hormones localized in various neuronal cells may have been formed by similar intracellular mechanisms catalyzed by a highly specific enzyme.

If all the components of the renin-angiotensin system are present in neuronal cells, one may wish to know the mechanism which would regulate the rate of angiotensin formation. We found that both renin and angiotensin I-converting enzyme are largely in membrane bound forms. Thus, compartmentalization of these components seems to regulate the rate of angiotensin formation. It is also likely that an endogenous renin inhibitor may function as a regulator of brain renin (unpublished results).

THE TYPE OF RENIN CONTAINING CELLS AND ROLE OF ANGIOTENSIN II

The immunohistochemical staining of renin and angiotensin II and studies on neuroblastoma cells indicated that renin is present in neuronal cells and these neuronal cells are equipped with enzymes for the formation of angiotensin II. A large variety of neuronal cells seem to contain renin. Only a few of them seem to be well characterized with respect to their neurotransmitters and electrophysiological responses.

Cerebellar Purkinje cells and neuronal cells in deep cerebellar nuclei and medial fastigial nuclei are well known to contain gamma aminobutyric acid (GABA) as inhibitory transmitter (25). Thus, some of these well characterized renin containing cells are GABAergic neurons. Electrophysiological studies on hippocampal pyramidal cells have shown that angiotensin functions to disinhibit the response of these neurons to electrical stimulation (10). It is quite likely that cerebellar Purkinje cells and Deiters cells may be affected by angiotensin II in a similar way.

This finding adds a new facet to an already complex body of information concerning the electrophysiological mechanism of angiotensin II action on neuronal cells which include both excitatory (26) and inhibitory mechanisms (29).

CONCLUDING REMARKS

The existence of specific renin in the brain and adjacent neuroendocrine tissues is now established. Wide distribution of not only renin but also other components of the renin-angiotensin system indicate central functions of angiotensin II which may be of general physiological significance. Differences of the isoelectric points of brain renin and renal renin and the presence of inactive prorenin in the brain indicate the endogenous synthesis of renin. The coexistence of renin and angiotensin II in certain neuronal cells and neuroblastoma cells indicates intracellular mechanism of angiotensin formation in contrast to the extracellular mechanism in plasma. Although a preliminary observation suggests an intracellular source of angiotensinogen, the origin of brain angiotensinogen in the intracellular formation of angiotensin II is yet to be established. Also the exact control mechanism of angiotensin II formation remains to be clarified.

ACKNOWLEDGEMENT

This work was supported by NIH Research Grants HL-24112 and HL-14192 and by the Schweizerische Stiftung für Kardiologie, Zürcher Hochschulverein and Hartman Müller-Stiftung.

REFERENCES

1. AUGUSTIO-TOCCO G, SATO G (1969) Establishment of functional clonal lines of neurons from mouse neuroblastoma. Proc Nat Acad Sci USA 64:311-315

2. CELIO MR, INAGAMI T (1981) Angiotensin II immunoreactivity coexists with renin in the juxtaglomerular granular cells of the kidney. Proc Nat Acad Sci USA 78:3897-3900

3. CELIO MR, CLEMENS DL, INAGAMI T (1980) Renin in anterior pituitary, pineal and neuronal cells of mouse brain: immunohistochemical localization. Biomedical Res 1: 427-431

4. COHEN S, TAYLOR JM, MURAKAMI K, MICHELAKIS AM, INAGAMI T (1972) Isolation and characterization of renin-like enzymes from mouse submaxillary glands. Biochemistry 11:4286-4292

5. DAY RP, REID IA (1976) Renin activity in dog brain: enzymological similarity to cathepsin D. Endocrinology 99:93-100

6. FISCHER-FERRARO C, NAHMOD VE, GOLDSTEIN DJ, FINKIELMAN S (1971) Angiotensin and renin in rat and dog brain. J Exp Med 133:353-361

7. FUXE K, GANTEN D, HÖKFELT T, LOCATELLI V, POULSEN K, STOCK G, RIX E, TAUGNER R (1980) Renin-like immunocytochemical activity in rat and mouse brain. Neurosci Lett 18:245-250

8. GANTEN D, MARQUEZ-JULIO A, GRANGER P, HAYDUK K, KARSUNKY KP, BOUCHER R, GENEST J (1971) Renin in dog brain. Am J Physiol 221:1733-1737

9. GANTEN D, SPECK G (1978) The brain renin-angiotensin system: A model for the synthesis of peptides in the brain. Biochem Pharmacol 27:2379-2389

10. HAAS H, FELIX D, CELIO MR, INAGAMI T (1980) Angiotensin II stimulates hippocampal response to electrical stimulation. Experientia 36:1394-1395

11. HACKENTHAL E, HACKENTHAL R, HILGENFELDT U (1978) Purification and partial characterization of rat brain acid proteinase (isorenin). Biochim Biophys Acta 522:561-573

12. HIROSE S, NARUSE M, OHTSUKI K, INAGAMI T (1981) Totally inactive renin zymogen and different forms of active renin in hog brain tissues. J Biol Chem 256:5576-5592

13. HIROSE S, WORKMAN RJ, INAGAMI T (1979) Specific antibody to hog renal renin and its application to the direct radioimmunoassay of renin in various organs. Circ Res 45:275-279

14. HIROSE S, YOKOSAWA H, INAGAMI T (1978) Immunochemical identification of renin in rat brain and distinction from acid proteases. Nature 274:392-393

15. HIROSE S, YOKOSAWA H, INAGAMI T, WORKMAN RJ (1980) Renin and prorenin in hog brain: high concentration in the pituitary and pineal. Brain Res 91:489-499

16. INAGAMI T, CELIO MR,CLEMENS DL, LAU D, TAKII Y, KASSELBERG AG, HIROSE S (1980) Renin in rat and mouse brain: immunohistochemical identification and localization. Clin Sci 59:49s-51s

17. INAGAMI T, CLEMENS DL, CELIO MR, BROWN A, SANDRU L, HERSCHKOWITZ N, HOFFMAN LH, KASSELBERG AG (1980) Immunohistochemical localization of renin in mouse brain. Neurosci Lett 18:91-98

18. INAGAMI T, YOKOSAWA H, HIROSE S (1978) Definitive evidence for renin in rat brain by affinity chromatographic separation from protease. Clin Sci Mol Med 55:121s-123s

19. MATOBA T, MURAKAMI K, INAGAMI T (1978) Rat renin: purification and characterization. Biochim Biophys Acta 526:560-571

20. MICHELAKIS AM, YOSHIDA H, MENZIE J, MURAKAMI K, INAGAMI T (1974) A radioimmunoassay for the direct measurement of renin in mice and its application to submaxillary gland and kidney studies. Endocrinology 94:1101-1105

21. MINNA J, GLAZER D, NIRENBERG M (1972) Genetic dissection of neural properties using somatic cell hybrids. Nature New Biology 235:225-231

22. NARUSE K, TAKII Y, INAGAMI T (1982) Immunohistochemical localization of renin in rat pituitary LH gonadotrophs. Proc Nat Acad Sci USA 80

23. OKAMURA T, CLEMENS DL, INAGAMI T (1981) Renin, angiotensin and angiotensin-converting enzyme in neuroblastoma cells: evidence for intracellular formation of angiotensins. Proc Nat Acad Sci USA 79:in press

24. OSMAN MY, SEMBY RR, SEN S (1979) Separation of dog brain renin-like activity from acid protease activity. Hypertension 1:52-60

25. OTSUKA M, OBATA K, MIHATA Y, TANAKA Y (1970) Measurement of gamma-aminobutyric acid in isolated nerve cells of cat central nervous system. J Neurochem 18:287-293

26. PHILLIPS MI, WEYHENMEYER J, FELIX D, GANTEN D, HOFFMAN WE (1979) Evidence for an endogenous brain renin angiotensin system. Fed Proc 38:2260-2266

27. REID IA (1977) Is there a brain renin-angiotensin system? Circ Res 41:147-152

28. SCHUBERT D, HEINEMAN S, CARLISLE W, TARIKAS H, KIMES B, PATRICK J, STEINBACK JH, CULP W, BRANDT BL (1974) Clonal cell lines from the rat central nervous system. Nature 249:224-227

29. SIMMONET G, BIOULAC B, RODRIQUEZ F, VINCENT J (1980) Evidence of a direct action of angiotensin II on neurons in the septum and in the medial preoptic area. Pharm Biochem Behav 13:359-363

30. SLATER EE, DEFENDINI R, ZIMMERMAN E (1980) Wide distribution of immunoreactive renin in nerve cells of human brain. Proc Nat Acad Sci USA 77:5458-5460

31. STERNBERGER LA (1979) Immunocytochemistry, 2nd Ed. WILEY, New York, pp 50-53

32. TOMOZAWA Y, SUEOKA N (1978) In vitro segregation of different cell lines with neuronal and glial properties from a stem cell line of rat neuroblastoma RT 4. Proc Nat Acad Sci USA 75:6305-6309

Isolation and in Vivo Activity of Brain Renin from Mice and Men

G. A. Speck, T. Unger, R. E. Lang, and D. Ganten

INTRODUCTION

The discovery of renin-like enzymatic activity in brain tissue of dogs and rats ten years ago (5,6,7) lead to the postulation of a complete intrinsic brain renin-angiotensin system (6,8).

At this time no pure renin preparation was available and the separation of renin from other enzymes which were able to split angiotensin I (ANG I) from angiotensinogen had not been achieved.

When it was found that some renin preparations from brain tissue exhibited cathepsin D-like activity (4,11), it was suggested that the renin activity in brain could be due to cathepsin D-like enzymes and it was questioned whether these enyzmes would be active under physiological conditions and thus be responsible for brain ANG formation in vivo.

The purification of renin from brain tissue of man (8,9), rat (12,16) and pig (13) as well as the demonstration of their enzymatic specificity and similarity with kidney renin were important steps to solve this problem. We report here the separation of brain renin from cathepsin D-like activity in mouse and human brain tissue by a three step procedure and the inhibition of renin activity by monospecific antibodies raised against submandibular gland renin (20). This antibody also inhibited kidney and submandibular gland renin from mouse which were purified in the same way.

The biological activity of purified brain renin was tested under physiological conditions by injection of the enzyme into the lateral brain ventricles of rats. The ANG I which was split from the angiotensinogen in the cerebrospinal fluid (CSF) is then converted into ANG II by converting enzyme, which is also present in the CSF. The effects of the so formed ANG II on blood pressure and water intake are described.

Experimental Brain Research, Suppl. 4

MATERIAL AND METHODS

Estimation of Renin (E.C. 3.4.23) and Angiotensin

Renin was estimated by incubation with angiotensinogen from plasma of nephrectomized sheep (for human renin) or nephrectomized rats (for mouse renin). The angiotensinogen was partially purified by precipitation with 2.7 M ammonium sulfate, followed by gel filtration on Sephadex G 100 and ion exchange chromatography on Whatman DEAE 52 cellulose.

Samples of 100 µl were incubated at 37°C with 600 µl 0.1 M citrate phosphate buffer pH 7.5 containing 1 mg angiotensinogen. Angiotensinase inhibitors were added in a 50 µl inhibitor cocktail consisting of 12 µmol phenylmethylsulfonylfluoride (PMSF), 16.1 µmol 8-hydroxyquinoline hemisulfate salt and 2.5 µmol dimercaptopropanol. To stop the reaction the incubation mixture or aliquots of it were boiled for 5 min, centrifuged at 10 000 x g for 10 min and ANG I was measured in the supernatant by specific radioimmunoassay (9,24).

If the sample was to be pretreated with antibodies against submandibular gland renin, 50 µl were preincubated for 24 h at 4°C with 50 µl mouse submandibular gland renin antibodies (diluted 1:50) or 50 µl preimmune serum in the same dilution at pH 7.5. Renin activity is expressed as mol ANG I generated per mg protein or volume and hour. The specificity of ANG I formation and measurement was verified by the peptides heat stability, biological effects and high pressure liquid chromatography, using microbondapak C 18 reversed phase columns (see also this volume Hermann et al.).

Estimation of Angiotensinase Activity

Angiotensin degrading enzymes (angiotensinases) were measured by incubation of ^{125}I-labelled ANG I with the sample for 1 h at 37°C and estimation of the remaining intact ANG I-like immunoreactive material by radioimmunoassay (9,24).

Measurement of Acid Protease Activity

Cathepsin D-like acid protease activity was estimated by a modified version of the method of Anson (1). Acid denatured hemoglobin substrate (HbS) was prepared fresh daily from lyophilized, purified ox hemoglobin. Two hundred µl HbS were incubated with 50 µl sample. After an incubation time of 1 h at 37°C the reaction was stopped by addition of 125 µl 0.6 M trichloroacetic acid (TCA). The non precipitated peptides in the supernatant were determined by the method of Lowry et al. (19) using bovine serum albumin (BSA) as standard. Cathepsin D-like acid protease is expressed as mol BSA equivalents generated per hour and mg protein.

Preparation of Acetone Powder

Adult male albino mice (NMRI strain) of 12 - 15 g bodyweight were used. Brains from intact mice or 16 h after nephrectomy were removed, weighed and frozen at -20°C until use. The brain tissue was homogenized with an ultra turrax (Janke & Kunkel, Staufen) under continuous cooling with dry ice in 3 vols. of precooled -80°C acetone. After centrifugation at -20°C for 10 min at 3000 x g, the supernatant was rejected and the acetone insoluble material was resuspended 3 times in 3 vols of cold acetone. The acetone insoluble material was dryed in vacuo at room temperature and stored dry at -20°C until use.

Acetone powder from the submandibular gland and kidney was prepared in the same way. For the preparation of human brain acetone powder human brains from the department of Pathology, not older than 24 h post mortem, were used.

Gel Filtration

The acetone powder was suspended (1:10; w/v) in 0.1 M sodium acetate buffer, pH 5.9, containing 15 mM NaN_3, 25 mM EDTA and 23 μM PMSF, stirred for 1 h and centrifuged at 50,000 x g for 30 min at 4°C.

The supernatant was applied to a Sephadex G 100 column or to an AcA 44 column (5 x 100 cm) and developed with the same buffer. Renin, cathepsin D-like activity, and angiotensinases were estimated in the 10 ml fractions as described above, proteins were monitored at 280 nm. Fractions with the highest renin activity were pooled, concentrated, and dialyzed against 20 mM phosphate buffer pH 7.5, by ultrafiltration on a PM 10 membrane with a TCF 10 Amicon system.

Chromatography on Blue Agarose Column

The dialyzed fractions from gel filtration were further purified by chromatography on Matrex gel blue A (Amicon, Witten), an agarose with the triazinyl dye Cibachron Blue F3GA as a ligand. The column (1.5 x 9 cm) was developed with 20 mM phosphate buffer, pH 7.5. In this buffer, different additives were used: ATP, NAD and NADH, and a linear gradient from 0 to 1 M NaCl followed by 6 M urea.

In the case of human brain renin the peak fractions from the gel filtration were dialyzed against 20 mM phosphate buffer pH 6.5 and the blue agarose column was developed at the beginning with the same buffer till no protein was eluted from the column. Then the elution was continued with the same buffer but at pH 7.5. The fractions with the highest renin activity were pooled, concentrated and dialyzed against water on Amicon filter.

Estimation of Molecular Weights

Protein standards (ovalbumin, chymotrypsinogen A, and myoglobin) for the molecular weight determination by sodium dodecyl sulfate (SDS) gel electrophoresis were purchased from Serva, Heidelberg.

SDS polyacrylamide gel electrophoresis was carried out in glass tubes (10,5 cm x 5 mm inner diameter), the polymerized 7.5% polyacrylamide gel having a lenght of 8 cm. The running conditions were 1 mA/tube at room temperature.

The samples in a volume of 100 μl were denatured by boiling for 5 min with a 25 μl aliquot of 30% sucrose, 4% beta mercaptoethanol and 1 μg bromphenol blue to obtain a final volume of 125 μl. The proteins were fixed in the gel with a solution of 2-propanol/acetic acid/water (25%/10%/65% v/v), stained with coomassie blue (Serva, Heidelberg) and destained with acetic acid/methanol/water (7%/40%/53% v/v). The stained gels were scanned for proteins at 600 nm with a Beckman gel scanning device for the Beckman Model Acta MVI absorbance monitor. The molecular weights were estimated by comparing their Rf values with those of the standard proteins.

Determination of the Isoelectric Points

Carrier ampholytes (Servalyte pH 4 - 6 and pH 4 - 9, Serva, Heidelberg) were used in granulated agarose (Sephadex IEF, Pharmacia Fine Chemicals, Freiburg). The isoelectrical focusing was carried out with the flat bed focusing apparatus from LKB, Karlsruhe. The sample was included in the original gel slurry from which the gel bed was prepared (110 ml total volume), containing 4% carrier ampholytes.

The running conditions were 8 Watts constant power, 6°C cooling temperature, and the focusing time was between 12 - 14 h. The separated zones were collected from the gel plate using a fractionation grid. The gel sections were transferred to small columns and eluted four times with 500 μl destilled water. In these fractions the pH and renin activity were determined.

Animal Experiments

Male Sprague-Dawley rats of 250 - 300 g bodyweight were implanted with a chronic cannula into the lateral brain ventricle (i.c.v.) as described elsewhere (25). Correct placement of the cannula was verified by injection of Evans blue after sacrifice. On the day of testing, a catheter was implanted into the femoral artery under light ether anaesthesia and tunneled under the skin to exit through the scruff of the neck. After recovery from surgery, blood pressure was recorded continuously in the freely moving rats. Enzyme preparations were dialyzed against 0.9% NaCl and injected i.c.v. in vols of 2 - 10 μl via the brain cannula. For CSF collection, the rats were mounted in a stereotaxic apparatus. The membrana atlanto-occipitalis was dissected and a 20-gauge record needle, connected to polyethylene tubing no. 20, was inserted into the fourth brain ventricle with a micromanipulator (Leitz, Wetzlar). CSF was drained by gravity and collected on ice into plastic vials containing angiotensinase inhibitor cocktail. The blood free samples were quickly boiled for 5 min, centrifuged at 3000 x g for 10 min at 4^{0}C, and ANG I and ANG II were determined by radioimmunoassay.

For the drinking experiments, rats were implanted with a brain cannula only. Water intake was determined by weighing the water bowls at time intervals before and after i.c.v. enzyme injection.

RESULTS

The purification data for mouse brain and mouse submandibular gland renin are summarized in Table 1. The mouse tissue acetone powder extracts from brain and submandibular gland had specific activities of 6.35 and 6.39 pmol ANG I/mg protein and h, respectively. The enzymes remained stable for more than two years if stored at -20^{0}C. Gel filtration of mouse brain acetone powder extract on Sephadex G 100 or AcA 44 lead to a separation of renin from the bulk of other proteins and angiotensinases, but not from cathepsin D-like acid proteases. Gel filtration of mouse kidney and mouse submandibular gland acetone powder extract was less effective in separating other proteins, but angiotensinases were also eliminated. Chromatography on blue agarose lead to the separation of renin from cathepsin D-like acid proteases. The acid proteases eluted within the void volume while renin was eluted by the NaCl-gradient. The elution profiles of mouse brain (Fig. 1A), mouse kidney (Fig. 1B), and mouse submandibular gland renin (Fig. 1C) are shown in Fig. 1. Acid protease activity eluted in all cases in the void volume. Nucleotides such as ATP, NAD, and NADH in 1 mM concentrations which generally elute proteins from blue columns were ineffective to elute renin.

The human brain acetone powder extract had a specific activity of 1.05 pmol ANG I/mg protein and h and was stable if stored dry at -20^{0}C for more than 4 years. Gel filtration also separated inactive proteins and angiotensinases from renin as reported for mouse brain renin. The separation from acid proteases could be achieved on a blue agarose column with a saline gradient, but the elution as described for mouse brain renin did not lead to a separation of human brain renin from acid proteases. Therefore a change in pH was carried out before starting the saline gradient. Cathepsin D-like acid proteases eluted in three peaks: in the void volume, after pH change and immediately after the start of the NaCl gradient. Renin also eluted

Table 1. Purification data for mouse brain and submandibular gland renin. Renin activity is expressed in pmol ANG I/mg protein and h. For details see text

Purification step	mouse brain renin	purification	mouse submandibular gland renin	purification
1) Tissue homogenate	0.034	1	30	1
2) Acetone powder extract	6.35	187	6394	206
3) Gel filtration	378	11130	22076	713
4) Ultrafiltration	777	22840		
5) Blue agarose	2530	74460	104609	3382

partially in the void volume and after pH change from 6.5 to 7.5. But with the saline gradient renin eluted later than the cathepsin D-like activity (Fig. 2). Human brain renin was purified 12,000 fold by this method.

The molecular weight as determined by gel filtration was 40,000 dalton. The molecular weights of the mouse tissue renins were 40,000 as determined by SDS polyacrylamide gel electrophoresis (Fig. 3). The influence of pH on ANG I formation of mouse brain, kidney and submandibular gland renin gave pH optima within the physiological pH range at 7.0, 8.5 and 7.0 and 9.0, respectively. The pH optimum for the reaction of human brain renin with sheep angiotensinogen was between 7.5 and 9.5. Differences between mouse tissue renins were also found by isoelectric focusing, as demonstrated in Fig. 4. Mouse brain renin had a single pI at 5.65 (Fig. 4A); mouse submandibular gland renin exhibited two pI at 5.6 and 6.4 (Fig. 4C); mouse kidney renin was heterogenous in this respect the activity being distributed between pH 4 and 8, with the highest activity at pI 5.8 and 6.74 (Fig. 4B).

The biochemical data are summarized in Table 2. All purified renins were free from angiotensinase and converting enzyme activity and were stable at 4^{0}C and after lyophilization. The mouse brain renin, mouse submandibular gland renin and mouse kidney renin could be inhibited by monospecific antibodies against mouse submandibular gland renin. Human brain renin could be inhibited by substrate analogs, as described earlier (8,9).

Renins from mouse and man specifically formed ANG I upon incubation with plasma angiotensinogen and with tetradecapeptide renin substrate.

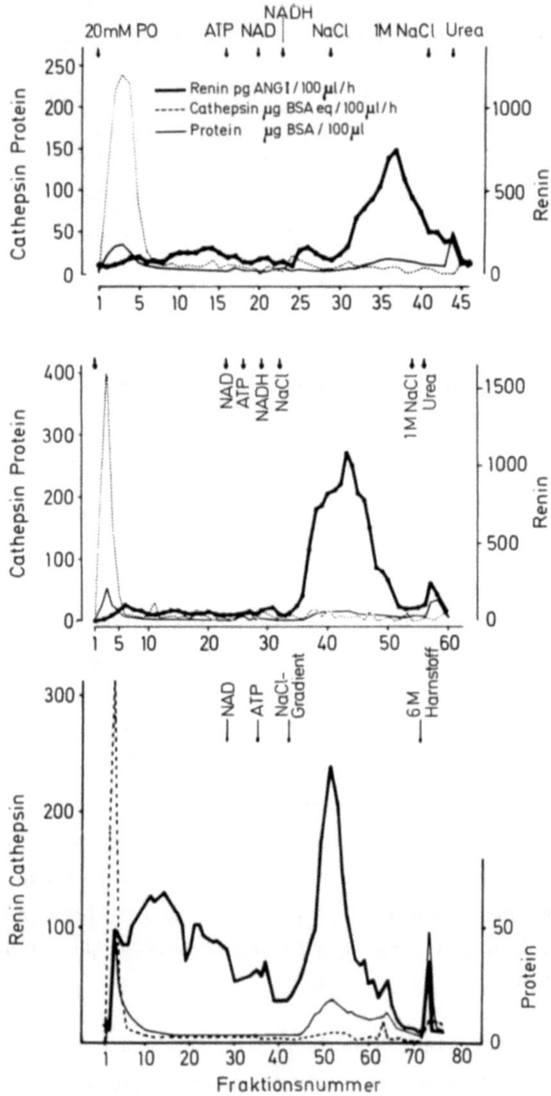

Fig. 1. Separation of renin and cathepsin D-like acid proteases on blue agarose column. (A) indicates mouse brain, (B) mouse kidney, and (C) mouse submandibular gland samples

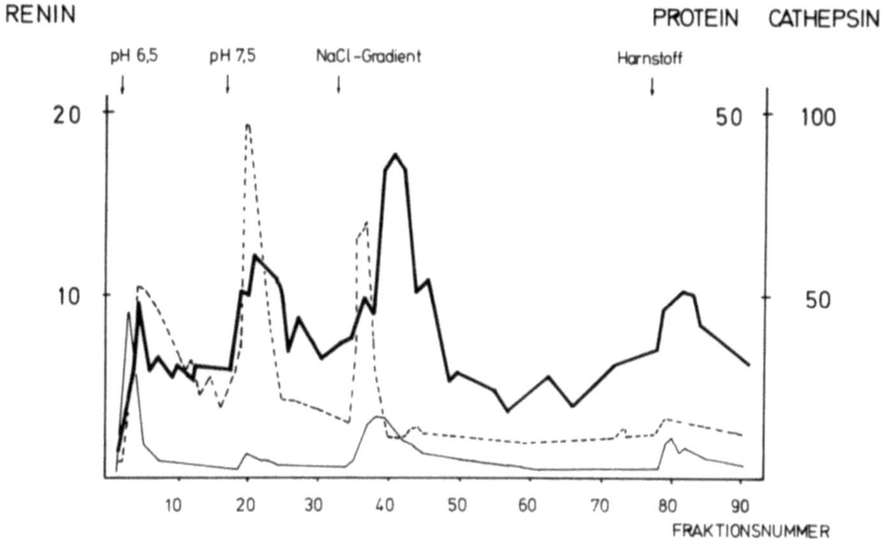

Fig. 2. Purification of human brain renin by chromatography on blue agarose column. Note that the separation of renin from cathepsin D-like proteases occurs with the NaCl gradient

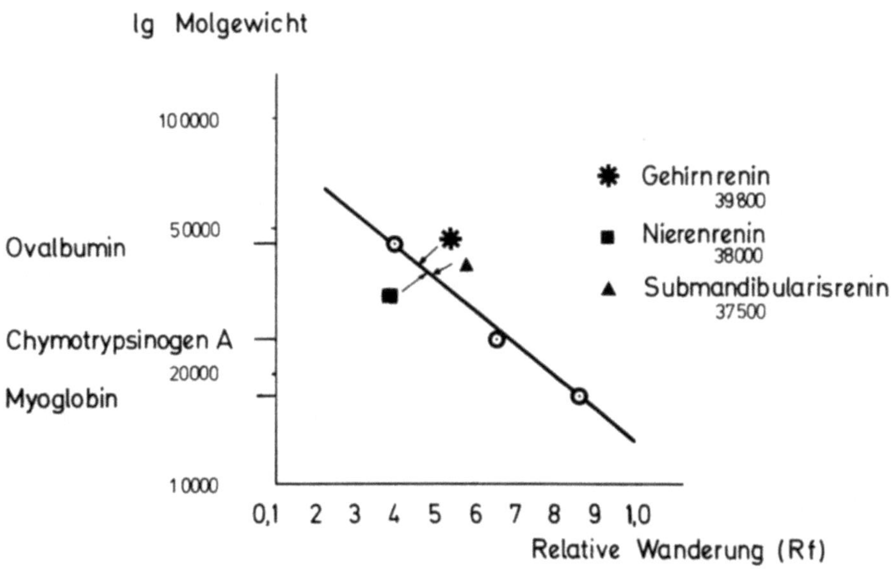

Fig. 3. Molecular weight estimation of mouse brain, kidney and submandibular gland renin by SDS polyacrylamide gel electrophoresis. The abszissa indicates the Rf values, the ordinate the molecular weights in logarithmic scale

Table 2. Biochemical characteristics of the purified renins from mouse and man

Renin source	Molecular weight	pH optimum		Isoelectric point	Specific ANG I formation
Mouse brain	39 800	7.0	(1)	5.65	yes
Mouse kidney	38 000	8.5	(1)	5.8/6.74	yes
Mouse submandibular gland	37 500	7.0/9.0	(1)	5.8/6.4	yes
Human brain	40 000	7.5-9.5	(2)	not determined	yes

(1) Reacted with rat plasma angiotensinogen. (2) Reacted with sheep plasma angiotensinogen

The in vivo activity of different renins was tested by bolus injections of the enzymes into the lateral brain ventricles of rats. After 15 min the CSF was collected and ANG I and ANG II determined by radioimmunoassay. The results after injection of mouse submandibular gland renin, mouse brain renin, human brain renin, hog kidney renin and tonin are demonstrated in Fig. 5. All renins lead to a dramatic increase of the ANG I concentration in the CSF, while tonin did not cause changes in CSF ANG I concentration but the ANG II concentration increased to high levels.

The i.c.v. injection of renin from mouse brain, kidney and submandibular gland as well as of human brain renin lead to a long lasting increase of blood pressure, which could be blocked by either the ANG II receptor antagonist saralasin or the angiotensin converting enzyme inhibitor captopril as reported before (22,23,24). A dose response curve of mean blood pressure increases with different concentrations of renin was constructed with i.c.v. injected mouse kidney renin (Fig. 6). The half maximal blood pressure increase of 43 mmHg was observed after injection of 0.3 μg kidney renin (purification step 5). The data on the in vivo activity of the purified renins are summarized in Table 3.

The i.c.v. injection of purified mouse brain renin stimulated water intake. After an application of 40 μg mouse brain renin from purification step 5, rats drank 22 $\pm$ 2 ml (n:4) within 90 min. This drinking response could be inhibited by saralasin (24).

Fig. 5. ANG I formation after injection of renins into the lateral brain ventricles of rats determined in the CSF 15 min after injection. Note logarithmic scale and that tonin forms ANG II only and no ANG I; levels indicated after tonin injection are endogenous CSF ANG I baseline concentrations ▷

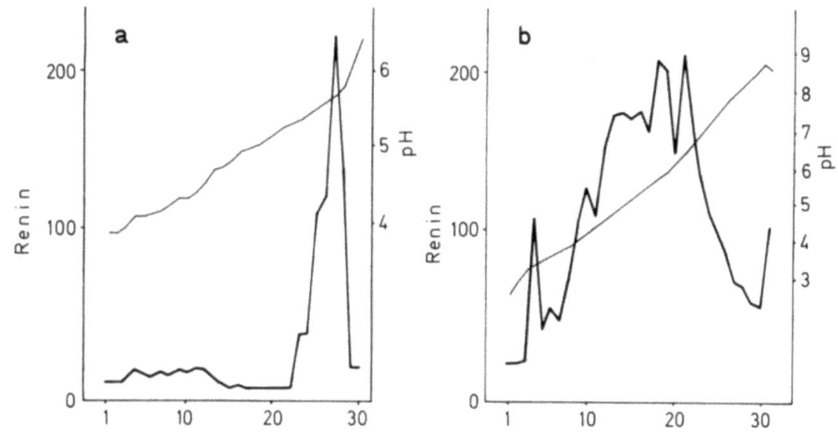

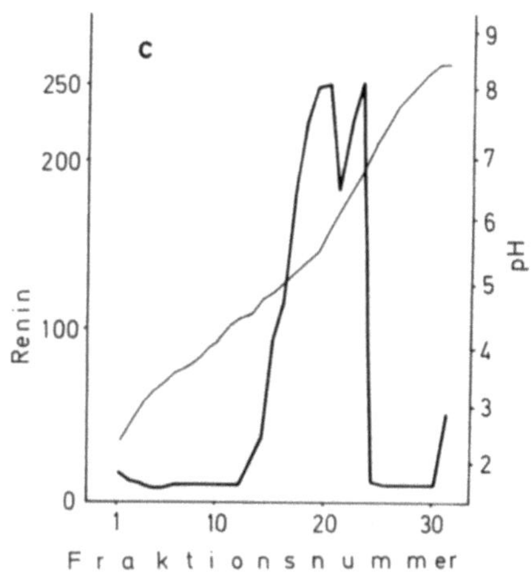

Fig. 4. Isoelectric focusing profiles of mouse brain (A), kidney (B), and submandibular gland renin (C)

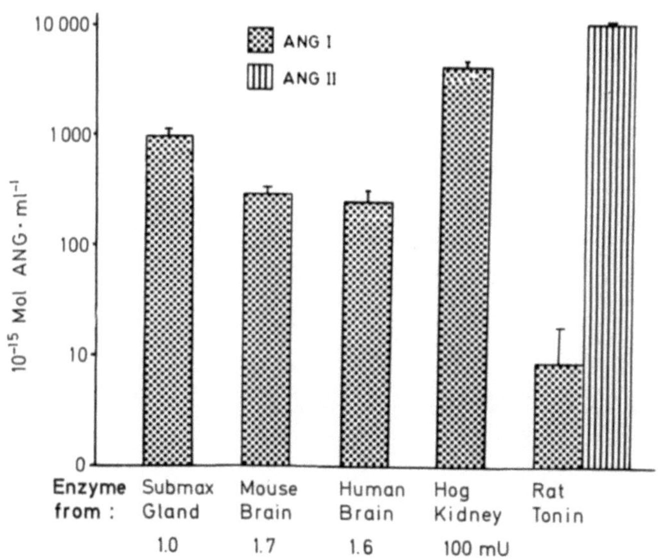

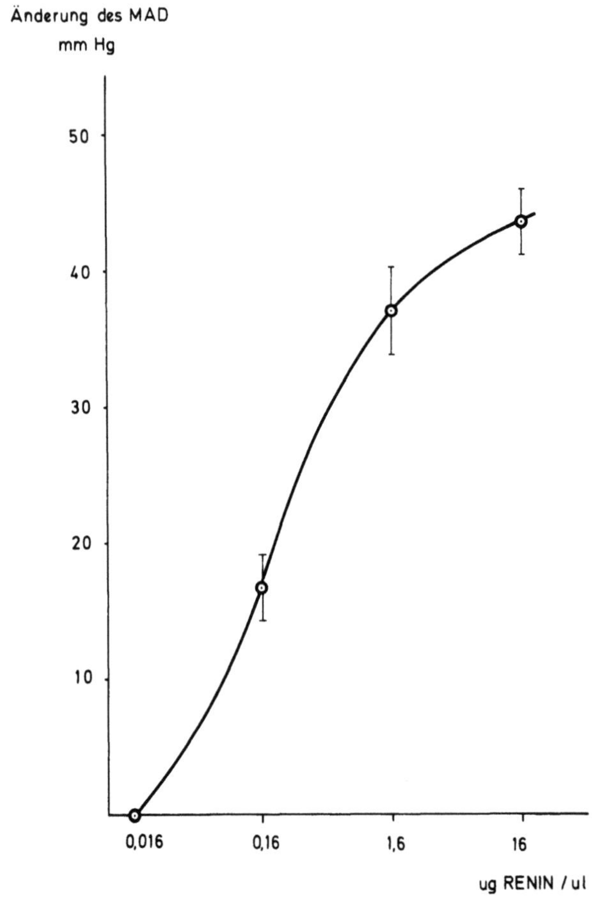

Fig. 6. Dose response curve of mean arterial blood pressure (ordinate) following injections of different concentrations of mouse kidney renin (abszissa) into the lateral brain ventricles of rats

Table 3. Biological effects of purified renins from mouse brain, kidney and submandibular gland and from human brain after injection into the lateral brain ventricles of rats

Renin source	i.c.v. injection of the enzyme caused a blood pressure increase	stimulation of water intake	formation of ANG I in the CSF (1)
Mouse brain	yes	yes	11 ± 2
Mouse kidney	yes	yes	23243 ± 4456
Mouse submandibular gland	yes	yes	533 ± 107
Human brain	yes	yes	9250 ± 1187 (2)

(1) In pmol ANG I/ml CSF after injection of 1 μg enzyme preparations as indicated in Table 1, step 5. (2) See (8,22)

DISCUSSION

The results demonstrate that brain renin is similar to kidney and submandibular gland renin with respect to its biochemical characteristics and its in vivo activity.

We have demonstrated that renin from human brain, mouse brain, kidney and submandibular gland are active under in vivo conditions, raising blood pressure, inducing thirst, and forming ANG I from brain angiotensinogen.

The main problem with the purification of brain renin as compared to kidney or mouse submandibular gland renin is due to the relatively low concentrations of the enzyme in this tissue and the fast degradation by autolytic processes. The considerable increase in brain renin activity in the acetone powder extract could indicate that degrading processes were stopped under this condition, that the enyzme was activated or that an inhibitor was lost by this purification step. Our results with the blue agarose column differ from the experience of other authors with human kidney renin,where only weak binding of active and stronger binding of inactive renin on blue agarose was reported (17,26). In the dog it was found that brain renin did bind to blue agarose (14) but eluted together with cathepsin D-like acid proteases.

In our hands the blue agarose column was a powerful procedure for the separation of renin from cathepsin D-like acid proteases in human brain, mouse brain, mouse kidney and mouse submandibular gland. The fact that nucleotides did not elute renin from the blue agarose column may indicate an unspecific electrostatic binding to the gel rather than a specific interaction with the ligand. The pH optimum of the reaction of renin with angiotensinogen has been reported to be between 4 and 9, depending on species, source and reaction conditions (8,13,23). In this study the pH optima of mouse and human renin were all in the physiological pH range when reacted with purified angiotensinogens from rat and sheep plasma.

We had earlier observed a more acidic pH optimum of another human brain renin preparation but it was nevertheless found to be active under in vivo conditions and to form ANG I in the rat enzyme test paradigm in which cathepsin D was inactive (9,22). The lower pH optimum may have been due to contaminating enzymes of the less pure enzyme preparations or to an alteration in enzyme characteristics following different purification procedures and incubation conditions.

Differences between the three mouse renins following exactly the same purification procedure were found for the pH optima and isoelectric points. Cohen et al. (2) also found two pI for mouse submandibular gland renin at pI 5.4 and 5.62. The pI for kidney renin is reported to be between 4.5 and 5.4 (3,15,18,21). Yokosawa et al. (27) reported a pI at 5.7 for human kidney renin.

The rat in vivo enzyme assay is based on the high specific angiotensinogen concentrations in CSF and the presence of angiotensin converting enzyme and ANG II receptors in the brain. This assay in rat and dog has been described in detail (22,23). Our results clearly indicate that human brain renin as well as renin from mouse brain, mouse submandibular gland, and mouse kidney are capable of ANG I generation under physiological conditions. It is also evident that the biological effects on blood pressure and drinking are caused by angiotensin II generation because they could be inhibited by ANG II receptor blockade with saralasin and by inhibition of the converting enzyme with captopril.

In conclusion we have shown that renin from brain, submandibular gland and kidney can be separated from cathepsin D-like activity by chromatography on blue agarose column. Purified brain renin is physicochemically, immunologically, and biologically similar to kidney and submandibular gland renin. All purified enzymes are active under in vivo conditions and lead to specific ANG I formation, blood pressure increase, and drinking.

SUMMARY

Brain renin from mice and men was purified by a three step procedure: acetone powder extraction, gel filtration, and chromatography on blue agarose. The latter efficiently separated renin from cathepsin D-like acid proteases.

The 75 000-fold purified mouse brain renin had a pH optimum at 7.0 and an iso-electrical point at 5.65. The 12 000-fold purified human brain renin showed its pH optimum between pH 7.5 and 9.5. Brain renin from mouse and man had a molecular weight of 40 000, they were free of angiotensin degrading enzymes and converting enzyme activity.

Mouse brain renin was inhibited by monospecific antibodies raised against mouse submandibular gland renin. Both brain renins from mouse and man were inhibited by synthetic substrate analogs.

The in vivo activity of mouse brain and human brain renin was tested by injection into the lateral brain ventricles of rats. This resulted in the formation of angiotensin I from endogenous angiotensinogen, in a long lasting blood pressure increase and in a stimulation of water intake. These effects are due to the formation of angiotensin II under physiological conditions in vivo and could be blocked by injection of the angiotensin II-receptor antagonist saralasin and also by the converting enzyme inhibitor captopril. Renins purified from mouse submandibular gland and mouse kidney by the same method were also active under in vivo conditions. They stimulated water intake and caused long lasting blood pressure increase.

Brain renin of men and mice thus is similar to kidney renin and is active under in vivo conditions but probably not identical.

ACKNOWLEDGEMENT

We thank Clarlies Bayer and Friedlinde Bles for technical assistance. Dr. Knud Poulsen, Copenhagen, has kindly provided us with antibodies against submandibular gland renin and the late Dr. Roger Boucher, Montreal, supplied us with the pure tonin.

REFERENCES

1. ANSON ML (1937) The estimation of cathepsin with hemoglobin and the partial purification of cathepsin. J Gen Physiol 20:565-574

2. COHEN S, TAYLOR JM, MURAKAMI K, MICHELAKIS AM, INAGAMI T (1972) Isolation and characterization of renin-like enzymes from mouse submaxillary glands. Biochem 11: 4286-4293

3. CORVOL P, DEVAUX C, ITO T, SICARD P, DUCLOUX J, MENARD J (1977) Large purification on hog renin. Physicochemical characterization. Circ Res 41:616-622

4. DAY RP, REID IA (1976) Renin activity in dog brain: enzymological similarity to cathepsin D. Endocrinology 99:93-100

5. FISCHER-FERRARO C, NAHMOD VE, GOLDSTEIN DJ, FINKIELMAN S (1971) Angiotensin and renin in rat and dog brain. J exp Med 133:353-361

6. GANTEN D, MARQUEZ-JULIO A, GRANGER P, HAYDUK K, KARSUNKY KP, BOUCHER R, GENEST J (1971) Renin in dog brain. Am J Physiol 221:1733-1737

7. GANTEN D, MINNICH JL, GRANGER P, HAYDUK K, BRECHT HM, BARBEAU A, BOUCHER R, GENEST J (1971) Angiotensin-forming enzyme in brain tissue. Science 173:64-65

8. GANTEN D, SPECK G (1978) The brain renin-angiotensin system: A model for the synthesis of peptides in the brain. Biochem Pharmacol 27:2379-2389

9. GANTEN D, SPECK G, MEYER D, LOOS HE, SCHELLING P, RETTIG R, UNGER TH (1980) Biochemical and functional aspects of brain renin. In: GROSS F, VOGEL HG (eds) Enzymatic release of vasoactive peptides. Raven Press, New York, pp 31-46

10. GANTEN D, SPECK G, MANN JFE, UNGER TH (1981) Role of the brain renin-angiotensin system in central mechanisms of blood pressure control. In: Frontiers in hypertension research. LARAGH JH, BÜHLER FR, SELDIN DW (eds) Springer, New York Heidelberg Berlin, pp 268-273

11. HACKENTHAL E, HACKENTHAL R, HILGENFELDT U (1978) Purification and partial characterization of rat brain acid proteinase (Isorenin). Biochem Biophys Acta 522:561-573

12. HIROSE S, YOKOSAWA H, INAGAMI T, (1978) Immunochemical identification of renin in rat brain and distinction from acid proteases. Nature 274:392-393

13. HIROSE S, YOKOSAWA H, INAGAMI T, WORKMAN RJ (1980) Renin and prorenin in hog brain: Ubiquitous distribution and high concentration in the pituitary and pineal. Brain Research 191:489-499

14. HUSAIN A, SEN S, SMEBY RR (1980) Affinity chromatography in the purification of dog brain isorenin. Fed Proc 39:369

15. INAGAMI T, MURAKAMI K (1977) Pure renin. Isolation from hog kidney and characterization. J Biol Chem 252:2978-2983

16. INAGAMI T, YOKOSAWA H, HIROSE S (1978) Definitive evidence for renin in rat brain by affinity chromatographic separation from protease. Clin Sci Molec Med 55:121s-123s

17. INAGAMI T, YOKOSAWA N, YOKOSAWA H, TAKII Y (1979) Isolation of human plasma prorenin by affinity chromatography and studies of its activation. Circulation 59/60 (Suppl. II):132

18. LAURITZEN M, DAMSGAARD JJ, RUBIN I, LAURITZEN E (1976) A comparison of the properties of renin isolated from pig and rat kidney. Biochem J 155:317-323

19. LOWRY OH, ROSEBROUGH NJ, FARR AL, RANDALL RJ (1951) Protein measurement with the Folin phenol reagent. J Biol Chem 193:265-275

20. MALLING C, POULSEN K (1977) A direct radioimmunoassay for plasma renin in mice and its evaluation. Biochem Biophys Acta 491:532-541

21. MATOBA T, MURAKAMI K, INAGAMI T (1978) Rat renin:Purification and characterization. Biochem Biophys Acta 526:560-571

22. RETTIG R, SPECK G, SIMON W, SCHELLING P, FAHRER A, GANTEN D (1978) In vivo enzyme activity of purified human brain renin. Klin Wochenschr 56 (Suppl I): 43-45

23. SCHELLING P, SPECK G, UNGER TH, GANTEN D (1980) The brain renin angiotensin system: Biochemistry,localization and functional aspects. In: PARVEZ S, PARVEZ H (eds) Advances in Experimental Medicine. A centenary tribute to CLAUDE BERNARD. Elsevier, Amsterdam, pp 243-288

24. SPECK G, POULSEN K, UNGER TH, RETTIG R, BAYER C, SCHÖLKENS B, GANTEN D (1981) In vivo activity of purified mouse brain renin. Brain research 219:371-384

25. UNGER TH, KAUFMANN-BÜHLER I, SCHÖLKENS B, GANTEN D (1981) Brain converting enzyme inhibition: a possible mechanism for the antihypertensive action of captopril in spontaneously hypertensive rats. European J Pharmacol 70:467-478

26. YOKOSAWA N, TAKAHASHI N, INAGAMI T, PAGE DL (1979) Isolation of complete inactive plasma prorenin and its activation by kallikreins. Biochem Biophys Acta 569: 211-219

27. YOKOSAWA N, HOLLADAY LA, INAGAMI T, HAAS E, MURAKAMI K (1980) Human renal renin: Complete purification and characterization. J Biol Chem 255:3498-3502

Isolation and Regulation of Brain Renin

V. J. Dzau

INTRODUCTION

The simultaneous discoveries of an enzyme in the brain capable of generating angiotensin I from renin substrate by Ganten et al. (13) and Fischer-Ferraro et al. (10) have opened new frontiers in brain research. Since their initial reports, other components of the renin-angiotensin system have been reported to be present in the brain-angiotensinogen, angiotensin converting enzyme, angiotensin II, angiotensin II receptors, and angiotensinases (12,27). Despite these observations, many controversial issues remain to be settled. Among these, three important questions are:
a) Does renin exist in the brain independent of the circulating and renal enzyme?
b) If so, what are the factors that regulate brain renin synthesis and activity? and
c) What is the function of the brain renin-angiotensin system?

EVIDENCE FOR THE EXISTENCE OF BRAIN RENIN

Angiotensin I generating enzymes have been demonstrated in extracts of rat, dog, hog, and human brain (12,27). Renin is present in all brain regions studied, but highest in the pineal gland. The brain renin-like activity is a thousandth of renin activity (13) in the kidney and its pH optimum has been reported by some groups to be 4.7 (15). Several investigators reported that the angiotensin I generating activity of whole brain extracts decreased markedly above pH 5.0 and was barely detectable at pH 7.4 (3,12). These data have led to the suggestion that the renin-like activity may be due to nonspecific acid proteases released during homogenization of brain tissue. Indeed, Day and Reid (4) isolated from the dog brain an enzyme with angiotensin I generating activity that resembled cathepsin D. Hackenthal et al. (15) reported that purified rat brain "isorenin" had acid protease activity, low pH optimum and other properties consistent with those of cathepsin D. On the other hand, several investigators reported that fractionation of brain extracts yielded fractions with angiotensin I generating activity in low concentration which is distinct from acid proteases. Osman et al. (24) utilizing carboxy methyl cellu-

Experimental Brain Research, Suppl. 4

lose chromatography separated a neutral protease with angiotensin I generating activity from acid proteases. Using different separation techniques, other investigators (5,14) made similar observations. Identity of these angiotensin generating enzymes which are active at neutral pH was not established in these studies, however. They may be nonrenin neutral proteases. Indeed, Osman et al. were unable to inhibit angiotensin I generating activity of the isolated enzyme with antirenin antibody (24). Immunoreactive brain renin was first demonstrated by Hirose et al. (17). Using α-casein-Sepharose chromatography, they separated an enzyme that resembled renal renin immunochemically from acid proteases in the rat brain. Subsequently, these investigators have been successful in detecting renin in various regions of the brain using a direct radioimmunoassay of renin. These observations were confirmed in our laboratory using renin specific-antibody raised against pure renal renin (8) and two different types of affinity chromatography, namely, pepstatin-Sepharose and antirenin-Sepharose chromatography (6). We observed that the major angiotensin I generating activity, as fractionated by these columns, is not inhibitable by renin specific antibody, has a low pH optimum (pH of 4), and contains acid proteolytic activity, similar to cathepsin D. In addition, we isolated fractions containing neutral proteases which were capable of generating angiotensin I from angiotensinogen as well as cleaving hemoglobin as substrate. The activity of these fractions were also not inhibitable by antirenin antibody (Fig. 1). Finally, we also isolated an enzyme whose activity resembled that of renal renin in that (a) the activity can be measured at pH 7.4, (b) the activity is completely inhibited by renin-specific antibody, (c) the pH optimum is 6.0, like that of renal renin, and (d) it has no proteolytic activity on C^{14}-hemoglobin at either acid or neutral pH (Table 1). Thus, the differences in the literature can be reconciled. Our results suggest that brain extracts contain more than one enzyme capable of generating angiotensin I from renin substrate. These enzymes include acid and neutral proteases as well as an enzyme with immunochemical identity to renal renin.

Since a mixture of enzymes capable of angiotensin I generation are present in the brain homogenate, the condition of the assay is an important consideration in the interpretation of data. The angiotensin I generating activity of brain homogenate at pH 6.0 was ten fold greater than at pH 7.4. However, the activity of brain extract at pH 6.0 was not significantly affected by antirenin antibody. On the other hand, the activity at pH 7.4, although small, was largely inhibited by antirenin antibody. Thus, renin activity is masked by nonrenin proteases when assayed at acidic pH. In all our experiments, brain renin activity was determined as the angiotensin I generating activity at pH 7.4 using nephrectomized dog or rat plasma as substrate (6).

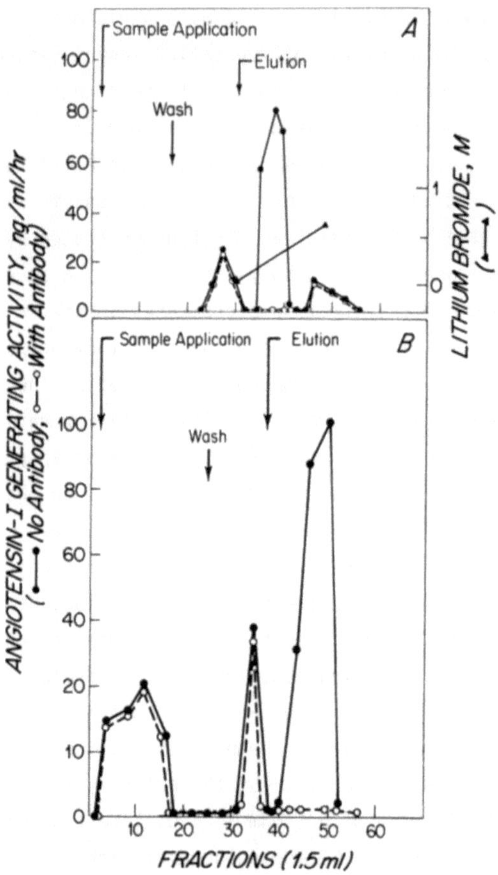

Fig. 1. Elution profile of (A) pepstatin affinity chromatography and (B) immuno-affinity chromatography. •——• represents angiotensin I generating activity of various fractions, o----o represents activity of fractions which were incubated with renin specific antisera prior to assay. Elution of (A) was effected by linear concentration of lithium bromide ▲——▲ in 0.1 M Tris-HCl buffer, pH 7.5. Renin was eluted from (B) by 1 M acetic acid, pH 3.0 in this experiment

Table 1. Properties of angiotensin I generating enzymes separated by affinity chromatography

Affinity chromatography	Proteolytic activity		Angiotensin I generating activity		
	pH 7.0	pH 3.0	pH optimum	Specific activity ANG I/hr/mg protein	% inhibition of renin activity
Immunoaffinity					
Nonbound fraction	no	yes	4.0	12.5	5
Eluate	no	no	6.0	115.5	100
Pepstatin					
Wash	yes	no	---	18	0
Eluate I	no	no	6.5	141	100
Eluate II	no	yes	4.0	104	0
α-casein					
Nonbound fraction	yes	no	6.5	18	83
Eluate	no	yes	4.0	116	0

DEFINITION OF BRAIN RENIN

Since several brain enzymes are capable of angiotensin I generation, a definition of brain renin becomes necessary.

1) Substrate specificity: brain renin releases angiotensin I from angiotensinogen but is unable to hydrolyze hemoglobin at acid or neutral pH.
2) pH requirements: the activity of brain renin is detectable at neutral pH and its pH optimum is not very acidic (pH ≙ 5.0).
3) Immunoidentity: it is detected by the direct radioimmunoassay using antirenin antibody and/or its activity is neutralized by antirenin antibody.

This definition of brain renin does not preclude the existence of other enzymes which resemble renin but do not completely fulfill all of the above criteria.

PURIFICATION OF BRAIN RENIN

Rapid progress has been made in the past decade in the purification and characterization of renal renin. Complete purification of hog, dog, rat, and human renins (3,9,11,22,28,29) have become possible by the introduction of affinity chromatography and the use of protease inhibitors during the purification procedures. In our laboratory, dog renal renin was purified to homogeneity using an eight step procedure. The introduction of lithium bromide gradient (9), a chaotropic agent, for the elution of dog renin from the pepstatin-Sepharose column, improved the recovery and stability of the enzyme. This has subsequently been proven to be an important modification for the human renin purification (7). Other modifications of our original protocol for human renin purification have yielded a product with higher specific activity (Table 2b). These have included the introduction of alpha-casein-Sepharose affinity chromatography early in the purification for the removal of non-renin acid proteases and an additional DEAE-cellulose chromatographic procedure in the final step (7).

Attempts to purify brain renin have been limited by the small quantity of enzyme in brain tissues. Brain renin appears to be more stable than renal renin during purification. To avoid contamination of the brain extracts with plasma renin, our initial approach (6) included using rats which were bilaterally nephrectomized for 48 hours. Under ether anesthesia, the chest was opened and a needle was inserted into the beating left ventricle while the right atrium was split open. Two hundred ml of phosphate buffered saline was infused into each animal until the drainage from the atrium was free of any visible blood. The brain was then removed and washed extensively with extraction buffer. Based on our subsequent observation that

Table 2a. Purification of rat brain renin (130 grams)

	Angiotensin I Generating Activity a. (ng/h)	Protein (mg)	Specific Activity (ng/mg)		Purification Fold	Recovery %
1A. Homogenate			1.45		1	
1B. Acetone powder	2.4×10^3	192.00	12.50		9	100
2. α casein-sepharose chromatography	1.0×10^3	42.00	23.80		160	42
3. Immunoaffinity chromatography	4.2×10^2	0.40	1050.00		7,240	18
4. Pepstatin-affinity chromatography	5.0×10^2	0.066	7576.00		52,000	21
5. Gel filtration AcA34 column	4.8×10^2	0.02	24000.00		165,000	20
				(4.8 Gu/mg)[a]		

Pure rat renal renin, specific activity = 452-626 Gu/mg (Matoba et al., 1978)

Table 2b. Purification of human renal renin (8kg)

	Renin Activity a. (ng AI/h)	Protein	Specific Activity (ng/mg)	(Gu/mg)	Purification Fold	Recovery %
1A. Isolation of cortex				(0.002)		
1B. Crude extract	3.0×10^7	52.00 g	5.8×10^2	(0.120)	60	100
2. DEAE batch concentration	1.8×10^7	9.00 g	2.0×10^3	(0.400)	200	60
3. Acid-salt fractionation	1.2×10^7	3.75 g	3.2×10^3	(0.600)	320	40
Ammonium sulfate precipitation						
4. α casein-affinity chromatography	1.0×10^7	1.50 g	6.6×10^3	(1.300)	650	33
5. Pepstatin-affinity chromatography	8.0×10^6	82.00 mg	9.8×10^4	(19.600)	9,800	27
6. Sephadex G-100 gel filtration	5.0×10^6	22.00 mg	2.3×10^5	(46.000)	23,000	17
7. Carboxy methyl cellulose chromatography	4.0×10^6	4.00 mg	1.0×10^6	(200.000)	100,000	13
8. DE-52 chromatography	2.0×10^6	0.50 mg	4.0×10^6		400,000	6.6
				(800.0 Gu/mg)[a]		

a. Nephrectomized dog plasma used as substrate;1 GU = 5000 ng AI/h at pH 7.4

plasma renin contamination was not a problem, larger scale purification were performed on brains of intact rats. These brains were dissected free of meninges, visible blood vessels and then washed extensively with buffer.

We employed a 5 step purification protocol utilizing 3 affinity chromatographic steps alpha-casein-, pepstatin-, antirenin antibody-Sepharose columns. The pepstatin was coupled to aminohexyl Sepharose by the method of Murakami and Inagami (23). Alpha-casein and renin-specific immunoglobulin were coupled to Sepharose 4B using the cyanogen bromide activation method of Cuatrecases (2). The rationale for these affinity columns are as follows: alpha-casein binds nonrenin acid proteases but not renin or neutral proteases; the immunoaffinity column is specific for renin and should remove neutral proteases and other proteins. Pepstatin binds active renin but not inactive renin or neutral proteases. Indeed, renin activity and nonspecific proteolytic activity (at pH 7.0) were observed in the nonbound fractions of alpha-casein-Sepharose column. The eluate of the immunoaffinity columns contains both active and trypsin-activatable brain renin. The nonbound fraction of pepstatin column contains trypsin-activatable inactive brain renin. Elution with lithium bromide gradient separated active renin from nonrenin and proteases (Table 3).

130 gram whole rat brains (including pituitaries) were homogenized in 700 mls acetone at -10°C using a Virtus-60 homogenizer. The homogenate was centrifuged at 12,000 g for 30 mins. The precipitate was homogenized again with 200 mls cold acetone three times and the precipitate was dried. The acetone powder was resuspended in 200 mls of 50 mM sodium acetate pH 3.5 containing 0.25 mM EDTA, 0.1 M phenyl-methylsulfonyl-fluoride (PMSF), and 0.25 mM sodium tetrathionate, centrifuged and the supernatant was applied to an alpha-casein-Sepharose column (2.5 x 30 cm) at 30 ml/min. The nonbound fractions were dialysed against 50 mM potassium phosphate pH 7.5 with 150 mM NaCl containing the same mixture of inhibitors. This sample was applied to the antirenin immunoaffinity column (1.5 x 6 cm). Brain renin activity was eluted by 2 M lithium bromide in 0.1 M Tris-HCl pH 7.5 with protease inhibitors. After dialysis of this eluate against 5 mM sodium acetate pH 5.5, it was applied to a 1.5 x 6 cm pepstatin-aminohexyl-Sepharose column and eluted with 1 M lithium bromide in 0.1 M Tris-HCl pH 7.5. Final step was gel filtration chromatography on a 2.5 x 45 cm AcA-34 column equilibrated in 50 mM potassium phosphate - 150 mM NaCl pH 7.5.

A 53,000 fold purification with a yield of 20% was obtained. The final specific activity was 4.8 Gu/ml. A comparison with renal renin purification is shown in Table 2.

Active renin in rat brain has an apparent molecular weight of 42,000 as estimated by gel filtration chromatography. Interestingly, SDS gel electrophoresis of the final product yielded three major protein bands corresponding to apparent molecular weights of 60,000, 49,000, and 38,000 (Fig. 2). Recently, Hirose et al. (16) demonstrated that hog brain renin existed in 3 MW forms based on gel filtration chromatography. They reported that hog brain tissues contain a 50,000 MW inactive form of renin, a 43.000 MW active renin which can be converted to a higher MW form (60,000) when the brain extract was treated with thiol blocking agents. All our purification procedures were performed in the presence of sodium tetrathionate except for the last gel filtration step. The finding of three major MW forms on SDS gel electrophoresis thus supports the data of Hirose et al. (16).

The properties of purified rat brain renin are compared to those of renal renin (20,22) and plasma renin (1) in Table 3. Active renin in rat brain has a pH optimum of 6.0 when reacted with rat substrate. This is similar to renal renin and plasma renin. The Michaelis constant Km of brain renin with the rat substrate is 0.9×10^{-6}M whereas that of renal renin is 1×10^{-6}M (20) to 2.2×10^{-6}M (our observation) and that of plasma renin has been reported to be 3×10^{-6}M (1). Isoelectrofocusing of brain renin demonstrated marked heterogeneity with peak activity at pH 4.82 similar to that of hog brain renin (18). Indeed, the isoelectric point of brain renin from both species was significantly more acidic than those of the kidney and plasma renins. The activity of this highly purified rat brain renin was completely neutralized by renin-specific antibody (1:100). Brain renin did not cleave C^{14}-hemoglobin at any pH.

Rat brain renin has many physiochemical properties which are similar to those of renal and plasma renin such as molecular weight (of active renin) and pH optimum. In addition, these renins exhibit antigenic homology. However, differences do exist between the brain and the renal enzyme. The brain enzyme appears to have lower specific activity and lesser affinity for plasma angiotensinogen. Since several investigators have reported that brain angiotensinogen differed from plasma angiotensinogen (25,26), the higher Km of brain renin for plasma angiotensinogen might reflect a lower specificity of brain enzyme for the plasma substrate than for brain renin substrate itself. Finally, the isoelectric point of brain renin is more acidic than plasma or renal renin. This may be due to differences in glycosylation of the respective enzymes.

As will be discussed in a later section, brain renin activity remains unchanged in conditions which are associated with marked increases in plasma and renal renin activity, (e.g., renal ischemia) or total suppression of plasma renin activity (e.g., nephrectomy). These observations argue strongly against contamination of brain extract by plasma renin as a factor in the purification and support the existence of an independent enzyme in the brain.

Table 3. Comparison of the properties of brain and renal renins

	Rat Renal Renin (Matoba et al. 1978)	Rat Brain Renin	Hog Renal Renin (Hirose et al. 1980) (Inagami et al 1977)	Hog Brain Renin (Hirose et al. 1980)
Molecular Weight				
Gel filtration	41,000 (active)	41,000 (active)	42,000 (active)	43,000 (active)
				50,000
				60,000
SDS-PAGE	35,000 - 37,000	38,000	36,400	
		49,000		
		60,000		
PH optimum	6.0[a]	6.0[a]	5.5 - 7.0[a]	6.0[a]
Isoelectric point	5.05 - 5.22	4.82	5.05 - 5.2	4.48 - 4.94
Km	2.2×10^{-6} [a]	8×10^{-6} [a]		
Inactive renin	?	Present	Present	Present
Glycoprotein	Yes	?	Yes	Yes

[a] Nephrectomized rat plasma or purified rat angiotensinogen as substrate

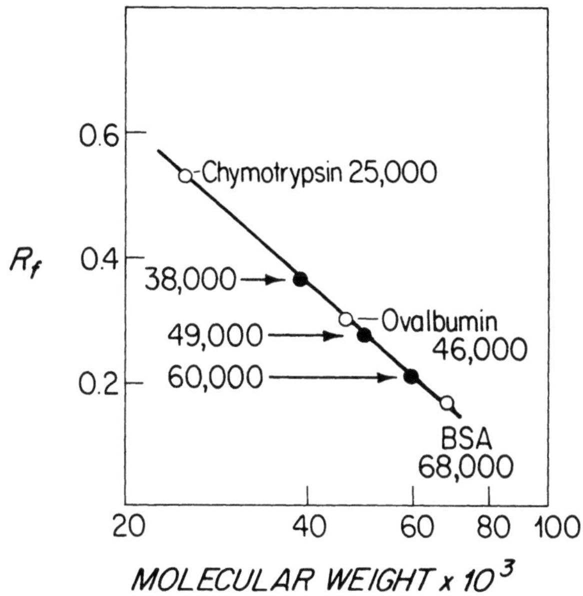

Fig. 2. SDS polyacrylamide gel electrophoresis (11%) of brain renin step 5. Three bands are observed at estimated molecular weights of 60,000 (A), 49,000 (B) 38,000 (C); molecular weight markers are bovine serum albumin (68,000), ovalbumin (46,000) and chymotrypsin (25,000).

Logarithem of molecular weight is plotted against retardation factor (RF)

INACTIVE RENIN IN RAT BRAIN TISSUES

Hirose et al. (18) have observed widespread distribution of active renin in hog brain. The pineal gland was found to contain the highest activity and concentration of renin followed by the anterior pituitary and the choroid plexus. Many regions also contain trypsin and acid activatable renin. Lewicki et al. (21) reported widespread distribution of angiotensinogen in the rat brain. These data predict that brain renin should be distributed in many areas in the rat brain. We did not examine the detailed regional distribution of rat brain renin. Instead, we compared the renin activity of whole brain to those of the anterior pituitary and pineal gland. Indeed, the renin activity of pineal gland was higher than the anterior pituitary (10.9 ± 0.2 vs 5.4 ± 0.1 ng ANG I/mg). Whole rat brain contained only 1.78 ± 0.13 ng ANG I/mg protein. In order to determine the presence of activatable renin, samples were treated with trypsin according to the method of Hirose et al. (18). 10 μl of trypsin (Worthington) mg/ml was added to 100 μl samples and incubated for 30 min at 25^{0}C. The reaction was stopped by the addition of 450 μM of soyabean trypsin inhibitor (Worthington) and 3 mM PMSF. The resultant mixture was assayed for renin activity.

Inactive renin was present in extracts of pineal and anterior pituitary glands. A 46% and 20% increase in renin activity were respectively observed after trypsin treatment of the two extracts. This trypsin activatable renin activity was completely neutralized by renin-specific antibody. Activatable renin was not observed in whole brain extracts. Furthermore, no increase in renin activity was seen when brain renin, as isolated by alpha-casein Sepharose chromatography, was treated with trypsin. This may mean that inactive renin was retained by the alpha-casein column. A more likely explanation is that the acidic conditions of alpha-casein fractionation have resulted in the total activation of inactive renin in brain extracts.

When anterior pituitary gland extracts were subjected to pepstatin-Sepharose chromatography, trypsin-activatable renin was recovered in the nonbound fractions (Table 4). This property of inactive renin in the brain resembles that of plasma inactive renin. Hirose et al. (16) speculated that this inactive renin was the precursor of brain renin.

EFFECTS OF VARIOUS PHYSIOLOGIC STATES ON BRAIN RENIN ACTIVITY

To examine the factors which regulate brain renin biosynthesis, we studied brain renin activities in various physiologic states. Whole brains and anterior pituitary glands were obtained from a) normal rats, b) those with suppressed plasma renin activity, c) rats with elevated plasma and renal renin activities, d) animals treated with sympatholytic agents, and e) rats with various endocrine-metabolic disorders. Brain renin was isolated from brain extracts by alpha-casein-Sepharose chromatography as described by Hirose et al. (18). Separation of inactive renin from active renin in brain extracts was performed by using pepstatin-Sepharose chromatography. Active brain renin is defined as the angiotensin I generating activity detected in the nonbound fraction of alpha-casein Sepharose. This activity was invariably neutralized (83-100%) by antirenin antibody. Inactive brain renin is defined as the increase in renin activity after trypsin treatment.

Renin was measured in the following experimental groups:

a) Normal rats - whole brains (not including anterior pituitary) and anterior pituitary glands were extracted separately in appropriate buffers as described earlier (n=11). Mean PRA in these rats was 2.2 $\pm$ 0.8 ng/ANG I/ml/hr.

b) Suppressed plasma renin activity - this consisted of two groups of rats, i.e., bilaterally nephrectomized for 48 hours (n=4) and those injected with antirenin antibody intravenously (n=4). Rats from the latter group were sacrificed 3 hours after antibody injection. The plasma renin activities were suppressed to undetectable levels in both groups.

c) Elevated plasma renin activity - three groups of animals with elevated plasma renin activity were examined. One group received continuous renin infusion (4.5 ng ANG I/hr per min) for 3 days by the Alza osmotic pump implanted intraperitoneally (n=5). A second group of animals (n=3) received intraperitoneal captopril injection (10 µg/g body weight twice daily) for 10 days. In a third group of animals (n=4) right renal ischemia was created by ligation (90% narrowing) of the abdominal aorta above the right renal artery and below the left.

The animals were studied 24 hours after surgery. PRA was elevated in all 3 groups (8.8 $\pm$ 1.9, 5.3 $\pm$ 1.3, and 12.6 $\pm$ 2.9 ng ANG I/ml/hr respectively).

d) Inhibition of sympathetic nervous activity - this was achieved by administration of reserpine (0.125 mg/rat for 4 days) in 10 rats, and propranolol (20 mg/100 g body weight daily in 2 divided doses for 5 days) in 5 rats.

e) Endocrine-metabolic disorders-brain renin activities were studied in thyroidectomized (n=5) and hyperthyroid rats (n=5). The latter state was achieved by daily subcutaneous injection of L-thyroxine (10 µg/100 g body weight) for 7 days to normal rats.

Streptozotocin induced diabetic rats (n=2)(generous gift of Dr. Hostetter) and hypophysectomized rats (n=5) (Charles River Laboratories, Mass) were also studied.

Table 4. Effects of various physiologic states on brain renin activity

	Whole brain		Anterior pituitary		
	Total angiotensin I-generating activity (ng ANG I/h/mg protein)	Active brain renin activity (ng ANG I/h/mg)	Total angiotensin I-generating activity (ng ANG I/h/mg)	Active brain renin activity (ng ANG I/h/mg	Inactive renin (ng ANG I/h/10 mg)
a. Normal	0.10	1.78 ± 0.13	2.8 ± 0.16	5.4	0.39 ± 0.06
b. Suppressed PRA					
Nephrectomy	0.15	1.9 ± 0.24	2.9 ± 0.88	2.0	0.9
Antirenin infusion	0.10	1.85 ± 0.35	2.4		
c. Elevated PRA					
Renin infusion	0.12	1.95 ± 0.14			0.45 ± 0.1
Captopril	0.10	2.4 ± 0.43		4.0	0.45 ± 0.1
Renin ischemia	0.12	1.62 ± 0.1			0.6 ± 0.02
d. Sympatholysis					
Reserpine	0.15	1.18 ± 0.005 ($p \lesssim 0.005$)	3.5 ± 0.7	2.0	0.24 ± 0.01 ($p \lesssim 0.05$)
Propranolol	0.13	1.0 ± 0.1 ($p \lesssim 0.01$)			
e. Endocrine disorders					
Thyroidectomy	0.14	1.38 ± 0.19	3.0	4.6	
Hyperthyroidism	0.11	1.47 ± 0.15	4.2 ± 0.93	4.6	
Diabetes	0.12	1.28 ± 0.03		5.0	
Hypophysectomy	0.14	1.44 ± 0.11			

These results are summarized in Table 4. Total angiotensin I generating activity of whole brain and anterior pituitary extracts did not change appreciably with most of the physiologic states examined. Brain renin activity remained unchanged despite marked elevation or suppression of PRA. Similarly, brain or pituitary renin activities were not affected by changes in endocrine-hormonal states such as diabetes, hypo- or hyperthyroidism and hypopituitarism. However, when rats were treated with reserpine (depletion of central catecholamine stores) or propranolol (beta-adrenergic antagonist), whole brain and anterior pituitary renin activity fell significantly suggesting a causal relationship between sympatholysis and decreased brain renin activity.

Inactive renin in the anterior pituitary glands could be separated from active renin by pepstatin-affinity chromatography. This trypsin activatable pituitary renin activity did not change with alterations in plasma renin activity except in the rats treated with reserpine. In these rats, inactive renin decreased accompanying the fall of active renin.

The above data suggest that the total brain or pituitary renin activities are not under the same general influence and regulation as plasma and renal renins. However, the central catecholaminergic system appears to influence whole brain and pituitary renin activity. Depletion of catecholamine by reserpine and blockade of beta-adrenergic receptors by propranolol resulted in a significant decrease in whole brain and pituitary renin activities. The exact nature of the interaction is not clear. One intriguing possibility is that central catecholamines play a role in the regulation of the synthesis of brain renin which may in turn regulate the production of the neuropeptide angiotensin II. Since we did not examine brain renin activities in various regions, we were unable to address the question of whether there were changes in regional distribution of brain renin under different physiologic states. Taken together, our data lends strong support to the existence of an independent brain enzyme. The exact function of this enzyme is as yet undefined.

SUMMARY

An enzyme with immunologic identity, similar molecular weight, and pH optimum as renal renin has been partially purified from the rat brain using three affinity chromatographic steps. This enzyme exists in possibly three molecular weight forms. The active form (MW 42,000 estimated by gel filtration chromatography) has lower specific activity and higher Km for rat plasma angiotensinogen when compared to renal renin. Inactive brain renin is present in anterior pituitary tissue and can be separated from active renin by pepstatin affinity chromatography.

Whole brain and pituitary renin activities remain essentially unchanged in many physiologic states. Most factors which stimulate or suppress renin release from the kidney have no influence on brain renin activity. These observations lend strong support to the existence of an independent brain renin. On the other hand, reserpine and propranolol treatment resulted in significant decreases in brain renin activity. This is further evidence indicating that a close relationship exists between the central catecholaminergic system and the brain renin angiotensin system. The interaction of these two systems may play an important role in the central regulation of systemic blood pressure.

ACKNOWLEDGEMENTS

The author is indebted to Drs. Richard Pratt, Gordon Williams, and Howard Hermann for many helpful suggestions and consultations. I wish to thank Elizabeth Ross, Laurie Sturzenbecker, and Kenneth Sands for excellent technical assistance. I thank D. Malone for expert secretarial help.

This research was supported by NIH grants, 5 K08 00750 and HL 19259, and a grant from R. J. Reynolds Industry. Dr. V. J. Dzau is the recipient of a NHLBI Clinical Investigator Award from the National Institutes of Health.

REFERENCES

1. BARRETT JD, EGGENA P, SAMBHI M (1981) Activation of rat plasma renin. Endocrinology 108:778-785
2. CUATRECASES P (1970) Protein purification by affinity chromatography. Derivation of agarose and polyacrylamide beads. J Biol Chem 245:3059-3065
3. CORVOL P, DEVAUX C, ITO T, SICARD P, DUCLOUX J, MENARD J (1977) Large scale purification of hog renin. Circ Res 41 (Suppl II) 616-622
4. DAY RP, REID IA (1976) Renin activity in dog brain: Enzymological similarity to cathepsin D. Endocrinology 99:93-100
5. DWORSHAK RT, GREGORY TJ, PRINTZ MP (1978) Identification of neutral pH renin-like activity from brain homogenate by chromatography of phenylamine-Sepharose. Fed Proc 37:1385 (abstract)
6. DZAU VJ, BRENNER A, EMMETT N, HABER E (1980) Identification of renin and renin-like enzymes in rat brain by a renin specific antibody. Clin Sci 59:45s-47s
7. DZAU VJ, HABER E (in preparation) Human renal renin purification.
8. DZAU VJ, KOPELMAN RI, BARGER AC, HABER E (1980) Renin specific antibody for study of cardiovascular homeostasis. Science 207:1091-1093
9. DZAU VJ, SLATER E, HABER E (1979) Complete purification of dog renal renin. Biochemistry 18:5224-5228

10. FISCHER-FERRARO C, NAHMOD VE, GOLDSTEIN DJ, FINKIELMAN S (1971) Angiotensin and renin in rat and dog brain. J Exp Med 133:353-361

11. GALEN FX, FEVAUX C, GYNENNE T, MENARD J, CORVOL P (1979) Multiple forms of human renin . Purification and characterization. J Biol Chem 254:4848-4855

12. GANTEN D, FUXE K, PHILLIPS I, MANN JFE, GANTEN U (1978) The brain isorenin-angiotensin system: Biochemistry, localization and possible role in drinking and blood pressure regulation. In Frontiers in Neuroendocrinology, Vol. 5, edited by W. F. GANONG, L. MARTINI. New York, Raven Press, p 61-69 (Review)

13. GANTEN D, MARQUEZ-JULIO A, GRANGER P, HAYDUK K, KARSUNKY KP, BOUCHER R, GENEST J (1971) Renin in dog brain. Am J Physiol 221:1733-1737

14. GANTEN D, SPECK G, MEYER D, LOOS HE, SCHELLING P, RETTIG R, UNGER T (1980) Biochemical and functional aspects of brain renin. In: Enzymatic release of vasoactive peptides, edited by F. GROSS, G. VOGEL. New York, Raven Press. p 31-49

15. HACKENTHAL E, HACKENTHAL R, HILGENFELDT U (1978) Purification and partial characterization of rat brain acid proteinase (isorenin). Biochim Biophys Acta 522:561-573

16. HIROSE S, NARUSE M, OHTSUKI K, INAGAMI T (1981) Totally inactive renin zymogen and different forms of active renin in hog brain tissues. J Biol Chem 256:5572-5576

17. HIROSE S, YOKOSAWA H, INAGAMI T (1978) Immunochemical identification of renin in rat brain and distinction from acid proteases. Nature 274:392-393

18. HIROSE S, YOKOSAWA H, INAGAMI T, WORKMAN R (1980) Renin and prorenin in hog brain: ubiquitous distribution and high concentration in the pituitary and pineal. Brain Research 191:489-499

19. INAGAMI T, MURAKAMI K (1977) Pure renin: Isolation from hog kidney and characterization. J Biol Chem 252:2978-2983

20. LAURITZEN M, DANSGAARD JJ, RUBEN I, LAURITZEN E (1976) A comparison of properties of renin isolated from pig and rat kidney. Biochem J 155:317-323

21. LEWICKI JA, FALLON JH, PRINTZ MP (1978) Regional distribution of angiotensinogen in rat brain. Brain Research 158:359-371

22. MATOBA T, MURAKAMI K, INAGAMI T (1978) Rat renin: purification and characterization. Biochim Biophys Acta 526:560-571

23. MURAKAMI K, INAGAMI T (1975) Isolation of pure and stable renin from hog kidney. Biochem Biophys Res Commun 62:757-763

24. OSMAN MY, SMEBY RR, SEN S (1979) Separation of dog brain renin-like activity from acid protease activity. Hypertension 1:52-60

25. PRINTZ MP, PRINTZ JM, LEWICKI JA, GREGORY TJ (1976) Resolution of multiple forms of human angiotensinogen. Isoelectric focusing heterogeneity profile analysis. Circ Res 41 (Suppl II):37-43

26. PRINTZ MP, LEWICKI JA, WALLIS CJ (1980) Brain angiotensinogen: Origin and evidence for an influence by adrenal corticosteroids. In: Enzymatic release of vasoactive peptides, Raven Press, New York, p 193-207

27. REID IA (1977) Is there a brain renin-angiotensin system. Circ Res 41:147-152 (Review)

28. SLATER EE, COHN RC, DZAU VJ, HABER E (1978) Purification of human renal renin. Clin Sci Mol Med 55:117S-119S

29. YOKOSAWA H, HOLLADAY L, INAGAMI T, HAAS E MURAKAMI K (1980) Human renal renin. Complete purification and characterization. J Biol Chem 255:3498-3502

In Vitro and in Vivo Evidence for Angiotensin Generation in the Rat Brain by Endopeptidase Activity Distinct from Cathepsin D

J. A. D. M. Tonnaer, V. M. Wiegant, and W. de Jong

INTRODUCTION

Evidence for renin-like activity in the brain has been provided as early as 1971 (7, 11). During recent years, using more sophisticated techniques, renin-like enzymes have been identified in, and extracted and purified from brain tissue (9,10,15). The functional activity of these enzymes may be deduced from the angiotensin I (ANG I) generation from synthetic substrate and from natural plasma substrates (12).

It is questionable, however, whether such estimation of renin activity reflects its physiological activity, since the endogenous ANG I precursor within the brain is not yet known. Plasma-like angiotensinogen is present in the cerebrospinal fluid (CSF) and in brain tissue (17,19,25) and this precursor is the most likely endogenous brain renin substrate.

The use of tetradecapeptide for the screening of renin-activity has the advantage that renin has a high affinity for this synthetic substrate. The specifity of this substrate, however, has been debated since the peptide is a substrate for renin, but additionally can be attacked by ANG I-converting enzyme (4) and the lysosomal enzyme cathepsin D (5). On the other hand, cathepsin D also has been reported to be capable of releasing ANG I from natural angiotensinogen (5,20).

In the present paper, a high-performance liquid chromatographical (HPLC) technique for the analysis of angiotensins and tetradecapeptide cleavage products is described. Using this HPLC technique, the ANG I generation from tetradecapeptide by brain enzymes is studied. The conditions for the activity of the brain enzymes for ANG I generation have been compared to those required for the activity of purified renal renin and cathepsin D. In addition, the characteristics of enzyme activity involved in the in vivo release of ANG I from tetradecapeptide and from natural brain angiotensinogen are described.

Experimental Brain Research, Suppl. 4

MATERIAL AND METHODS

HPLC Separation of Peptides

Protein-free peptide mixtures were fractionated using an HPLC equipment (Waters Ass., Milford, U.S.A.) which consisted of two pumps (model 6000 A), a programmer for gradient elution (model 660) and an universal liquid chromatograph injector (model U 6K). The peptides were separated using a reversed-phase µBondapak C18 column (30 cm x 0.39 cm I.D.; 10 µm particle size; Waters Ass.). Elution was performed at room temperature using a 45-min linear methanol gradient from 20 - 75% in 0.01 M ammonium acetate, pH 4.15. The flow rate was 2 ml/min. The HPLC eluate was monitored continuously at 210 nm, using a variable wave-length UV monitor (model 450, Waters Ass.).

Enzymatical Cleavage of Tetradecapeptide

Tetradecapeptide-cleavage by purified enzymes or proteolytic activity from brain tissue was studied as described previously (28). Tetradecapeptide (Beckman Bioproducts, Geneva, Switzerland) (20 µM) was incubated with the enzyme preparations in 500 µl of 0.1 M citrate-phosphate buffer (pH 4.4 or pH 7.4) at 37°C for 30 min. Incubations with non-purified enzyme preparations were performed in the presence of 5 mM EDTA, 5 mM 8-hydroxyquinoline (8-HQ) and 3.5% ethanol. The incubations were stopped by 10 min boiling. Denatured proteins were removed by centrifugation and the supernatant subjected to HPLC fractionation as described above. Tetradecapeptide-cleavage products were quantified by measuring their peak heights in the HPLC chromatograms.

Enzyme Sources

Cathepsin D (E.C.3.4.23.5) was commerciably available from Sigma Chemical Co., St. Louis, U.S.A. Purified renal hog renin (280 GU/mg protein) was a gift of Dr. J. Ménard, Paris, France. The enzyme preparation of rat brain was obtained by subcellular fractionation: rat brain homogenate was centrifuged for 10 min at 1000xg, and the supernatant recentrifuged for 20 min at 10,000xg. The resulting pellet was hypo-osmotically lysed and the particulate fraction pelleted by centrifugation for 60 min at 100,000xg. The supernatant contained the proteolytic activity studied.

Animal Studies

The lateral cerebral ventricle of male Wistar rats (Cpb: WU, TNO Zeist, The Netherlands), weighing about 200 g, were cannulated as described previously (29). Intracerebroventricular (ICV) administration of enzymes, saralasin (Beckman Bioproducts) and N-acetyl-pepstatin (NAP) (gift of Dr. M. Miyazaki, Ohtsu, Japan), in a volume of 3 μl, was done with a 10 μl syringe with an adapted needle (1). Subsequently, the animals were replaced in their home cages and water was provided. Water intake was measured from calibrated reservoirs after 20 min. Animals were used once only.

RESULTS

HPLC of Peptides

Closely related peptides were rapidly separated with high resolution in one single step by HPLC. Table 1 summarizes the retention times (t_r) for tetradecapeptide, ANG I, ANG II, ANG III and the tetrapeptide Leu-Val-Tyr-Ser (the 11-14 sequence of tetradecapeptide). UV absorbance of the various peptides, as measured from the peak heights in the HPLC chromatogram, was linear with the quantity of peptide loaded on the column over a range of 0.1 - 10 nmoles.

Table 1. Separation of peptides by HPLC

Peptide	t_r (min)
Tetradecapeptide	37.2
Angiotensin I	30.2
Angiotensin II	23.0
Angiotensin III	22.0
Leu-Val-Tyr-Ser	6.6

Peptides were separated by reversed-phase HPLC on a μBondapak C18 column. A 45-min linear methanol gradient from 20 - 75% in 0.01 M ammonium acetate buffer, pH 4.15, was used.

In Vitro Cleavage of Tetradecapeptide

When incubated with cathepsin D or purified renal renin, tetradecapeptide was specifically cleaved at the Leu-Leu peptide bond, as indicated by the accumulation of both ANG I and the complementary tetrapeptide Leu-Val-Tyr-Ser (29). A similar pattern was observed upon incubation of tetradecapeptide with the enzyme preparation of brain tissue. However, during this incubation the minor accumulation of some other, unidentified, peptides was observed. The formation of ANG I by the three enzyme preparations was highly pH dependent. Table 2 shows that cathepsin D was active at acidic pH only, whereas purified kidney renin exerted maximal activity at pH 7.4. The rat brain preparation maximally generated ANG I at the acidic pH, but also displayed considerable activity at pH 7.4.

Table 2. In vitro ANG I formation from tetradecapeptide

Enzyme	pH	ANG I generated (nmoles)
Cathepsin D	4.4	6.2 $\pm$ 0.2
Cathepsin D	7.4	n.d.
Renal renin	4.4	0.6 $\pm$ 0.1
Renal renin	7.4	1.9 $\pm$ 0.1
Brain preparation	4.4	10.1 $\pm$ 0.2
Brain preparation	7.4	1.2 $\pm$ 0.1

Cathepsin D (1 μg), renin (0.2 μg) or brain enzyme preparation (0.1 mg) was incubated with 20 μM tetradecapeptide at 37°C for 30 min. Incubations with the brain preparation occured in the presence of EDTA and 8-HQ. Samples were analysed by HPLC fractionation. Mean values $\pm$ S.E.M. are given (n = 3). n.d. = not detectable.

In Vivo Studies

ICV injection of tetradecapeptide in rats resulted in a pronounced water intake during 20 min. This water intake was dose-dependently inhibited by ICV pretreatment of the animals with saralasin (Table 3). The dipsogenic effect of tetradecapeptide also was significantly reduced by pretreatment with NAP (Table 3).

Table 3. Inhibition of tetradecapeptide-induced drinking

Pretreatment	Dose (nmol)	Water intake (ml)	Inhibition
Vehicle	-	10.3 $\pm$ 0.7	
Saralasin	0.1	6.6 $\pm$ 0.9	36%$^{++}$
Saralasin	1.0	3.1 $\pm$ 0.8	70%$^{+++}$
Vehicle	-	8.2 $\pm$ 0.5	
NAP	75.0	3.8 $\pm$ 1.2	54%$^{+}$

Saralasin was given 2 min prior to 250 pmol tetradecapeptide, and NAP 5 min prior to 25 pmol tetradecapeptide. Mean values $\pm$ S.E.M. are presented (n = 5-7). $^{+}P<0.05$; $^{++}P<0.01$; $^{+++}P<0.001$ (Student's t-test)

Cathepsin D was not dipsogenic, as the water intake upon ICV injection of this enzyme was not significantly different from the water intake following vehicle injection (Table 4). Injection of purified renin, on the other hand, resulted in a marked water intake. This disogenic effect was dose dependent (Table 4). The latency to the onset of drinking and the total drinking time were longer than those upon ICV injection of angiotensins (not shown).

Table 4. Dipsogenic activity of cathepsin D and renin

Treatment	Dose	Water intake (ml)
Vehicle	-	0.4 $\pm$ 0.2
Cathepsin D	0.6 μg	1.5 $\pm$ 0.6
Cathepsin D	3.0 μg	1.4 $\pm$ 0.6
Renal renin	4 ng	5.0 $\pm$ 1.4^{++}
Renal renin	20 ng	10.1 $\pm$ 2.6^{++}

Water intake was measured 20 min after ICV injection of the enzymes. Mean values $\pm$ S.E.M. are given (n = 6). Values significant different from control are indicated by $^{++}P<0.01$ (Student's t-test).

DISCUSSION

The separation of angiotensins and related peptides, i.e. tetradecapeptide, in the past has been hampered by the very close physicochemical properties of the peptides. During the last decade polyacrylamide gelelectrophoresis (3), isoelectric focussing (3), thin-layer chromatography (2) and paper chromatography (18,24) have been employed, but did not always lead to complete separation. The development of HPLC, however, opened the possibility of separating closely related compounds with high resolution. Employment of this technique for the analysis of angiotensin has been reported already in 1978 (14) and during recent years more applications have been described (13,23,27). The reversed-phase HPLC system used in the present study has extensively been described in a previous paper (28). The separation of angiotensins, tetradecapeptide, and tetradecapeptide cleavage products is performed in one single, rapid step. The separation time can be shortened to 20 min without impairment of the resolution. The system also has proved its applicability for the separation of a variety of angiotensin fragments, and thus enables an accurate study of angiotensin metabolism in various tissues (Tonnaer et al., in preparation).

In the present study tetradecapeptide cleavage products, generated upon incubation of this substrate with purified renin, cathepsin D and with enzymes present in rat brain tissue, were analysed by HPLC fractionation of the incubation mixtures. Incubation of the renin substrate with the purified enzymes resulted in the exclusive cleavage of the Leu-Leu peptide bond. Cathepsin D was inactive at pH 7.4, as reported previously (29), whereas renin was most active at this pH. The rat brain preparation also contained Leu-Leu cleaving activity, as indicated by the accumulation of both ANG I and the complementary tetrapeptide. The high activity of this preparation at pH 4.4 is explained by the presence of cathepsin D, which was released from the lysosomes upon lysis of the 10,000xg pellet. The activity at pH 7.4 might be due to renin-like enzymes, and is suggested to consist partly of renal renin-like activity (31).

The presence of in vivo tetradecapeptide cleaving activity in the brain is illustrated by the dipsogenic activity of this peptide. The inhibition of this dipsogenic activity by the ANG II antagonist saralasin clearly demonstrates the involvement of ANG II receptors. Tetradecapeptide lacks intrinsic activity, as its dipsogenic effect is blocked by the converting enzyme inhibitor SQ 20881 (8,16). Therefore, in order to stimulate central ANG II receptors, tetradecapeptide has to be converted into ANG II. The route of this conversion in the brain has been disputed. It has been suggested that this conversion should merely be due to the action of ANG I-converting enzyme (21,22,26). However, the renin inhibitor isovaleryl pepstatin has been reported to reduce tetradecapeptide induced drinking (6,8), indicating the involvement of a renin-like enzyme in the dipsogenic mechanism. This is reinforced in the present study, where the dipsogenic effect of the synthetic renin substrate was reduced by the water-soluble renin inhibitor NAP. Although the in vivo contribution of solely ANG I-converting enzyme in the dipsogenic mechanism of tetradecapeptide cannot be excluded, it is concluded that a renin-like enzyme is, at least partly, involved in tetradecapeptide induced drinking.

The endogenous in vivo generation of ANG I in the brain may be dependent on both the local availability of an ANG I precursor, and the substrate specificity of the ANG I releasing enzyme. Although direct evidence for a functional participation in the central renin-angiotensin system is not yet present, the in brain tissue and in CSF occuring natural angiotensinogen (17,19,25) is a very likely endogenous ANG I precursor. Cathepsin D has been reported to release ANG I from CSF-angiotensinogen in vitro at acidic pH (20). The in vivo studies presented in this paper, however, do not support that cathepsin D significantly contributes to the release of ANG I at physiological pH. ANG I was not generated from the precursor molecule upon injection of the enzyme directly into the CSF of rats. It might be suggested that ANG I can be released from angiotensinogen by cathepsin D within the lysosomes. As suggested by Morris and Reid (20), however, such generation would not expected to be functional, as ANG I would be rapidly degraded by proteolytic activities within these organelles.

ICV injection of renin, in contrast, resulted in a dose dependent ANG I release in the CSF, as indicated by the stimulated drinking activity of the animals. The pattern of water intake was different from that following ICV injection of angiotensins (30) in that both the latency to renin induced drinking and the total drinking time were longer. This indicates the generation of ANG I from endogenous renin substrate.

These data thus point to the involvement of endopeptidases in the generation of ANG I from tetradecapeptide in vitro as well as in vivo. Cathepsin D and renin both are capable of cleaving the Leu-Leu peptide bond of this substrate in vitro, under selective conditions. It is suggested that a renin-like enzyme, rather than cathepsin D, is responsible for the in vivo ANG I generation from synthetic renin substrate and from natural brain angiotensinogen. Such renin-dependent ANG I formation in the brain might have physiological significance with respect to angiotensin-mediated functions as the regulation of blood pressure and water intake.

SUMMARY

The in vitro and in vivo release of angiotensin I from the synthetic renin substrate tetradecapeptide, and the in vivo angiotensin I release from brain angiotensinogen was studied. Purified cathepsin D, renin and non-purified enzymes from brain origin were capable of specifically cleaving the Leu-Leu peptide bond in tetradecapeptide, as analysed by high-performance liquid chromatography of the generated cleavage products. Renin and the brain enzyme preparation, but not cathepsin D, were active at pH 7.4. The in vivo angiotensin I generation from the synthetic substrate by brain proteases was inhibited by N-acetyl-pepstatin. Central administration of renin caused prolonged drinking, whereas cathepsin D was inactive. It is concluded that a renin-like enzyme is responsible for angiotensin I generation in the brain.

ACKNOWLEDGEMENTS

The authors are grateful to Drs. M. Miyazaki (Ohtsu, Japan) and J. Ménard (Paris, France) for the gifts of NAP and purified renin. This study was supported by the Foundation for Medical Research FUNGO which is subsidized by the Netherlands Organization for the Advancement of Pure Research (Z.W.O.).

REFERENCES

1. BRAKKEE JH, WIEGANT VM, GISPEN WH (1979) A simple technique for rapid implantation of a permanent cannula into the rat brain ventricular system.Lab Anim Sci 29:78-81

2. COHEN S, TAYLOR JM, MURAKAMI K, MICHELAKIS AM, INAGAMI T (1972) Isolation and characterization of renin-like enzymes from mouse submaxillary glands. Biochemistry 11:4286-4293

3. CORVOL P, RODBARD J, DROUET J, CATT K, MÉNARD J (1973)Monoiodinated angiotensins: preparation and characterization by polyacrylamide gel electrophoresis and isoelectric focussing. Biochem Biophys Acta 332:392-400

4. DORER FE, KAHN JR, LENTZ KE, LEVINE M, SKEGGS LT (1975) Formation of angiotensin II from tetradecapeptide renin substrate by angiotensin-converting enzyme. Biochem Pharmacol 24:1137-1139

5. DORER FE, LENTZ KE, KAHN JR, LEVINE M, SKEGGS LT (1978) A comparison of the substrate specificities of cathepsin D and pseudorenin. J Biol Chem 253:3140-3142

6. EPSTEIN AN, FITZSIMONS JT, JOHNSON AK (1974) Peptide antagonists of the renin-angiotensin system and the elucidation of the receptors for angiotensin-induced drinking. J Physiol(Lond) 238:34p-35p

7. FISCHER-FERRARO C, NAHMOD VE, GOLDSTEIN DJ, FINKIELMAN S (1971) Angiotensin and renin in rat and dog brain. J Exp Med 133:353-361

8. FITZSIMONS JT, EPSTEIN AN, JOHNSON AK (1978) Peptide antagonists of the renin-angiotensin system in the characterization of receptors for angiotensin-induced drinking. Brain Res 153:319-331

9. FUXE K, GANTEN D, HÖKFELT T, LOCATELLI V, POULSEN K, STOCK G, RIX E, TAUGNER R (1980) Renin-like immunocytochemical activity in the rat and mouse brain. Neuroscience Lett 18:245-250

10. GANTEN D, SPECK G (1978) The brain renin-angiotensin system: a model for the synthesis of peptides in the brain. Biochem Pharmacol 27:2379-2389

11. GANTEN D, MINNICH JL, GRANGER P, HAYDUK K, BRECHT HM, BARBEAU A, BOUCHER R, GENEST J (1971) Angiotensin-forming enzyme in brain tissue. Science 173:64-65

12. GANTEN D, MARQUEZ-JULIO A, GRANGER P, HAYDUK K, KARSUNKY KP, BOUCHER R, GENEST J (1971) Renin in dog brain. Am J Physiol 221:1733-1737

13. GUY MN, ROBERSON GM, BARNES LD (1981) Analysis of angiotensins I, II, III and iodinated derivatives by high-performance liquid chromatography. Anal Biochem 112:272-277

14. HANCKOCK WS, BISHOP CA, PRESTIDGE RL, HARDING DRK, HEARN MTW (1978) Reversed-phase, high-pressure liquid chromatography of peptides and proteins with ion-pairing reagents. Science 200:1168-1170

15. HIROSE S, YOKOSAWA H, INAGAMI T (1978) Immunochemical identification of renin in rat brain and distinction from acid proteases. Nature 274:392-393

16. HOFFMAN WE, SCHELLING P, PHILLIPS MI, GANTEN D (1976) Evidence for local angiotensin formation in brain of nephrectomized rats. Neuroscience Lett 3: 299-303

17. LEWICKI JA, FALLON JH, PRINTZ MP (1978) Regional distribution of angiotensinogen in rat brain. Brain Res. 158:359-371

18. MENDELSOHN FA, JOHNSTON CI (1971) A radiochemical renin assay. Biochem J 121: 241-244

19. MORRIS BJ, REID IA (1978) The distribution of angiotensinogen in dog brain studied by cell fractionation. Endocrinology 103:492-500

20. MORRIS BJ, REID IA (1978) A "renin-like" enzymatic action of cathepsin D and the similarity in subcellular distributions of "renin-like" activity and cathepsin D in the midbrain of dogs. Endocrinology 103:1289-1296

21. RAMSAY DJ, REID IA, BROWN C (1979) Mechanism of the dipsogenic action of tetradecapeptide renin substrate in dogs. Endocrinology 105:947-951

22. RAMSAY DJ, BROWN C, REID IA, KEIL LC (1979) Effect of tetradecapeptide renin substrate on blood pressure, plasma renin activity, and vasopressin and adrenocorticotropin concentrations. Endocrinology 105:1406-1409

23. SEIDAH NG, DENNIS M, CORVOL P, ROCHEMONT J, CHRÉTIEN M (1980) A rapid high-performance liquid chromatography purification method of iodinated polypeptide hormones. Anal Biochem 109:185-191

24. SEMPLE PF, BOYD AS, DAWES PM, MORTON JJ (1976) Angiotensin II and its heptapeptide (2-8), hexapeptide (3-8), and pentapeptide (4-8) metabolites in arterial and venous blood of man. Circ Res 39:671-678

25. SERNIA C, REID IA (1980) Release of angiotensinogen by rat brain in vitro. Brain Res 192:217-225

26. SIMPSON JB, REID IA, RAMSAY DJ, KIPEN H (1978) Mechanism of the dipsogenic action of tetradecapeptide renin substrate. Brain Res 157:63-72

27. SPECK G, POULSEN K, UNGER TH, RETTIG R, BAYER C, SCHÖLKENS B, GANTEN D (1981) In vivo activity of purified mouse brain renin. Brain Res 219:371-384

28. TONNAER JADM, VERHOEF J, WIEGANT VM, DE JONG W (1980) Separation and quantification of angiotensins and some related peptides by high-performance liquid chromatography. J Chromatogr 183:303-309

29. TONNAER JADM, WIEGANT VM, DE JONG W (1981) Angiotensin generation in the brain and drinking: indications for the involvement of endopeptidase activity distinct from cathepsin D. Brain Res 226:in press

30. TONNAER JADM, WIEGANT VM, DE JONG W, DE WIED D (1982) Central effects of angiotensins on drinking and blood pressure: structure activity relationships. Brain Res:in press

31. TONNAER JADM, WIEGANT VM, DE JONG W (submitted) Subcellular localization in rat brain of angiotensin I generating endopeptidase activity distinct from cathepsin D

Properties and Subcellular Localization of Brain Renin and Its Substrate

R. R. Smeby, A. Husain, and R. C. Speth

INTRODUCTION

During the past twenty years the octapeptide angiotensin II has been shown to produce a variety of effects in the central nervous system (CNS) such as increase of blood pressure (3), induction of thirst (8), vasopressin release (19), and induction of sodium appetite (1). These effects may be caused either by peripherally generated angiotensin II (7) or by the peptide generated within the brain tissue itself (10). Recently a neutral protease with renin-like activity (11,20) has been demonstrated in crude brain extracts. However, neither the site at which this enzyme acts to form angiotensin I nor the tissue compartment from which it originates have been established. Renin activity of crude extracts might be derived from three different tissue compartments: first, the parenchyme of the brain itself; second, plasma trapped in the brain at the time of extraction; third, brain microvessels.

Until recently, perfusion of brain has commonly been used to remove blood contamination before preparation of tissue extracts. However, Printz and Gregory (21) in their studies on renin substrate in the brain, have shown that saline perfusion cannot be relied upon to remove all blood, particularly in those areas of the CNS possessing dense vascularity or are devoid of a blood brain barrier. Further, the time period required to wash out the blood during the perfusion procedure may cause cell disruption because of sustained hypoxia.

Recently we have conducted experiments to allow rapid separation of brain renin from plasma renin by chromatographic methods thus avoiding the need for perfusion. Furthermore we have localized the subcellular compartment containing the renin activity in brain tissue and eliminated the possibility that microvessels represent a major source of this enzyme (14).

We have previously demonstrated (20) separation of plasma renin from the brain enzyme using CM-cellulose in dogs, but this method is time consuming and not suitable for use with either large volumes of extract or large numbers of samples. Since active plasma renin is not retained on Affi-gel blue (4) and brain renin is retained (13), we have utilized this absorbant to remove plasma renin activity from brain re-

Experimental Brain Research, Suppl. 4

nin activity in crude extracts. For this separation, a column of Affi-gel blue (0.9 x 6 cm,containing 5 ml of settled gel) was equilibrated with 0.02 M phosphate buffer pH 7.4 and the crude brain extract containing protein from a maximum of 5 g of brain tissue was introduced into the column in the same buffer. The column was washed with 25 ml buffer and then washed with 25 ml of the same buffer containing 1% NaCl. Active plasma renin was not retained on this column and was eluted in the void volume, while brain renin activity was eluted with the salt containing buffer. Thus, this procedure provides a simple, rapid method to remove plasma renin activity from crude brain extracts and to eliminate interference by the plasma enzyme.

After Affi-gel blue treatment, dog brain extract was divided into two equal portions. One half was subjected to chromatofocusing without further treatment and 3 peaks renin activity were observed (Fig 1). The major peak had a pI of 7.92 and 2 other peaks had pI's of 7.73 and 7.42. The other half of the Affi-gel blue treated brain extract was subjected to the usual acid activation procedure by dialysis to pH 3.3 and then dialysis back to pH 7.4 (22). No increase in renin-like activity was observed at this stage. However, when this acid activated extract was dialyzed to pH 9.4 against ethanolamine-acetic acid buffer (0.025 M) and then subjected to chromatofocusing a marked increase in the 2 major peaks of renin activity (pI 7.92 and 7.73) was observed (Fig 2). Further, there was a marked decrease in total protein eluted from the column.

The range of isoelectric points we observed for dog brain renin is markedly different from those reported for renin in plasma and other tissues of species other than dog (Table 1).

Table 1. Isoelectric points of renin from various sources

Renin	pI	Ref
Human plasma (MW 45 to 55K)	5.1, 5.3	6
Rat plasma (MW >150K)	5.3	2
Rat plasma (MW 44K)	4.71, 4.89, 5.05, 5.19$^+$	2
Rat kidney (MW 38K)	4.8, 5.0, 5.3$^+$, 5.5	2
Rat aorta (MW 44K)	5.0, 5.3	2
Hog kidney (MW 42K)	5.2	15
Hog pituitary (MW 43K)	4.48, 4.85, 5.0, 5.14	12
Dog brain (MW 40K)	7.42, 7.73, 7.92$^+$	

$^+$Major peak

Hirose et al. (11) have studied the possible existence of renin and its precursors in the highly vascular regions of hog brain, anterior pituitary, pineal and cerebellum. They observed activation of brain renin activity by both acid and trypsin treat-

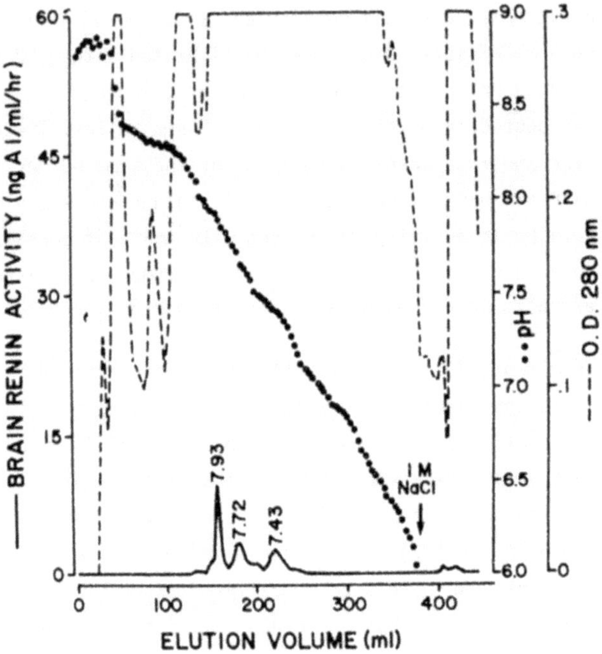

Fig. 1. Fractionation of partially purified dog brain proteins by chromatofocusing; 150 mg soluble brain proteins partially purified on Affi-gel blue dialysed against 0.025 M ethanolamine-acetic acid buffer pH 9.4 was applied to a polybuffer exchanger PBE 94 (0.9 x 60 cm) column pre-equilibrated with the same buffer. Proteins with pI above pH 6 were eluted by a self-forming pH gradient with polybuffer 96. Proteins with pI lower than 6 were eluted with 1 M NaCl

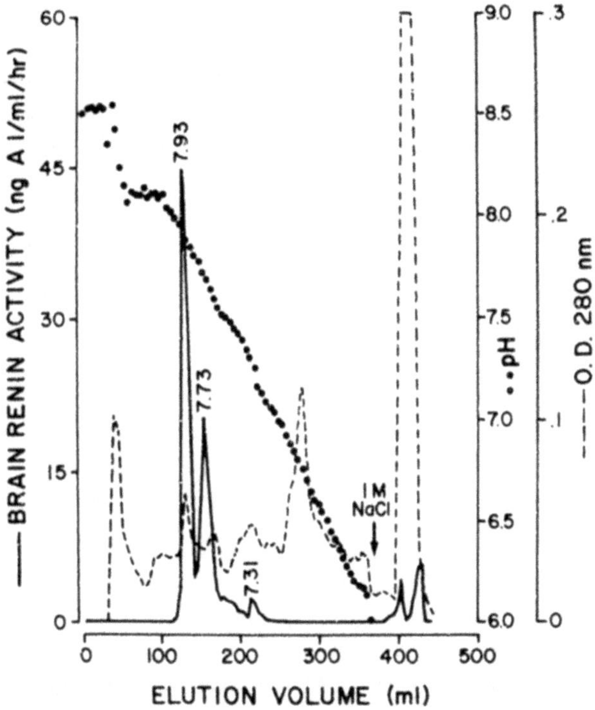

Fig. 2. Chromatofocusing of dog brain proteins as in Fig 1 but following acid activation

ment. Since saline perfusion used in their study does not always remove all blood, there is a possibility of plasma contamination. Because of the perfusion difficulties and the similarity of pI's of hog plasma and brain renin, it is difficult to interpret the observations of Hirose et al. (11). However, in our studies brain renin has been clearly separated from plasma renin and it has a markedly different pI. Therefore, renin in dog brain is a different protein from renin in plasma and it can be activated. Furthermore, Gallagher et al. have shown that inactive plasma renin cannot be demonstrated by acid activation in the dog (9). This lends support to our earlier observation (20) that dog brain activity is poorly inhibited by anti-hog kidney renin antibody at a dilution that inhibited dog kidney renin.

To determine if brain activity is associated with neural elements and to determine where brain renin is localized at the subcellular level, we prepared biochemically characterized synaptosomes from rat brain (14). The brains of 6 female rats (200 - 250 g) were removed and homogenized in 10 volumes (w/v) of 0.32 M sucrose using a loose fitting (0.25 mm clearance) glass teflon homogenizer. The combined homogenates were subjected to subcellular fractionation as described by DeRobertis (5). The distribution of renin and the renin substrate in fractions obtained by differential centrifugation is shown in Table 2. The crude mitochondrial pellet (P_2) contained 81% of the activity, while renin substrate (77%) was found in the soluble fraction as observed by Morris and Reid previously (17). Since very little renin activity was observed in P_1, cerebral vascular renin can be eliminated as a significant contributing source of this enzyme. The P_2 fraction was further fractionated on a discontinuous sucrose gradient, again as described by DeRobertis (5). The majority of the renin activity (59%) after sucrose density gradient separation was located in the synaptosomal fractions C and D (Table 3). The renin activity of the soluble fraction (fraction A) was probably due to particle rupture during the handling, and the renin activity of fraction B may be due to synaptosomal contamination.

From these experiments the following conclusions can be drawn: 1) Renin activity of dog brain tissue can be easily and quickly separated from plasma renin activity using Affi-gel blue. However, the brain enzyme has a different isoelectric point than plasma renin. It is therefore clearly different from the renin found in dog plasma.
2) Renin in the brain exhibits microheterogeneity as has been observed for renin in other tissues and renin activity of brain extracts is significantly increased following acid activation. 3) Renin activity is associated with synaptosomal fractions which suggest that it is present in nerve terminals. Immunohistochemical localization studies have also demonstrated that brain renin activity is associated with neural cells (16,23). 4) Vascular renin of rat brain can probably be excluded as contributing to renin-like activity in brain extracts. This conclusion is in agreement with the demonstration that the majority of enzyme activity from this source is probably cathepsin D (17). 5) Renin substrate is found in the soluble

Table 2. Differential centrifugation of a homogenate from whole rat brain

Fraction (g/min)	Ultrastructur	% Renin Activity	% Renin Substrate Activity
P_1 ($9x10^3$)	Cellular debris	6	3
P_2 ($230x10^3$)	Crude mitochondrial pellet	81	12
P_3 ($3x10^6$)	Microsomes	11	7
S_3	Soluble	2	77

Table 3. Brain renin activity of discontinuous sucrose density fractions from the crude mitochondrial pellet

Fraction	Ultrastructure[+]	Sucrose Conc (M)	Brain Renin (%)
A	Myelin	<0.8	18
B	Membranes, synaptic debris, some synaptosomes	0.8 - 1.0	16
C	Cholinergic nerve endings	1.0 - 1.2	33
D	Non-cholinergic nerve endings	1.2 - 1.4	26
E	Free mitochondria	>1.4	7

[+]As described by DeRobertis et al. (5)

portion of brain homogenates; from a more detailed investigation of substrate distribution, Morris and Reid (18) concluded that the substrate is entirely located within the brain extracellular fluid space. Thus, renin may be physically separated from its substrate in brain by a membrane barrier, and this raises a most interesting question as to how these two molecules interact in vivo to act as a functional brain renin-angiotensin system.

ACKNOWLEDGEMENT

Supported in part by NHBLI Grants HL6835 and HL27164

REFERENCES

1. AVRITH D, FITZSIMONS JT (1978) Intracranial angiotensin II stimulates sodium appetite. J Physiol (Lond) 282:40P

2. BARRETT JD, EGGENA P, KRALL JF, SAMBHI MP (1981) A comparison of physical characteristics of active renin isolated from aorta, plasma and kidney of the rat. Clin Sci in press

3. BICKERTON RK, BUCKLEY JP (1961) Evidence for a central mechanism in angiotensin-induced hypertension. Proc Soc Exp Biol 106:834-836

4. CARLSON EJ, HSUCH WA, LUETCHER JA (1977) Separation of active and inactive renins in human plasma. Circulation 58:II-250

5. DeROBERTIS E, PELLEGINO de IRALDI A, RODRIQUEZ de LORES ARNAIZ G, SALGANICOFF L (1961) Cholinergic and non-cholinergic nerve endings in rat brain. J Neurochem 9:23-25

6. EGGENA P, BARRETT JD, WIEDMAN CE, SAMBHI MP (1979) Studies on renin activation in normal human plasma. Hypertension 1:523-528

7. FERRARIO CM, DICKINSON CJ, McCUBBIN JW (1970) Central vasomotor stimulation by angiotensin. Clin Sci 39:239-245

8. FITZSIMONS JT, SIMONS BJ (1969) The effect of drinking in the rat of intravenous infusions of angiotensin, given alone or in combination with other stimuli of thirst. J Physiol (Lond) 203:45-47

9. GALLAGHER JF, LARAGH JG, ATLAS SA, SEALEY JE (1980) Effect of trypsin or acid treatment on dog plasma renin activity measurements. Endocrinology 107:147-154

10. GANTEN D, HUTCHINSON JS, HAEBARA H, SCHELLING P, FISCHER H, GANTEN U (1976) Tissue isorenin. Clin Sci Mol Med 51:1175-1206

11. HIROSE S, YOKOSAWA H, INAGAMI T (1978) Immunochemical identification of renin in rat brain and distinction from acid proteases. Nature 274:392-393

12. HIROSE S, NARUSE M, OHTSUKI K, INAGAMI T (1981) Totally inactive renin zymogen and different forms of active renin in hog brain tissues. J Biol Chem 256:5572-5576

13. HUSAIN A, SEN S, SMEBY RR, BUMPUS FM (1981) Separation of dog brain isorenin from plasma renin and acid proteases by affinity chromatography. In: Sambhi M (Ed) Heterogeneity of renin and renin substrate. Elsevier, Amsterdam, in press

14. HUSAIN A, SMEBY RR, KRONTIRIS-LITOWITZ J, SPETH RC (1981) Brain renin: localization in rat brain synaptosomal fractions. Brain Research 225:in press

15. INAGAMI T, MURAKAMI K (1977) Pure renin. J Biol Chem 252:2978-2983

16. INAGAMI T, CLEMENS DL, CELIO MR, BROWN A, SANDRU L, HERSCHKOWITZ N, HOFFMAN LH, KASSELBERG AG (1980) Immunohistochemical localization of renin in mouse brain. Neurosci Letters 18:91-98

17. KOWALOFF H, GAVRAS H, BRECHER P (1980) Renin-like enzymatic activity in the cerebral microvessels of the rat. Am J Physiol 238:H384-H388

18. MORRIS BJ, REID IA (1978) The distribution of angiotensin in dog brain studied by cell fractionation. Endocrinology 103:492-500

19. MOUW D, BONJOUR J, MALVIN RL, VANDER AJ (1971) Central action of angiotensin stimulated ADH release. Am J Physiol 220:237-242

20. OSMAN MY, SMEBY RR, SEN S (1979) Separation of dog brain renin-like activity from acid protease activity. Hypertension 1:53-60

21. PRINTZ MP, GREGORY TJ (1981) Brain angiotensinogen: Evidence for an independent and functional central angiotensin system. In: BUCKLEY JP, FERRARIO CM (eds) Central nervous system mechanisms in hypertension. Raven Press, NY 311-326

22. SKINNER SL (1967) Improved assay methods for renin "concentration" and "activity" in human plasma. Circ Res 20:391-402

23. SLATER EE, DEFENDINE R, ZIMMERMAN EA (1980) Wide distribution of immunoreactive renin in nerve cells of human brain. Proc Natl Acad Sci, USA 77:5458-5460

Immunocytochemical Demonstration of Renin and Converting Enzyme in Rat and Mouse Brain

E. Rix, D. Ganten, G. Stock, and R. Taugner

INTRODUCTION

Renin-like activity has been demonstrated in a variety of different tissues besides the kidney e.g. in the submaxillary gland, in the uterus, and in the brain. In 1971, Ganten and colleagues (10) and Fischer-Ferraro et al. (6) first described renin-like activity in the dog and rat brain. Recently, brain renin could clearly be distinguished from other angiotensin I generating acid proteases, like cathepsin D (12,13). Many biological effects of angiotensin, like blood pressure elevation, drinking behaviour, vasopressin release or neuronal excitation, have been observed to follow intraventricular or microiontophoretical application of ANG II or other components of the RAS (for ref. see 7). Furthermore, it has been shown that these effects could be abolished by simultaneously applied ANG II-receptor blocking agents like [Sar^1-Val^5-Ala^8]-angiotensin II. The presence of the other components of the brain-renin-angiotensin-system (RAS), namely angiotensinogen, angiotensin I and II (ANG I, ANG II), converting enzyme (CE), ANG II receptors and angiotensinases has later been shown in brain tissue or cerebrospinal fluid (CSF). Some of the RAS components have been localized by immunocytochemical methods in the central nervous system (8,9,11,16). The cellular localization is an important step to elucidate the physiological function of the RAS in specific brain areas and to study the interaction of the components. We therefore investigated the distribution of renin and converting enzyme, two key components of the RAS in brain by immunocytochemical techniques.

MATERIAL AND METHODS

Specific antibodies against purified submaxillary mouse renin had been raised in rabbits (for details see 17). Antibodies to converting enzyme had been raised in goat against rabbit lung CE (5); these antibodies fully crossreact with rat CE but also with that of some other species (4).

Male mice (NMRI, 40 g) and male rats (Wistar, 300 g) were fixed by retrograde aortic perfusion with a solution of 0.25% glutaraldehyde in formaldehyde-picric-acid (50%

Experimental Brain Research, Suppl. 4

of a saturated picric-acid solution containing 18.5% formaldehyde). The brains of the animals were embedded in paraffin and cut in 7 µm thick sections. These paraffin sections were stained by using the peroxidase-anti-peroxidase technique (PAP) (for details see 23).

Immunocytochemical staining of semithin sections or postembedding staining of ultrathin sections was performed on tissue fixed by perfusion with 0.5% glutaraldehyde in phosphate buffered saline (PBS) instead of formaldehyde-picric-acid solution. After ethanol-dehydration, this material was embedded in Epon 812 (Merck). Semithin and ultrathin sections were de-eponized with sodium methoxide and treated further in the same way as deparaffinized 7 µm sections. Preembedding staining was performed on tissue of rats perfused with an icecold solution of 0.25% glutaraldehyde in PBS. After removing the brain, the region of the choroid plexus was cut immediately in 50 µm sections in a vibratome. These sections were incubated with CE antibodies diluted with phosphate buffer containing 0.3% Triton X 100. After washing, sections were processed according to the PAP-technique, then postfixed in Karnovky's solution for 12 hours. Osmium-tetroxide treatment (0.2%, 15 min) and embedding in Epon 812 followed. After cutting, the ultrathin sections were inspected in the electron microscope without any counter-staining.

Specificity tests were carried out by preabsorbing the appropriate antibodies with purified renin and converting-enzyme, respectively, and by using preimmune sera instead of antisera. No crossreactivity of the renin-antibody with cathepsin D, extracted from spleen of rat and mouse, could be detected (Hackenthal, personal communication). The anti-submaxillary gland renin antibody fully crossreacted with mouse kidney renin (17) and, by about 3%, with rat kidney renin (for more details see 8,19).

DISTRIBUTION OF RENIN IN THE BRAIN

Renin-like immunoreactivity was detected in brains of both rats and mice. In both species, the strongest activity was located in the nucleus supraopticus (Fig. 1) and in the nuclei para- and periventricularis (Fig. 2). However, not all cells in these nuclei showed renin-like activity. Especially for the supraoptic nucleus our data could indicate that the renin-like material is present mainly in the neurosecretory oxytocin- and/or vasopressin-containing nerve cells. In the pituitary gland of the rat, strong renin-like immunoreactivity was also found especially in a number of fibres in the dorsal part of the posterior lobe of this gland. Previous studies have shown that the supraoptic and paraventricular nuclei contain immunocytochemically detectable ANG II-like-material (11) and specific ANG II-receptors (20). Furthermore, electrophysiological findings indicated that micro-

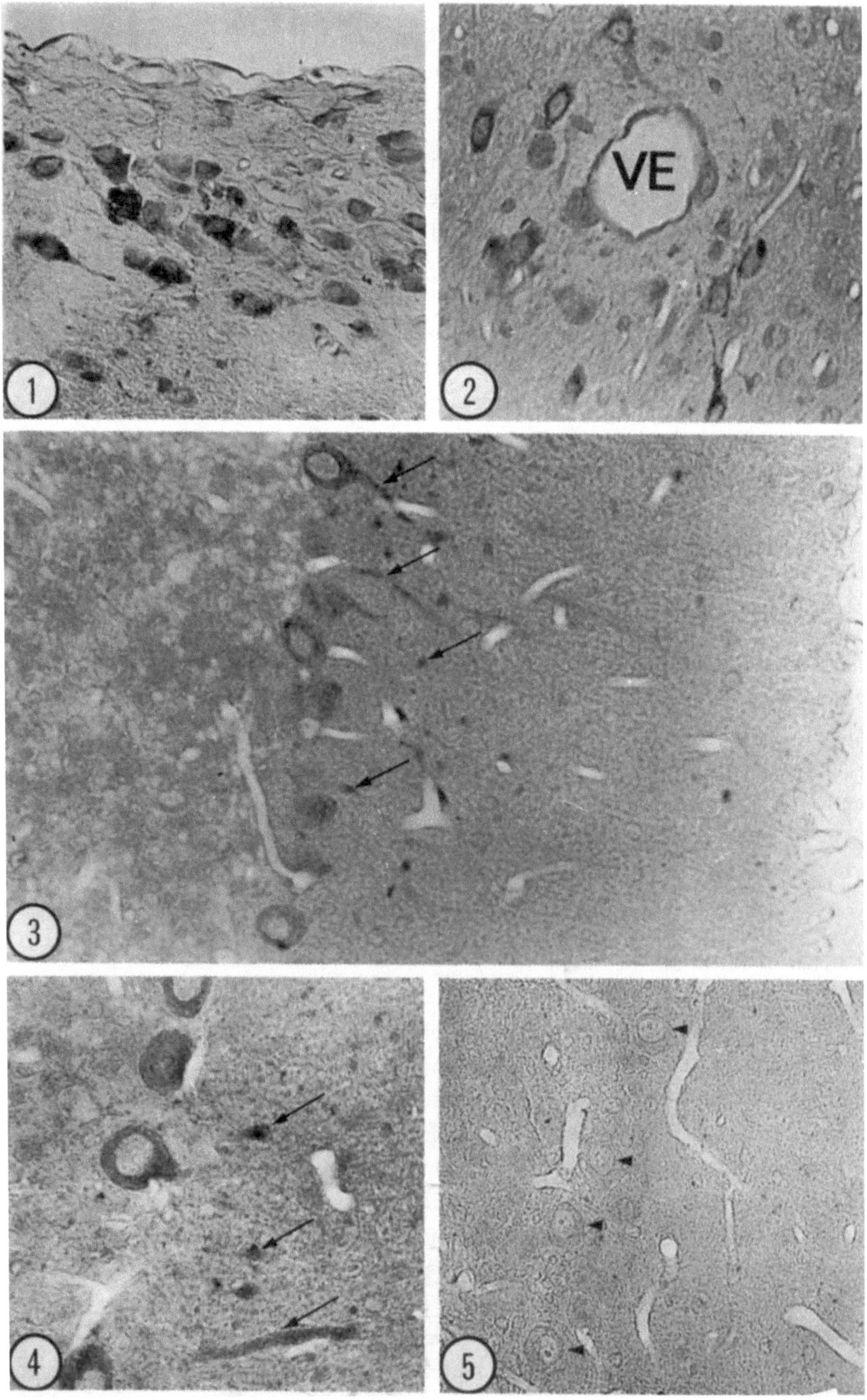
1
2
VE
3
4
5

iontophoretical application of ANG II to the supraoptic nucleus caused excitation of many neurons but, at the same time, reduced the firing rate of a small number of cells (18). In addition, cultured supraoptic cells have been reported to be excited by ANG II (7), others reported vasopressin secretion following ANG II applications (21). These effects could be antagonized by [Sar^1-Ile^8] -angiotensin II (1). ADH release and blood pressure increase following intracerebroventricular injection of ANG II in conscious rat have been reported by several authors (for ref. see 7). These responses were markedly less not only in rats with hereditary diabetes insipidus but also in normal animals after hypophysectomy (14). These data suggest together with our findings a possible close interaction of the brain-RAS, located in the periventricular region and the supraoptic nucleus, with the vasopressin system, with regard to the regulation of water and electrolyte balance and, therefore, blood pressure regulation.

In addition, we were able to demonstrate renin-like staining in mouse and rat cerebellar Purkinje cells (Figs. 3 and 4) and in the vicinity of the vestibular nuclei of the medulla oblongata. The absence of the reaction product from cerebellar Purkinje cells with preabsorbed renin antibodies (Fig. 5) supported the specificity of the immunocytochemical reaction of renin antibodies. At the present time it is difficult to find a functional correlate for the occurence of renin-like material in the Purkinje cells and in some neurons of the medulla oblongata. However, these data support earlier biochemical studies by which the presence of both renin and angiotensin-receptors in the cerebellum and midbrain had been established (11).

CONVERTING ENZYME

Converting enzyme (EC 3.4.15.1) is another important component of the RAS. The peptidyl-dipeptide-carboxyhydrolase transforms angiotensin I to angiotensin II, the ef-

◁ Fig. 1. Paraffin section of rat supraoptic nucleus stained by renin antibody. Antibody dilution 1:1000.(x250)

Fig. 2. Paraffin section of rat periventricular nucleus stained by renin antibody. Ve, Blood vessel. Antibody dilution 1:500.(x300)

Fig. 3. Paraffin section of mouse cerebellar cortex. Purkinje cells and their dendrites (arrows) contain renin-like activity. Antibody dilution 1:1000.(x450)

Fig. 4. Paraffin section of rat cerebellar cortex. Renin-like activity is detectable in Purkinje cells and their dendrites (arrows). Antibody dilution 1:500.(x600)

Fig. 5. Paraffin section of mouse cerebellar cortex. Preabsorption of renin antibodies causes a loss of staining of Purkinje cells (arrowheads); compare with Fig. 3. Antibody dilution 1:1000.(x400)

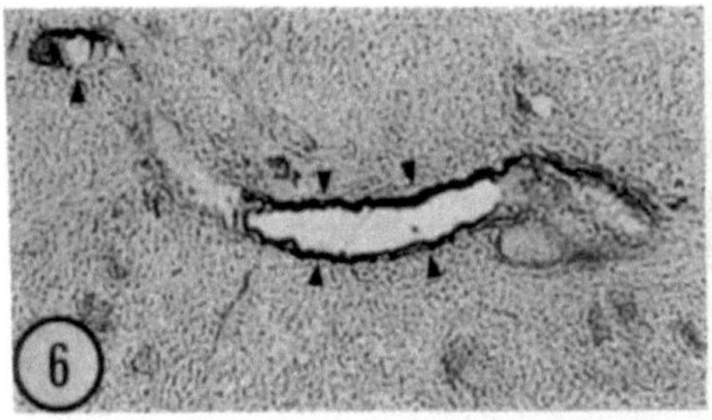
6

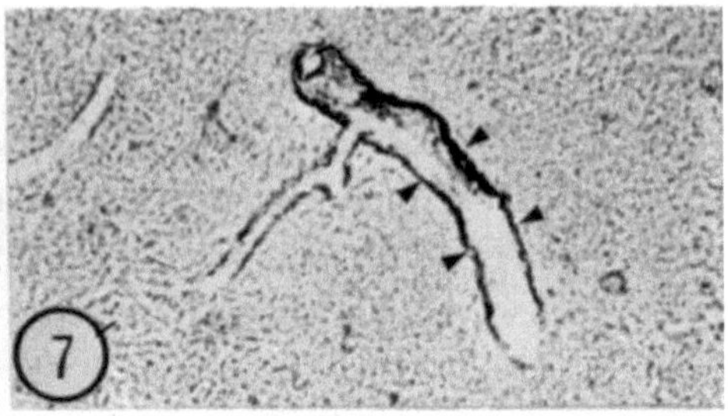
7

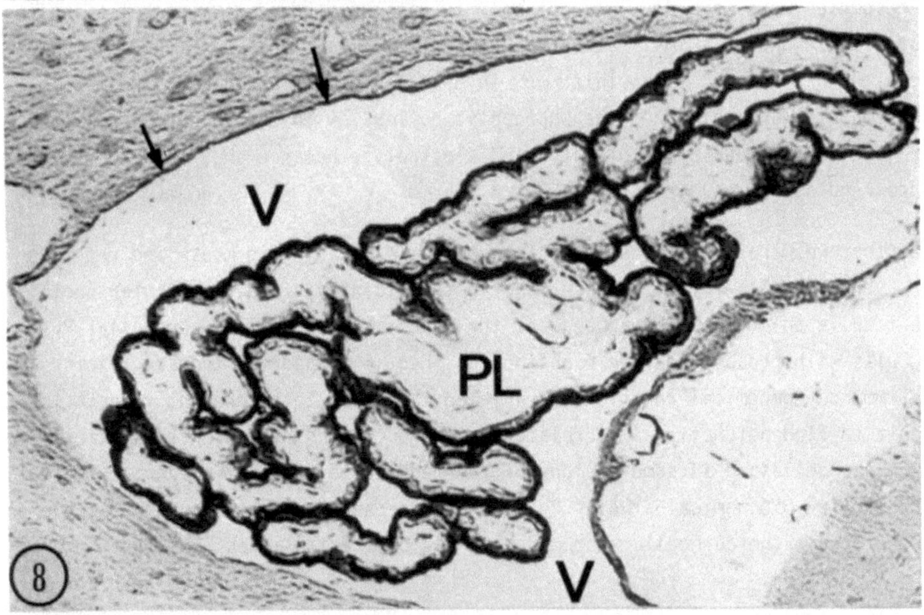
V
PL
V
8

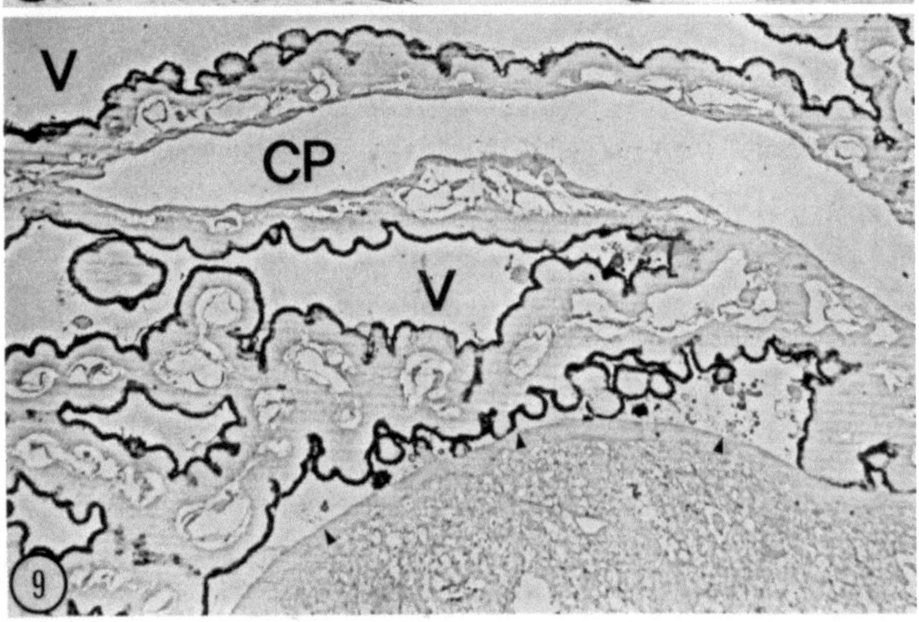
V
CP
V
9

fector peptide of the RAS. This enzyme is identical with kininase II which inactivates bradykinin. Besides its role in these two systems. CE has recently been shown to exhibit also some similarities to the enkephaline degrading enzyme (22). CE, first detected in blood and then found in the kidney, lung, and various other tissues (for ref. see 2) has been located by immunochemical methods in endothelial cells of lung and liver and in the proximal tubule cells of the kidney (2). Recently, the production and release of CE by cultured endothelial cells has been reported (3). In the rat brain, the occurence of CE and its regional distribution, measured by biochemical methods, has first been reported by Yang and Neff (26), The choroid plexus was described to contain the highest activity of CE (15). Other areas like the corpus striatum, caudate nucleus, substantia nigra, retina, and pituitary gland were also reported to contain fairly high concentrations of this enzyme.

The knowledge of cellular and, in addition, subcellular localization of CE in the brain could help to define the compartment in which interaction with its substrate peptides namely ANG I and bradykinin, respectively, take place. Especially for the choroid plexus with its numerous capillaries, the question is, whether the enzyme acts inside or outside the blood brain barrier. In all rats examined immunocytochemical staining for CE has been found in the choroid plexus (Fig. 8) as well as in the wall of some brain blood vessels (Fig. 6,7). On semithin sections of the choroid plexus, the immunocytochemical reaction product is restricted to a thin lining of the epithelial cells at the brain ventricular (luminal) side (Fig. 9). At the electron microscopic level, by pre-(Fig. 11) and postembedding (Fig. 10) immunocytochemical staining of CE could clearly be located on the surface of the brush border of the choroid plexus epithelial cells (Fig. 10,11). The absence of immunocytochemical staining product from epithelial cells and their microvilli with preabsorbed converting enzyme antibodies (Fig. 12 and 13) demonstrates the specificity of the

◁ Fig. 6. Paraffin section of rat cerebrum. Immunoreactivity of converting enzyme is detectable in the wall of a blood vessel (arrow heads). Antibody dilution 1:1000 (x 400)

Fig. 7. Paraffin section of rat cerebellum. Converting enzyme is present in the blood vessel wall (arrow heads). Antibody dilution 1:1000.(x 360)

Fig. 8. Paraffin section of the choroid plexus in the rat lateral ventricle. The reaction product of converting enzyme is detectable over the choroidal epithelium. The ependymal lining (arrows) is free of any staining product. PL, choroidal plexus; V, ventricle. Antibody dilution 1:5000. (x 50)

Fig. 9. Semithin section of the choroid plexus (lateral ventricle). Immunoreactivity of converting enzyme is detectable over the brush border of the choroidal epithelium. The ependymal cells (arrow heads) are free of any staining product. V, ventricle, CP, capillary; Antibody dilution 1:2000.(x 360)

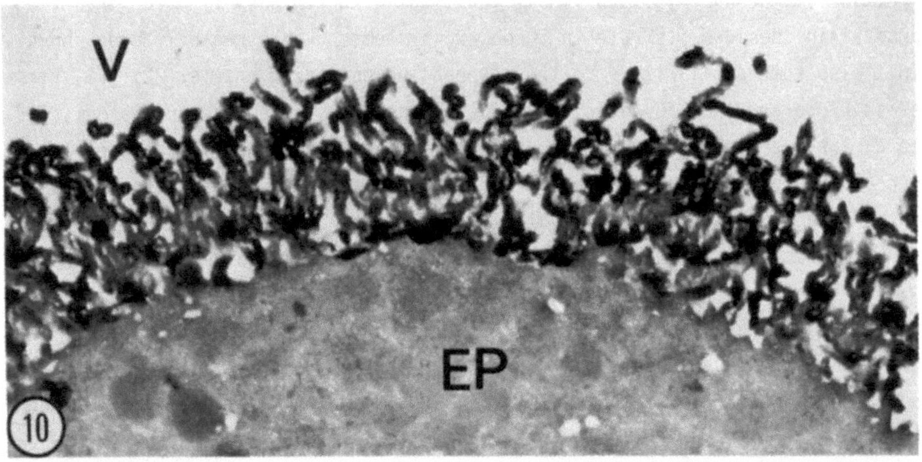
V
EP
10

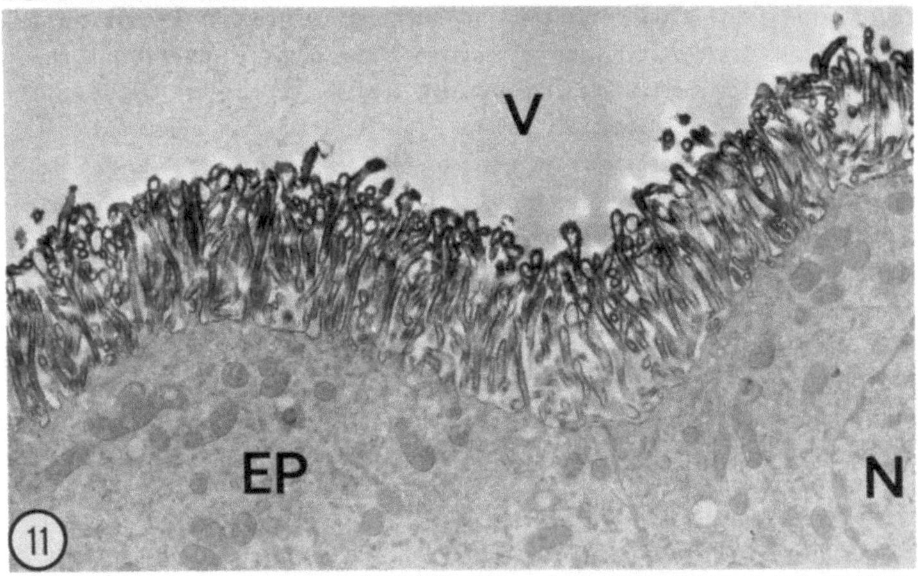
V
EP
N
11

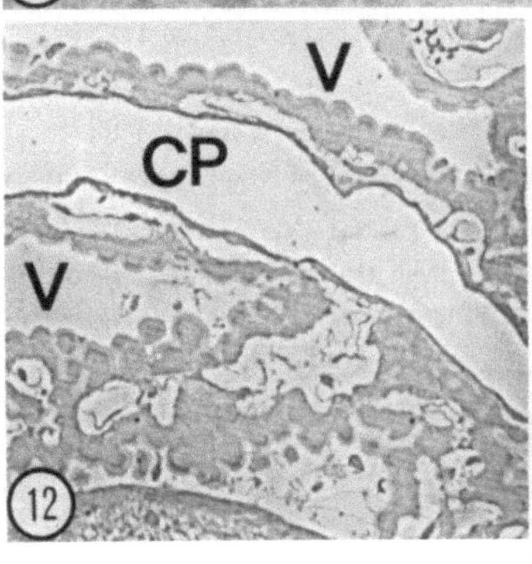
V
CP
V
12

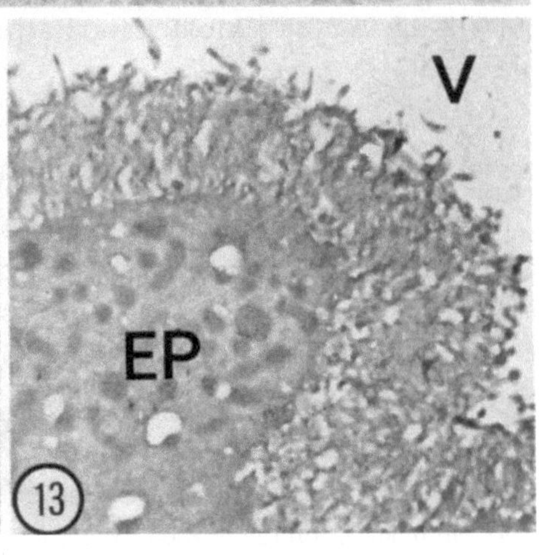
V
EP
13

immunocytochemical reaction of the non-absorbed converting enzyme antibodies. The large surface of the microvilli showing CE activity, is in direct contact with the CSF and thus enables a rapid conversion of ANG I to ANG II or degradation of other peptides like bradykinin or possibly enkephalins.

The localization of CE in brain blood vessels is shown in Fig. 6 and 7. The immunocytochemical reaction product is located over the wall of arterioles and/or capillaries. Although an extremely dense staining of the blood vessel walls was found in some brain areas, e.g. in the cerebellum, we were not able to demonstrate immunoreactivity in capillaries of the choroid plexus. In paraffin sections, the high immunocytochemical CE-activity in the epithelium might have obscured its presence in the nearby lying endothelial cells. On semithin sections negative staining of the endothelium might be due to the relatively high antibody dilutions and, especially, to the loss of antigenic sites by the different fixation media used for Epon embedding.

One can conclude, therefore, that the CE concentration on the surface of the microvilli of the choroid plexus cells is extraordinarily high. These interpretations are supported by biochemical data showing more than 1000-fold higher enzyme concentration in the choroid plexus compared to other brain regions (19). The importance of this localization for the generation or degradation of neuropeptides in the CSF and, therefore, for the peptide actions on their target organs in the periventricular structures remains to be studied. A possible biological role of CE in the brain ventricular system and adjacent brain regions not only for the RAS but also for the kinins and other brain peptides has recently been reported (24). On the other hand, CE could be speculated to have a similar function in the choroid plexus (viewed as a "miniature"-kidney) as in the renal tubules (25). Its influence on the generation

◁ Fig. 10. Electron microscopic micrograph of the choroidal epithelium. By using the postembedding procedure, converting enzyme heavily stains the microvilli. EP, choroidal epithelium; V, ventricle. Antibody dilution 1: 500.(x 9000)

Fig. 11. Converting enzyme immunoreactivity of the brush border of the choroidal epithelium observed in the electron microscope. Using the preembedding technique, staining product is detectable on the surface of the microvilli. EP, choroidal epithelium; N, nucleus; V, ventricle. Antibody dilution 1:1000.(x 8000)

Fig. 12. Semithin section of the choroid plexus. Using preabsorbed converting enzyme antibody, a complete loss of reaction product can be seen; compare to Fig. 9. CP, capillary; V, ventricle. Antibody dilution 1:1000.(x 300)

Fig. 13. By using the postembedding technique and observation in the electron microscope, preabsorbed converting enzyme antibody fails to show any staining product over the brush border of choroidal epithelium; compare to Fig. 10. EP, choroidal epithelium; V, ventricle. Antibody dilution 1:500.(x 6000) and degradation of oligopeptides could for example influence the secretory and resorptive processes and, thereby, the CSF composition.

In conclusion, using immunocytochemical techniques we were able to localize two components of the RAS in the brain. Renin was found especially in the para- and periventricular nuclei as well as in the supraoptic nucleus, and in nerve fibers of the neurohypophysis of rat and mouse brain. In addition, the cerebellar Purkinje cells and some neurons of the deep cerebellar nuclei contain immunocytochemically detectable renin-like material. Converting enzyme was found in high concentration on the brush border of the choroid plexus which is in direct contact to the CSF and, in addition, in the wall of brain arterioles and capillaries.

ACKNOWLEDGEMENTS

The generous gift of pure renin and renin antibodies from Dr. K. Poulsen, Copenhagen, and of pure converting enzyme and converting enzyme antibodies from Dr. R. L. Soffer, New York, and Dr. F. E. Dorer, Cleveland, is gratefully acknowledged. We thank Ms. Harlacher, Ms. Reb and Ms. Simmons for technical assistance.

These studies were supported by the Deutsche Forschungsgemeinschaft within the SFB 90 "Cardiovasculäres System".

REFERENCES

1. AKAISHI T, NEGORO H, KOBAYASI S (1980) Responses of paraventricular and supraoptic units to angiotensin II, Sar^1-Ile^8-angiotensin II and hypertonic NaCl administered into cerebral ventricle. Brain Res 188:499-511

2. CALDWELL PRB, SEEGAL BC, HSU KC, DAS M, SOFFER R (1975) Angiotensin-converting enzyme; vascular endothelial localization. Science 191:1050-1051

3. CHING SF, HAYES LW, SLAKEY LL (1981) Angiotensin-converting enzyme in cultured endothelial cells and growth medium. Biochem Biophys Acta 657:222-231

4. CONROY JM, HOFFMANN H, KIRK ES, HIRZEL HO, SONNENBLICK EH, SOFFER RL (1976) Pulmonary angiotensin-converting enzyme. J Biol Chem 251:4828-4832

5. DAS M, SOFFER RL (1976) Pulmonary angiotensin-converting enzyme anti-enzyme antibody. Biochemistry 15:5088

6. FISCHER-FERRARO C, NAHMOD VE, GOLDSTEIN DJ, FINKIELMAN S (1971) Angiotensin and renin in rat and dog brain. J exp Med 133:353-361

7. FITZSIMONS JT (1980) Angiotensin stimulation of the central nervous system. Rev Physiol Biochem Pharmacol 87:117-167

8. FUXE K, GANTEN D, HÖKFELT T, LOCATELLI V, POULSEN K, STOCK G, RIX E, TAUGNER R (1980) Renin-like immunocytochemical activity in the rat and mouse brain. Neurosci Lett 18:245-250

9. FUXE K, GANTEN D, HÖKFELT T, BOLME P (1976) Immunohistochemical evidence for the existence of angiotensin II containing nerve terminals in the brain and spinal cord of the rat. Neurosci Lett 2:229-234

10. GANTEN D, MINNICH JE, GRANGER P, HAYDUK K, BRECHT HM, BARBEAU A, BOUCHER R, GENEST J (1971) Angiotensin forming enzyme in brain tissue. Science 173:64-65

11. GANTEN D, FUXE K, PHILLIPS MI, MANN JFE, GANTEN U (1978) The brain isorenin-angiotensin system: biochemistry, localization and possible role in drinking and blood pressure regulation. In: GANONG WF, MARTINI L (eds) Frontiers in neuroendocrinology, vol 5, Raven Press, New York

12. GANTEN D, SPECK G, MEYER D, LOOS H-E, SCHELLING P, RETTIG R, UNGER T (1980) Biochemical and functional aspects of brain renin. In: GROSS F, VOGEL G (eds) Enzymatic release of vasoactive peptides. Raven Press, New York, p 31-49

13. HIROSE A, YOKOSAWA H, INAGAMI T (1978) Immunochemical identification of renin in rat brain and distinction from acid proteases. Nature 274:392-393

14. HUTCHINSON JS, SCHELLING P, MÖHRING J, GANTEN D (1976) Pressor action of centrally perfused angiotensin II in rats with hereditary hypothalamic diabetes insipidus. Endocrinology 99:819-823

15. IGIC RP, ROBINSON CJG, ERDÖS EG (1977) Angiotensin I converting enzyme activity in the choroid plexus and in the retina. In: BUCKLEY J and FERRARIO C (eds) Central actions of angiotensin and related hormones. Pergamon Press, New York, pp 23-27

16. INAGAMI T, CLEMENS DL, CELIO MR, BROWN A, SANDRU L, HERSCHKOWITZ N, HOFFMANN LH, KASSELBERG AG (1980) Immunohistochemical localization of renin in mouse brain. Neurosci Lett 18:91-98

17. MALLING C, POULSEN K (1977) A direct radioimmunoassay for plasma renin in mice and its evaluation. Biochem Biophys Acta 491:532-541

18. NICOLL RA, BARKER JL (1971) Excitation of supraoptic neurosecretory cells by angiotensin II. Nature N Biol 233:172-174

19. RIX E, GANTEN D, SCHÖLL B, UNGER TH,TAUGNER R (1981) Converting enzyme in the choroid plexus, brain and kidney: Immunocytochemical and biochemical studies in rats. Neurosci Lett 22:125-130

20. SAKAI KK, MARKS BH, GEORGE J, KOESTNER A (1974) Specific angiotensin II receptors in organ-cultured canine supraoptic nucleus cells. Life Sci 14:1337-1344

21. SLADEK C, JOYNT RJ (1980) Role of angiotensin in the osmotic control of vasopressin release by organ-cultured rat hypothalamo-neurohypophyseal system. Endocrinology 106:173-178

22. SWERTS JP, PERDRISOT R, MALFROY B, SCHWARTZ JC (1979) Is "enkephalinase" identical with "angiotensin-converting enzymes"? Europ J Pharmacol 53:209-210

23. TAUGNER C, POULSEN K, HACKENTHAL E, TAUGNER R (1979) Immunocytochemical localization of renin in mouse kidney. Histochemistry 62:19-27

24. UNGER TH,KAUFMANN-BÖHLER J, SCHÖLKENS B, GANTEN D (1981) Brain converting enzyme: A possible mechanism for the antihypertensive action of captopril in spontaneously hypertensive rats. Europ J Pharm 70:467-478

25. WARD PE, ERDÖS EG, GEDNEY CD, DOWBEN RB, REYNOLDS RC (1976) Isolation of membrane-bound renal enzyme that metabolizes kinins and angiotensin. Biochem J 157:643-650

26. YANG HYT, NEFF NH (1972) Distribution and properties of angiotensin converting-enzyme of rat brain. J Neurochem 14:2443-2450

Brain Converting Enzyme and Hypertension

T. Unger, B. Schüll, B. Schölkens, W. Rascher, R. E. Lang, and D. Ganten

INTRODUCTION

Angiotensin I (ANG I) is rapidly converted into angiotensin II (ANG II) when injected into the brain ventricles (icv) and induces a blood pressure increase, that is indistinguishable from the one seen after icv injection of ANG II. The same biological effects can also be elicited by injection of renin into the brain ventricles and can be prevented by simultaneous central administration of converting enzyme (CE) inhibitors and by ANG II receptor blockade. These and other results indicated the presence and in vivo activity of CE in the brain (6). Kinins given icv also produce blood pressure increases in conscious rats (13). These effects are potentiated by the CE inhibitor captopril, pointing to the identity of CE and the kinin degrading enzyme kininase II in the brain. The effect of icv pretreatment with the CE inhibitor captopril on the pressor responses to icv injected ANG I and bradykinin is shown in Fig. 1.

These data and reports demonstrating the presence of CE in the brain by biochemical methods (24) warranted an investigation into the role of brain CE in blood pressure control.

CENTRAL ADMINISTRATION OF THE CE INHIBITOR CAPTOPRIL IN SPONTANEOUSLY HYPERTENSIVE RATS

Having established that the brain CE activity can be inhibited by captopril in vivo, we addressed the question whether central CE blockade would have an effect on blood pressure in spontaneously hypertensive rats in which a stimulated brain RAS is believed to contribute to the maintenance of hypertension (7,14).

Captopril was injected at different doses into the lateral brain ventricle of conscious spontaneously hypertensive rats of the stroke-prone strain (SHRSP) and age- and sex matched normotensive Wistar Kyoto controls (WKY). In the SHRSP a

Experimental Brain Research, Suppl. 4

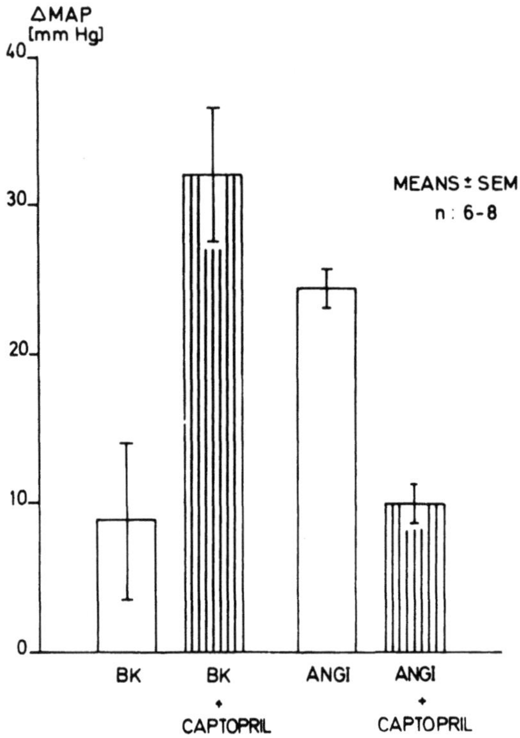

Fig. 1. Potentiation of the pressor responses to intracerebroventricularly (i.c.v.) injected bradykinin (2 μg) and reduction of the pressor responses to i.c.v angiotensin I (100 ng) by i.c.v. captopril (1 μg) in spontaneously hypertensive rats (SHRSP). BK: Bradykinin; ANG I: Angiotensin I; MAP: Mean arterial pressure

moderate decrease of blood pressure (-13 ± 1 mmHg) occurred at a dose of 5 µg which lasted about 20 min and was accompanied by a marked central inhibition of CE in vivo as indicated by a 68% reduction of the pressor responses to i.c.v. ANG I. The pressor responses to i.v. ANG I were not different from those in control SHRSP treated with i.c.v. saline instead of captopril. In the WKY captopril did not have any effect on blood pressure at this dose of 5 µg. When high doses of captopril (500 µg) were injected i.c.v., an initial pressor response was seen in SHRSP and in WKY which lasted about 30 min and which was then followed by a marked and long lasting blood pressure fall (-55 ± 6 mmHg) of more than 4-5 hours duration in SHRSP; in the WKY blood pressure returned to control levels and did not decrease further (Fig. 2). The in vivo CE activity was tested as described in the experiment above and found to be completely blocked centrally for up to 24 hrs., but also in the periphery for some hours. However, the blood pressure decreasing effect outlasted the peripheral CE inhibition by far: four hours after i.c.v. captopril injection, the pressor responses to i.v. ANG I had returned to 80% of the control values while blood pressure was still maximally decreased. Intravenous injection of the same dose of captopril (500 µg) was much less effective in lowering blood pressure in SHRSP than i.c.v. administration (-20.8 ± 6.1 mmHg) although the peripheral in vivo CE was almost totally inhibited. In WKY, 500 µg captopril i.v. did not change arterial blood pressure (for details of this study see 19).

From these experiments it can be concluded that the blood pressure fall in response to central captopril administration in SHRSP was due to a central and not to a peripheral inhibition of CE. This is indicated first by the lack of a peripheral CE inhibition following the blood pressure reducing central dose of 5 µg captopril, second by the markedly greater blood pressure lowering effect of 500 µg captopril injected i.c.v. when compared with the same dose given i.v., and third by the dissociation between blood pressure reduction and peripheral CE inhibition following 500 µg captopril i.c.v. Our findings confirm and extend those by Hutchinson et al. (9) and Stamler et al. (18) who observed similar blood pressure reductions in SHR following high i.c.v. doses of captopril but did not differentiate between central and peripheral CE blocking effects of the drug.

The initial blood pressure rise following the high i.c.v. dose of captopril in the hypertensive and normotensive animals, which has also been reported by Schölkens et al. (17) is difficult to explain. Unspecific effects such as changes in cerebrospinal fluid osmolality of pH were ruled out in control experiments. On the other hand, we have shown that captopril potentiated the pressor effects of exogenous bradykinin administered i.c.v. In view of these findings and on the basis of in vitro evidence, that CE inhibitors can inhibit brain enzymes that degrade peptides with central pressor actions, such as the enkephalins, substance P and kinins

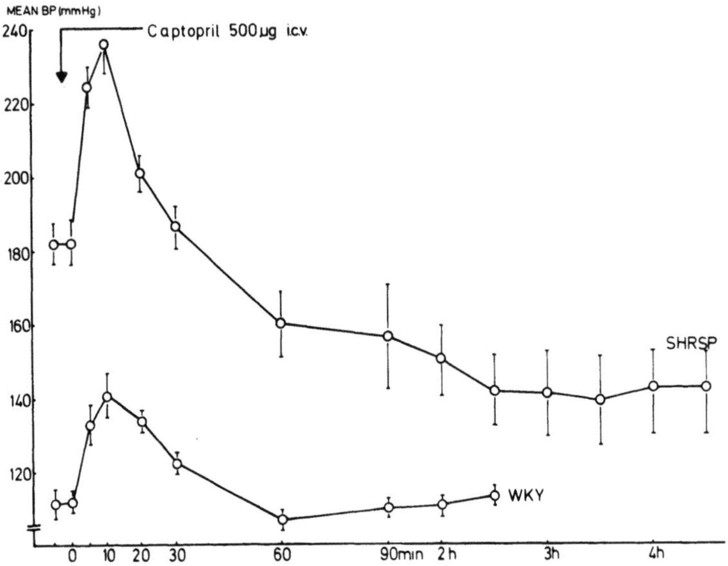

Fig. 2. Blood pressure (BP) responses to intracerebroventricular (i.c.v.) bolus injections of captopril (500 µg) in spontaneously hypertensive rats (SHRSP, n=14) and normotensive Wistar Kyoto controls (WKY, n=6). Shown are means ± S.E.M. Note the increases of BP in both SHRSP and WKY and the prolonged decrease of BP in SHRSP

(16,20,23) we speculate that the blood pressure increasing effects of high i.c.v. doses of captopril were brought about by an inhibition of CE-like peptidases in the brain which break down centrally acting endogenous pressor peptides.

Our observation of enhanced pressor responses to i.c.v. captopril in SHRSP would be in line with this interpretation, since in SHRSP a supersensitivity to the central pressor effects of enkephalins (15,16), bradykinin and substance P (20) has been demonstrated. The increase of substance P-like immunoreactive material in distinct brain areas following i.c.v. administration of CE inhibitors (8) is another argument in favour of this interpretation.

EFFECT OF ORAL TREATMENT WITH CE INHIBITORS ON BRAIN CE IN SHRSP

To test whether a central in vivo inhibition of CE would occur not only after i.c.v. but also after oral treatment with CE inhibitors, and whether this could contribute to the well-known but still poorly understood antihypertensive action of CE inhibitors in SHR (4,12), a number of experiments was carried out:

The CE inhibitors captopril and SA446 ((2R, 4R)-2-(2-Hydroxyphenyl)-3-(3-mercapto-propionyl)-4-thiazolidine-carboxylic acid) were given per os at a dose of 50 mg/kg/day to SHRSP for various periods of time. At the end of the respective treatment periods the rats were challenged with i.c.v. ANG I as an in vivo test of brain CE inhibition. In addition, renin concentrations and CE activities in blood pressure controlling brain regions (hypothalamus, medulla oblongata and neurohypophysis) were measured.

Four week oral treatment with captopril did not inhibit the pressor responses to i.c.v. ANG I. In contrast, the more lipophilic CE inhibitor SA446 produced slightly more central (21%) than peripheral (18%) converting enzyme inhibition in vivo; it enhanced the pressor responses to i.c.v. injected ANG II and increased the renin concentrations in the hypothalamus (22).

Inspite of the apparent lack of a brain CE inhibition in vivo after 4 weeks of oral treatment as tested by the i.c.v. injection of angiotensin, captopril induced an increase of renin concentration in the medulla oblongata after 2 and 6 months (Fig. 3) and in the hypothalamus and neurohypophysis after 6 months of oral treatment. In analogy to the well-known negative feedback mechanism between ANG II and renin in the periphery (2) this can be interpreted as an increase in tissue renin due to withdrawal of feedback inhibition following reduced local ANG II generation in the brain, this would suggest that central CE inhibition did occur after prolonged oral captopril treatment.

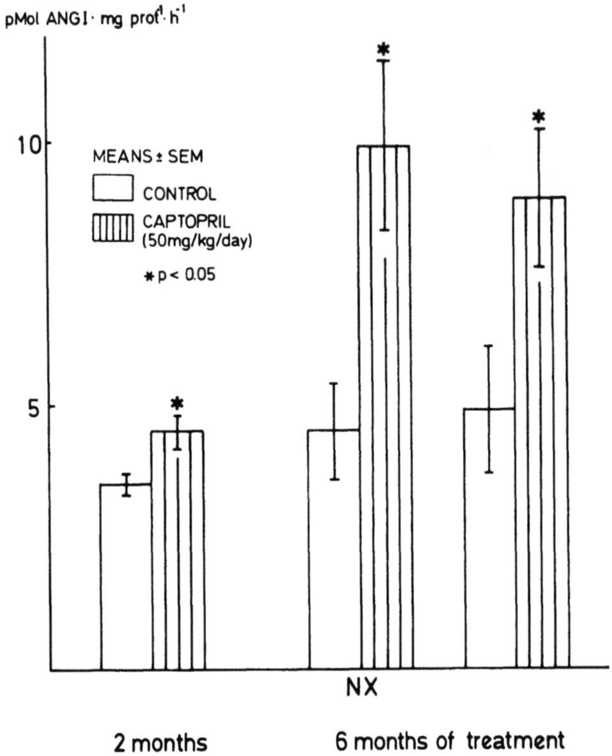

Fig. 3. Effects of chronic oral treatment with captopril on renin concentration in the medulla oblongata of spontaneously hypertensive rats (SHRSP).
NX: This group was nephrectomized 20 hours before sacrifice to exclude contamination of brain tissue with circulating renin of renal origin. It can be seen that renin was elevated in the treated animals irrespective of whether they had been nephrectomized or not

In vitro measurements of the CE activity performed with a fluorometric assay according to Depierre and Roth (3) revealed changes in CE activity in the neurohypophysis, medulla oblongata, plasma and lung after two and four months of oral captopril treatment. Surprisingly, increases of the CE activity were invariably found and not decreases as one might have expected in the presence of inhibited CE activities in vivo (Fig. 4). It has been reported that prolonged storage of plasma samples results in a loss of the inhibitory efficacy of captopril probably due to oxidation and dimerization of its sulfhydryl moiety (1,10,11). This prompted us to investigate whether such a phenomenon could explain the lack of CE inhibition observed in our long-term studies with captopril.

Plasma and tissue samples from rats treated acutely with captopril were stored on ice for different periods of time before CE was measured. We observed that the inhibition of CE was relatively stable for up to 2 hrs after sampling. With longer storage times, however, a progressive loss of the CE inhibition was noted (Fig. 5).

In our studies with chronic CE inhibitor treatment, CE activities were always analyzed within less than 2 hrs after sampling. Hence, it appears unlikely that inactivation of captopril during processing was responsible for the observed lack of CE inhibition. Moreover, this would not explain our finding that CE was increased to above control levels in the chronically treated animals (Fig. 4).

In another study (21) we found a good correlation between the pressor responses to i.v. ANG I and the CE activity in the plasma only after acute i.v. injections of captopril but not after chronic oral treatment: In these studies the ANG I pressor responses and the plasma CE activity were both inhibited by about 70% following 20 µg captopril i.v. After one week oral captopril treatment (50 mg/kg/day) the in vitro inhibition of CE in the plasma was less (23%) than the inhibition of the pressor responses to i.v. ANG I (84%), and after 4 months of oral treatment CE was markedly increased in the plasma while the pressor responses to i.v. ANG I were still inhibited by 70%.

Our findings and those of other investigators (1,5) therefore strongly suggest that chronic treatment with the CE inhibitor leads to an induction of CE,which gives rise to the dissociation between the in vivo and the in vitro measurements of the enzyme. Thus, in the presence of an increased CE concentration in the plasma and in tissue following chronic captopril treatment, the ANG I pressor assay still reflects the in vivo inhibition of CE activity despite increased enzyme concentrations, whereas in vitro assays have measured either increased concentration or reduced activity of CE (1).

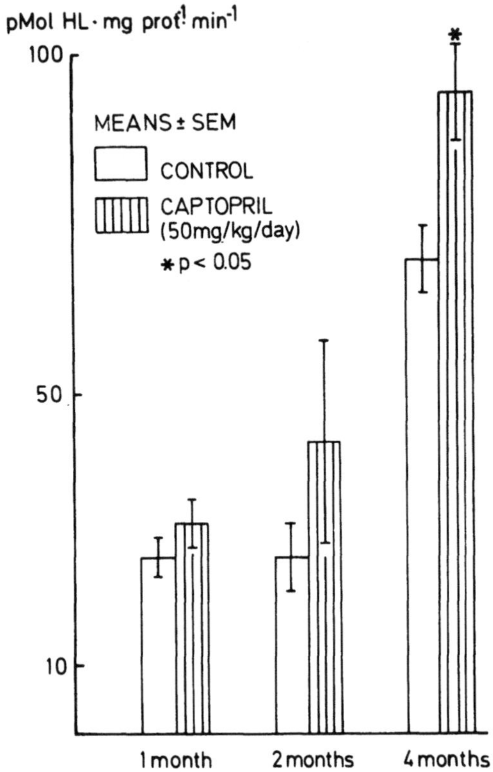

Fig. 4. Effects of chronic oral treatment with captopril on converting enzyme activity in the medulla oblongata of spontaneously hypertensive rats (SHRSP). Note that converting enzyme was increased above control levels in the chronically treated animals. HL: Histidyl-Leucine

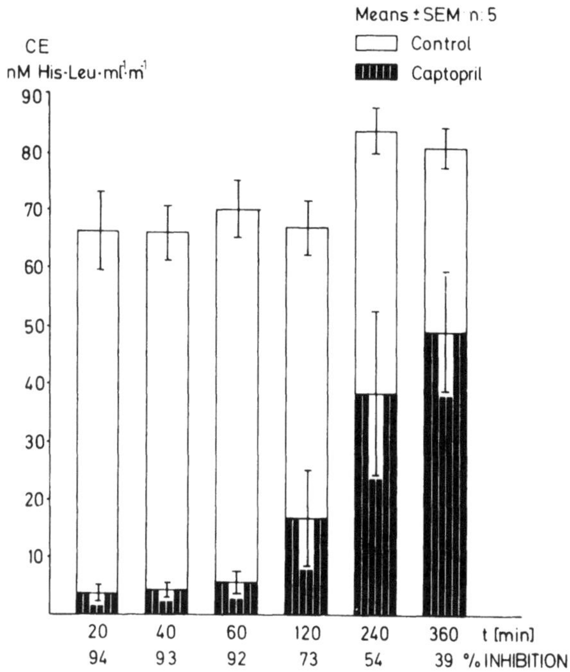

Fig. 5. Effects of cold storage on the inhibition of plasma converting enzyme activity by captopril. Rats were treated orally for one day with the converting enzyme inhibitor at a dose of 50 mg/kg. Samples were stored on ice, and aliquots were analyzed at different times after withdrawal as indicated

In conclusion, we have shown that centrally administered CE inhibitors such as captopril inhibit brain CE and lower blood pressure in SHRSP by a central action. Transient pressor effects after high doses of captopril injected i.c.v. may be due to the inhibition of enzymes involved in the inactivation of endogenous pressor peptides in the brain such as bradykinin, enkephalins or substance P. Chronic oral treatment of SHRSP with the CE inhibitors captopril and SA446 lowers blood pressure, produces characteristic signs of an inhibited brain CE activity and may lead to an induction of the enzyme in the brain and in the periphery. Interference with the stimulated brain RAS can contribute to the antihypertensive mechanisms of CE inhibitors in SHRSP.

REFERENCES

1. BOOMSMA F, DEBRUYN JHB, DERKX FHU, SCHALEKAMP MADH (1981) Opposite Effects of Captopril on Angiotensin I Converting Enzyme "Activity" and "Concentration"; Relation between Enzyme Inhibition and Long-Term Blood Pressure Response. Clin Sci 60:491-498

2. DAVIS JO, FREEMAN RH (1976) Mechanisms Regulating Renin Release. Physiol Rev 56: 1-56

3. DEPIERRE D, ROTH M (1975) Fluorimetric Determination of Dipeptidylcarboxy-peptidase (Angiotensin I-Converting Enzyme). Enzyme 19:65-70

4. FERRONE RA, ANTONACCIO MJ (1979) Prevention of the Development of Spontaneous Hypertension in Rats by Captopril (SQ14225). Europ J Pharmacol 60:131-137

5. FYHRQUIST F, FORSLUND T, TIKKANEN I, GRONHAGEN-RISKA C (1980) Induction of Angiotensin Converting Enzyme in Rat Lung with Captopril (SQ14225). Europ J Pharmacol 67:473-476

6. GANTEN D, SPECK G (1978) The brain renin-angiotensin system: A model for the synthesis of peptides in the brain. Biochem Pharmacol 27:2379-2389

7. GANTEN D, UNGER TH, ROCKHOLD R, SCHAZ K, SPECK G (1979) Central Peptidergic Stimulation: Its Possible Contribution to Blood Pressure Regulation. In: ALBERTINI A, da PRADA M, PESKAR BA (eds) Radioimmunoassay of Drugs and Hormones in Cardiovascular Medicine. Biomedical Press, p 33-40

8. HANSON GR, LOVENBERG W (1980) Elevation of Substance P-like Immunoreactivity in Rat Central Nervous system by Protease Inhibitors. J Neurochem 35:1370-1374

9. HUTCHINSON JS, MENDELSOHN FAO, DOYLE AE (1980) Blood Pressure Responses of Conscious Normotensive and Spontaneously Hypertensive Rats to Intracerebro-ventricular and Peripheral Administration of Captopril. Hypertension 2:546-550

10. LAI FM, TANIKELLA T, HERZLINGER H, CERVONI P (1980) Effects of the time interval between blood sampling and assay on serum angiotensin I converting enzyme activity from captopril-treated rats. Europ J Pharmacol 67:97-99

11. LEVENS NR, PEACH MJ, DARRACOTT VE, WEED UC, CAREY RU (1981) Responses of blood pressure and angiotensin-converting enzyme activity to acute captopril administration in normotensive and hypertensive rats. Endocrinology 108:536-544

12. MUIRHEAD EE, PREWITT RL, BROOKS B, BROSINS WL (1978) Antihypertensive action of the orally active converting enzyme inhibitor (SQ 14225) in spontaneously hypertensive rats. Circ Res 43 (Suppl.I) 53-58

13. LAMBERT GA, LANG WJ (1970) The effects of bradykinin and eledoisin injected into the cerebral centricles of conscious rats. Europ J Pharmacol 9:383-386

14. PHILLIPS MI (1980) Biological effects of angiotensin in the brain. In: GROSS F, VOGEL HG (eds) Enzymatic Release of Vasoactive Peptides. Raven Press, New York, 337-360

15. ROCKHOLD RW, CROFTON JT, SHARE L (1981) Vasopressin release does not contribute to pressor action of enkephalin in SHR. Hypertension 3:410-415

16. SCHAZ K, STOCK G, SIMON W, SCHLÖR KH, UNGER TH, ROCKHOLD R, GANTEN D (1980) Enkephalin effects on blood pressure, heart rate and Baro-receptor reflex. Hypertension 2:395-407

17. SCHÖLKENS BA, STEINBACH R, JUNG W (1979) Intracerebroventricular Injection of an Angiotensin I Converting Enzyme Inhibitor (SQ 14225) in Animals with Experimental Hypertension. Naunyn-Schmiedeberg's Arch Pharmacol 307 (Suppl) 184,R46

18. STAMLER JF, BRODY MJ, PHILLIPS MI (1980) The central and peripheral effects of captopril (SQ 14225) on the arterial pressure of the spontaneously hypertensive rat. Brain Res 186:499-503

19. UNGER TH, KAUFMANN-BÜHLER I, SCHÖLKENS B, GANTEN D (1981) Brain Converting Enzyme Inhibition: A possible mechanism for the antihypertensive action of captopril in spontaneously hypertensive rats. Europ J Pharmacol 70:467-478

20. UNGER TH, ROCKHOLD RW, YUKIMURA T, RETTIG R, RASCHER W, GANTEN D (1981) Role of kinins and substance P in blood pressure regulation of normotensive and spontaneously hypertensive rats. In: BUCKLEY JP, FERRARIO CM (eds) Central nervous system mechanisms in hypertension. Raven Press, New York, pp 115-127

21. UNGER TH, SCHÜLL B, HÜBNER D, YUKIMURA T, LANG RE, RASCHER W, GANTEN D (1981) Plasma-converting enzyme activity does not reflect effectiveness of oral treatment with captopril. Europ J Pharmacol 72:255-259

22. UNGER TH, YUKIMURA T, MARIN-GREZ U, LANG RE, RASCHER W, GANTEN D: SA446, a new orally active converting enzyme inhibitor: antihypertensive action and comparison with captopril in stroke-prone spontaneously hypertensive rats. Europ J Pharmacol (in press)

23. UNGER TH: In:Minireview:The brain renin angiotensin system 1982. This volume

24. YANG H-YT and NEFF NH (1972) Distribution and properties of angiotensin converting enzyme of rat brain. J Neurochem 19:2243-2250

Brain Angiotensin Converting Enzyme, Plasma Beta-Endorphin and Thyrotropin in New Zealand and Japanese Genetically Hypertensive Rats

J. S. Hutchinson, F. A. O. Mendelsohn, J. Csicsmann, R. DiNicolantonio, Y. Takata, and A. E. Doyle

INTRODUCTION

Angiotensin-containing neurones have been identified and mapped immunologically in the rat brain by several groups (7,17,20,23). Angiotensin II (ANG II) has central nervous actions on blood pressure (9,24,26), thirst (5,27), salt appetite (1,4,6), the release of vasopressin (25,31,32) and of ACTH (25). The brain ANG II forming system has been postulated to be involved in the pathogenesis of raised blood pressure (22). Since the levels of immunoreactive ANG II in the brain are very low and difficult to quantitate (12) it was decided to measure angiotensin converting enzyme (ACE) as a possible index of ANG II-formation in the brain of the Japanese spontaneously hypertensive rats (SHR) and their normotensive controls (WKY) Brain ACE levels were also investigated in the New Zealand strain of genetically hypertensive rats (GHR) and their normotensive controls (NT-NZ) as the hyperactivity of their brain ANG II system has been suggested to be involved in the maintenance of hypertension in GHR (16).

METHODS

Six female GHR and six female NT-NZ (28) bred from the same colony were studied at 17-22 weeks of age when their blood pressures were 164 ± 5.5 and 124 ± 3.6 mmHg (mean ± SEM) respectively. Twelve female SHR and twelve female WKY (21) bred from the same colony were studied at 20-22 weeks of age when their blood pressures were 173 ± 3.1 and 121 ± 2.7 mmHg. The animals were killed by decapitation, their brains dissected into eight different regions (11). ACE was measured as previously described using a fluorometric procedure and the synthetic substrate Hip-His-Leu (19); protein was measured by the Lowry method (18).

Blood pressures were measured in groups of hypertensive and normotensive rats of both strains by a tail cuff method in conscious rats. SHR, WKY, GHR and NT-NZ rats were decapitated, plasma samples obtained and immunoreactive ß-endorphin and thyrotropin estimated by radioimmunoassays (14,30).

Experimental Brain Research, Suppl. 4

RESULTS

As shown in Fig. 1, there were marked regional variations in ACE with highest levels in striatum, cerebellum, hippocampus and pituitary. ACE was significantly lower in medulla oblongata, hypothalamus and cerebral cortex of SHR compared to WKY. The ACE levels were not significantly different between the two groups at the other brain sites. ACE was lower in the cerebral cortex of GHR compared to NT-NZ but not different between the two groups for the other seven brain regions.

Plasma levels of immunoreactive β-endorphin are lower in SHR than in WKY ($p < 0.05$) but there is no statistically significant difference in plasma β-endorphin between GHR and NT-NZ. SHR have higher circulating thyrotropin levels than WKY ($p < 0.005$) but, again, there is no significant difference between the circulating immunoreactive thyrotropin of GHR and NT-NZ (Table 1).

Table 1. Plasma concentrations of immunoreactive β-endorphin and thyrotropin in hypertensive and normotensive rats of the New Zealand and Japanese strains

	NEW ZEALAND		JAPANESE	
	GHR	NT-NZ	SHR	WKY
n	10	10	7	7
Blood pressure (mmHg)	161 ± 6	123 ± 3	179 ± 3	117 ± 1.0
β-endorphin (ng/ml)	0.382 ± 0.07	0.363 ± 0.04	0.09 ± 0.03	0.19 ± 0.03
	n.s.		$p < 0.05$	
n	6	6	6	6
Blood pressure (mmHg)	158 ± 4	130 ± 4	154 ± 5	113 ± 3
Thyrotropin (µU/ml)	5.3 ± 1.4	8.8 ± 2.0	6.9 ± 0.6	2.0 ± 0.4
	n.s.		$p < 0.005$	

DISCUSSION

The regional variation of brain ACE is similar in both hypertensive and normotensive rats of both the Japanese and New Zealand strains and other strains of rats (33). The pattern of differences in brain ACE between the normotensive and hypertensive rats of the New Zealand strain is different from that observed for the Japanese strain where the SHR had lower levels of ACE in the medulla oblongata, hypothalamus and cerebral cortex. It is possible that the depression of ACE in cerebral cortex of both strains of hypertensive rats is related to hypertension, perhaps by effects on cerebral microvessels (3,10). However, the differences in medulla oblongata and hypothalamus are not seen in the GHR and may be a specific genetically determined phenomenon in the SHR.

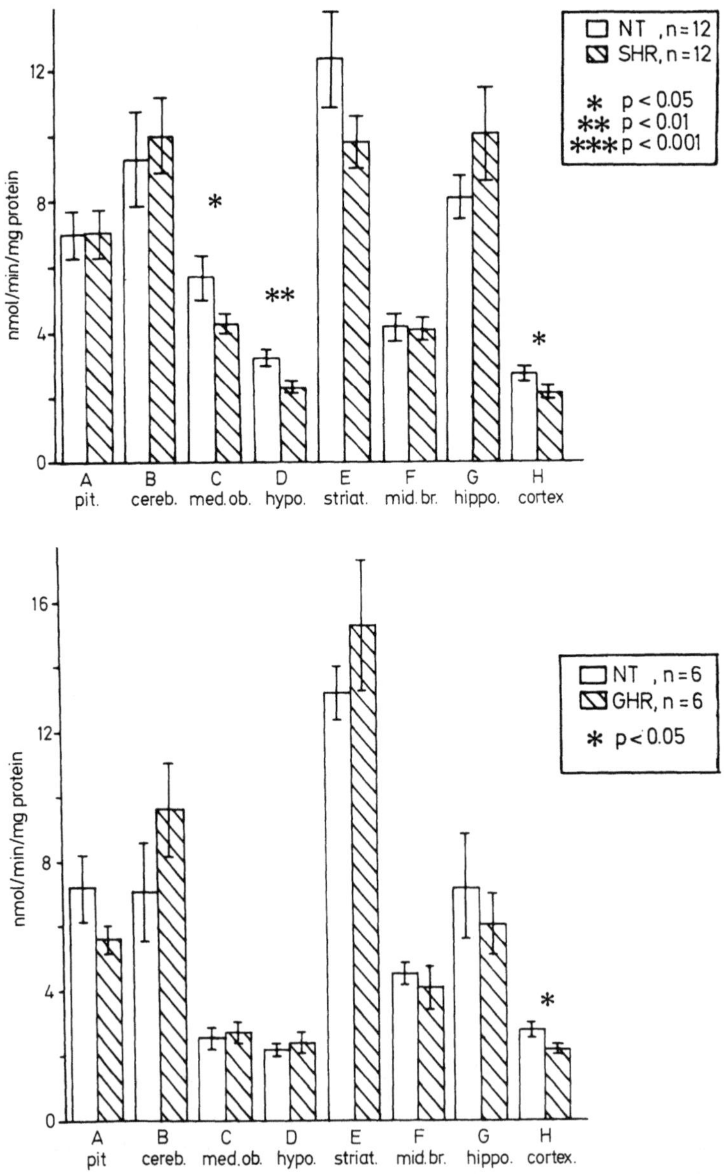

Fig. 1. Angiotensin converting enzyme levels in brain regions of Japanese and New Zealand strains of genetically hypertensive rats (SHR and GHR) and their normotensive (NT) controls

These results also suggest that if the brain ANG II system plays a role in hypertension as studies with central injections of angiotensin II antagonists suggest (22), then the GHR and SHR must have distinct abnormalities in the central neuropeptides controlled by ACE. It has been suggested that ACE may also act as an enkephalinase (2). Opioid levels in the brain of SHR and GHR may also be different. This is the case for β-endorphin where SHR have elevated levels in the neurointermediate lobe compared to WKY, but there are no differences in the levels found in the neurointermediate lobes of GHR and NT-NZ (15). Beta-endorphin, like several other neuroendocrine secretions appears to be under dopaminergic control: it is therefore of interest that infusions of bromocriptine decreased both blood pressure and the levels of β-endorphin in the neurointermediate lobe of SHR (13). The secretion of thyrotropin is also under dopaminergic control and elevated thyrotropin levels have been reported in SHR (29). Hypothalamic dopaminergic insufficiency reported in the SHR (8) may be the primary cause of the peptide abnormalities seen in this model. The GHR does not seem to possess this defect since plasma levels of β-endorphin and thyrotropin, whose secretion is under dopaminergic control appear to be normal (Table 1).

It is suggested while both SHR and GHR appear to be related models of genetic neurogenic hypertension, with an apparently similar overactivity of the brain angiotensin forming system, there are major differences between the two strains in the secretion of peptides under dopaminergic control and the regional variations in brain ACE.

SUMMARY

Rats of the New Zealand genetically hypertensive strain (GHR), the Japanese spontaneously hypertensive (SHR) strain and their respective normotensive controls (NT-NZ and WKY) were killed at 17-22 weeks of age. Their brains were removed and dissected into eight regions. Angiotensin converting enzyme activity (ACE) was measured in homogenates of the brain region using the substrate Hip-His-Leu as substrate. In all rats there were marked regional variations of ACE with the highest levels in striatum, cerebellum, pituitary and hippocampus. In the medulla oblongata, hypothalamus and cerebral cortex, ACE was lower in tissue from SHR compared to WKY. These region-specific differences in the SHR may lead to alterations in central neuropeptides involved in blood pressure regulation. ACE was lower in homogenates of cerebral cortex of GHR compared to NT-NZ but no significant differences were found in the other seven brain regions. These results show that the pattern of changes in brain ACE in GHR compared to NT-NZ are different from those seen in SHR compared to NT-WK and suggest different central nervous mechanisms in the two strains of hypertensive rats. Other neuroendocrine abnormalities such as increased pituitary β-endorphin and increased secretion of thyroid stimulating hormone indicate a decreased dopaminergic activity in the hypothalamus in SHR compared to WKY. No such neuroendocrine abnormalities were observed in GHR compared to NT-NZ. These observations support the suggestion that there are different central nervous mechanisms involved in these two strains of genetically hypertensive rats.

ACKNOWLEDGEMENTS

Supported by the National Health and Medical Research Council of Australia.

REFERENCES

1. AVRITH DB, FITZSIMONS JT (1980) Increased sodium appetite in the rat induced by intracranial administration of components of the renin-angiotensin system. J Physiol 301:349-364

2. BENUCK M, MARKS N (1979) Co-identity of brain angiotensin converting enzyme with a membrane bound dipeptidyl carboxypeptidase inactivating met-enkephalin. Biochem Biophys Res Comm 88:215

3. BRECHER P, TERCYAK A, GAVRAS H, CHOBANIAN AV (1978) Peptidyl dipeptidase in rabbit brain microvessels. Biochim Biophys Acta 526:537-546

4. BRYANT RW, EPSTEIN AN, FITZSIMONS JT,FLUHARTY SJ (1980) Arousal of a specific and persistent sodium appetite in the rat with continuous intracerebroventricular infusion of angiotensin II. J Physiol 301:365-382

5. FITZSIMONS JT (1979) The physiology of thirst and sodium appetite. Cambridge University Press, Cambridge, 288-297

6. FITZSIMONS JT (1980) Angiotensin in the control of water and sodium intake. In: BLOOM FE (ed) Peptides: integrators of cell and tissue function. Raven Press, New York, 99-106

7. FUXE K, GANTEN D, HÖKFELT T, BOLME P (1977) Immunohistochemical evidence for the existence of angiotensin II containing nerve terminals in the central nervous system of the rat. In: BUCKLEY JP, FERRARIO CM, LOKHANDWALA MF (eds) Central actions of angiotensin and related hormones. Pergamon, Oxford, 515-518

8. FUXE K, GANTEN D, JONSSON G, AGNATI LF, ANDERSSON K,HÖKFELT T, BOLME P, GOLDSTEIN M, HALLMAN H, UNGER TH, RASCHER W (1979) Catecholamine turnover changes in hypothalamus and dorsal midline area of the caudal medulla oblongata of spontaneously hypertensive rats. Neurosci Lett 15:283-288

9. GANTEN D, HUTCHINSON JS, SCHELLING P (1975) The intrinsic brain iso-renin angiotensin system in the rat: its possible role in central mechanisms of blood pressure regulation. Clin Sci 58:265_s-268_s

10. GIMBRONE MA Jr, MAJEAU GR, ATKINSON WJ, SADLER W, GRUISE SA (1979) Angiotensin-converting enzyme activity in isolated brain microvessels. Life Sci 25:1075-1083

11. GLOWINSKI J, IVERSEN LL (1966) Regional studies of catecholamines in the rat brain - 1. The disposition of ^{3}H dopamine and ^{3}H dopa in various regions of the brain. J Neurochem 13:655-669

12. HUTCHINSON J, CSICSMANN J (1978) The distribution of immunoreactive angiotensins I and II extracted from rat brain. Acta Med Acad Sci Hung 35:277-285

13. HUTCHINSON JS, DiNICOLANTONIO R, LIM A, CLEMENTS J, FUNDER JW (1981) Spontaneously hypertensive rats: effects of bromocriptine on blood pressure and ß-endorphin levels. Proceedings of the Eighth Meeting of the International Society of Hypertension, 191

14. HUTCHINSON JS, DiNICOLANTONIO R, TAKATA Y, VERONI M, JERUMS G, DOYLE AE (1981) Comparison of endocrine factors in Japanese and New Zealand strains of hypertensive rats. Proceedings of Endocrinology of Hypertension, Padua, Italy (in press)

15. HUTCHINSON JS, LIM A, DiNICOLANTONIO R, CLEMENTS JA, FUNDER JW (1981) Immunoreactive ß-endorphin levels in plasma and pituitary tissue from hypertensive and normotensive rats. Clin Exp Pharm Physiol 8 (in press)

16. HUTCHINSON JS, SCHELLING P, GANTEN D (1975) Effects of centrally administered angiotensin II and P 113 on blood pressure in conscious rats. Pfluegers Archiv 335:Suppl R28

17. KILCOYNE MM, HOFFMAN DL, ZIMMERMAN EA (1980) Immunocytochemical localization of angiotensin II and vasopressin in rat hypothalamus: evidence for production in the same neuron. Clin Sci 59:575-605

18. LOWRY OH, ROSENBROUGH NJ, FARR AL, RANDALL RJ (1951) Protein measurement with the folin phenol reagent. J Biol Chem 186:265-275

19. MENDELSOHN FAO, HUTCHINSON JS, CSICSMANN J (1980) Angiotensin converting enzyme in the brain of spontaneously hypertensive rats. Clin Exp Pharm Physiol 7: 551-554

20. NAHMOD VE, FINKIELMAN S, de GORODNER OS, GOLDSTEIN DJ (1977) On the neuronal localization and the physiological variations of brain angiotensin. In: BUCKLEY JP, FERRARIO CM, LOKHANDWALA MF (eds) Central actions of angiotensin and related hormones. Pergamon, Oxford, 573-579

21. OKAMOTO K, AOKI K (1963) Development of a strain of spontaneously hypertensive rats. Jap Circ J 27:282-293

22. PHILLIPS MI (1978) Angiotensin in the brain. Neuroendocrinology 25:354-377

23. PHILLIPS MI (1980) Biological effects of angiotensin in the brain. In: GROSS F, VOGEL G (eds) Enzymatic release of vasoactive peptides. Raven Press, New York, 335-363

24. PHILLIPS MI, HOFFMAN WE (1977) Sensitive sites in the brain for the blood pressure and drinking responses to angiotensin II. In: BUCKLEY JP, FERRARIO CM, LOKHANDWALA MF (eds) Central actions of angiotensin and related hormones. Pergamon, Oxford, 325-356

25. RAMSAY DJ, KEIL LC, SHARPE MC, SHINSAKO J (1978) Angiotensin II infusion increases vasopressin, ACTH, and 11-hydroxy-corticosteroid secretion. Am J Physiol 234 (1):R66-R71

26. SCHÖLKENS BA, JUNG W, RASCHER W, SCHÖMIG A, GANTEN D (1980) Brain angiotensin II stimulates release of pituitary hormones, plasma catecholamines and increases blood pressure in dogs. Clin Sci 59:53_s-56_s

27. SEVERS WB, DANIELS-SEVERS AE (1973) Effects of angiotensin on the central nervous system.Pharmacol Rev 25:415-449

28. SMIRK FH, HALL WH (1958) Inherited hypertension in rats. Nature 182:727-728

29. SOWERS JR, RESCH G, TEMPEL G, HERZOG J, COLANTINO M (1979) Hyperprolactinaemia in the spontaneously hypertensive rat. Acta endocrinol 90:1-7

Angiotensinogen, Angiotensin, Angiotensin Receptors

Evidence for a Functional, Independent Brain-Angiotensin System: Correlation Between Regional Distribution of Brain Angiotensin Receptors, Brain Angiotensinogen and Drinking During the Estrous Cycle of Rats

F. M. Chen, R. Hawkins, and M. P. Printz

INTRODUCTION

Historically, the finding that angiotensin II (ANG II) when injected into the vertebral artery of dogs in a classical cross-perfusion experiment resulted in an elevation of systemic arterial pressure led to the proposal of a central site of action of the peptide (1). This hypothesis stimulated studies by many laboratories which subsequently established that exogenous ANG II administered into the cerebro-ventricles of rats, rabbits, dogs and other mammals resulted in a triad of effects: elevation of sympathetic nervous system activity, elevated secretion of vasopressin (ADH) and ANG II-stimulated drinking. The net physiological result of these effects is an elevation of systemic arterial pressure and salt and water retention. It is essential to recognize that these net effects also result from peripheral administration of ANG II. However, the mechanisms by which ANG II elevates arterial pressure and induces salt and water retention differ depending on whether the peptide is given centrally or peripherally. This fact remains one of the most cogent arguments (2) in support of an independent central angiotensin system[+].

The concept of a renin-angiotensin system intrinsic to the brain was first advanced by Ganten et al. (3), based on his finding of apparent angiotensinogen and renin and angiotensin activity in canine brain tissue. The probable presence of angiotensinogen in the CNS was reinforced by (a) the finding that intraventricular

[+] The term "renin-angiotensin sytem" is traditionally used to refer to this putative central peptidergic system in analogy with historical terminology applied to the peripheral system. This is incorrect since it is not yet established rigorously either that renin is the important enzyme centrally or even that angiotensinogen yields ANG I (or ANG II) in vivo, although the evidence in favor of these possibilities is greatly reinforced by findings discussed in this volume. We will, therefore, refer to the putative central "angiotensinogen-renin-angiotensin system" as the "central angiotensin system"

Experimental Brain Research, Suppl. 4

administration of a crude renin preparation gave a central ANG II-like response in dogs (4) and (b) the demonstration and partial isolation of an angiotensin precursor (angiotensinogen)+ from extravascular rabbit brain tissue by Dr. John Lewicki in my laboratory (5) and from canine CSF by Reid and Ramsay (4). Further studies by members of my laboratory have demonstrated that CNS angiotensinogen is biochemically different from plasma angiotensinogen (2), that BRS is clearly in extravascular brain tissue (5,6,10), that BRS appears to be largely of brain origin and it is unlikely to derive from a plasma angiotensinogen pool (6,7), that BRS exhibits a unique topographical distribution in rats (8), that regional BRS levels are selectively regulated by adrenal glucocorticoids (6,9), and most recently that bilateral nephrectomy dissociates changes in levels of PRS from BRS (10).

This sequence of studies is continuing and has sought to establish, in a systematic manner, the validity and origin of a brain angiotensinogen. BRS, being a stable protein and the necessary precursor of the angiotensin peptides, should provide a reliable marker for a biosynthetic pathway for ANG II. However, it is essential to proceed cautiously in investigation of a brain ANG II system since plasma contamination of isolated brain tissue or even CSF is the rule and not the exception. Brain angiotensinogen could derive by three possible routes while a functional brain ANG II system could bypass local peptide formation if blood-borne ANG II could enter the CNS (Fig. 1).

Studies summarized above argue against path A, passage of PRS into a pool of "false BRS", as well as path B, the direct conversion of PRS to brain ANG II. Path C, where plasma ANG II acts on brain receptors (which would bypass central generation of ANG II), would be expected only with pharmacologic doses of ANG II or a specific transport pathway. Thus studies, to date, support path D from a centrally-derived brain angiotensinogen. Demonstration of a central angiotensin system must rest on more than demonstration of a precursor. However, one way to establish functionality of an intrinsic central ANG II system is to demonstrate a coupling between precursor and receptor and to define linkages between the peripheral and a central ANG II system. Studies in this direction are underway and summarized below. Our findings point to a relationship between BRS and ANG II receptors and a dissociation in the regulation of both peripheral and central receptors and precursors.

+ While angiotensinogen is the appropriate term for the precursor of the angiotensin peptides, operationally what we measure is a substrate of kidney renin. Thus, in this paper, we will interchangeably use the classical term for angiotensinogen, renin substrate, and refer to brain or plasma angiotensinogen as brain renin substrate (BRS) and plasma renin substrate (PRS) respectively

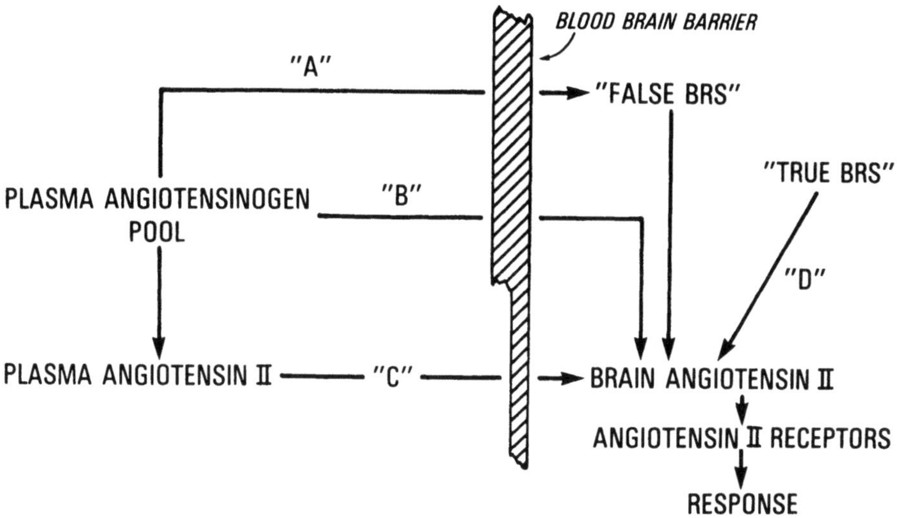

Fig. 1. Schematic illustration of pathways of origin of brain angiotensinogen

RADIOLIGAND BINDING STUDIES

ANG II receptors (ANG IIR) were identified by the quantitation of specific, saturable and high affinity binding of [^{125}I] ANG II by a rat brain membrane fraction (11). The methods employed were modifications of those described by Sirett et al. (12). However, we found binding to be very sensitive to how the tissue was handled, a fact not evident in the literature. Freezing and thawing brain tissue led to a marked decrease in binding, primarily of a high affinity class of receptors. If tissues were homogenized and membrane fractions were isolated from cytosol, before being subjected to freeze-thaw, specific binding was retained. The loss in specific binding in the intact tissue was not due to receptor solubilization and probably represents freeze-thaw release of tissue proteases or peptidases.

Dithiothreitol enhanced ANG II binding by decreasing K_D from 0.42 nM to 0.24 nM without affecting significantly receptor density (B_{max}) (11). In contrast to previous studies on calf cerebellum (13) we did not find that either PMSF or glucagon enhanced binding. Scatchard analysis of the HTSM (hypothalamus-thalamus-septum-midbrain) area gave a K_D of 0.24 $\pm$ 0.02 nM and a B_{max} of 537 $\pm$ 24 fmol/g tissue. The regional distribution of specific binding is shown in Fig. 2. The highest density of receptors was found in the septal region while cortex and hippocampus were consistently lowest among areas surveyed. Measurement of the IC_{50} for competition of [^{125}I -Ile^5]-ANG II by Val^5-ANG II yielded a similar value (0.22 $\pm$ 0.05 nM) for all areas except the pituitary, which was found to have an IC_{50} of 2.25 $\pm$ 0.63 nM.

The regional distribution of ANG IIR when compared with measured BRS levels in the same areas provided some interesting findings. Several areas where very low BRS is consistently found, also contain low receptor density, e.g., cortex and hippocampus. Such areas are probably either devoid of a central ANG II system, or if a system is present it is in a very localized, discrete region and not detectable by our procedures. Other areas contain significant BRS and ANG IIR, e.g., thalamus, hypothalamus, midbrain and brainstem. It is likely that such areas contain a complete ANG II system. Several regions contained high receptor density (septum, pituitary and olfactory bulb) but a comparatively low BRS content. This lack of correspondence may indicate that angiotensinogen has been converted to ANG II which is either utilized or stored. The spinal cord contained low receptor density but moderate levels of BRS.

Our results on the regional distribution of ANG IIR are in general agreement with data from other laboratories (Table 1) and indicate a non-uniform distribution of ANG IIR in rat brain. We find a correspondence between ANG IIR and BRS in several areas associated with either known ANG II responsiveness or neural pathways which

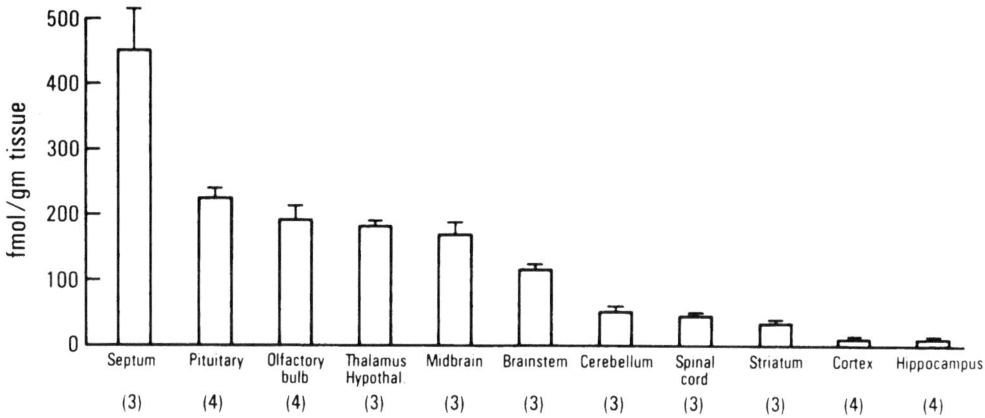

Fig. 2. Regional distribution of ANG II receptors in rat brain. Mean ± S.E. Concentration of [^{125}I] -ANG II was 0.1 nM

mediate ANG II-induced drinking or blood pressure changes. These correlations indicate that our experimental approach to simultaneous quantitation of ANG IIR and BRS may provide a sensitive probe for an endogenous brain ANG II system.

Table 1. Comparison of receptor density measurements

	Present Study	Snyder 1976	Sirett 1977	Sirett 1977	Snyder 1980
ANG II (nM)	0.1	0.05	0.2	0.3	0.3
Region	fmol/g tissue				
Septum	452	--	216	---	---
Pituitary	225	--	---	---	---
Olfactory bulb	193	--	---	---	---
Thalamus/Hypo.	182	49	145/ 92	220/ 134	225
Midbrain	169	44	140	---	158
Brainstem	115	21	71	113	86
Cerebellum	51	13	---	52	57
Spinal cord	45	--	---	---	---
Striatum	35	6	---	53	53
Cerebral cortex	11	10	15	33	51
Hippocampus	9	5	---	30	52

CORRELATION BETWEEN DRINKING AND RECEPTOR DISTRIBUTION

Previous studies from this laboratory established a marked potentation of ANG II-induced drinking 24 hours following bilateral nephrectomy (10). Concurrently, we found elevated BRS in brain areas which mediate dipsogenesis. This indicated a correspondence between ANG IIR which mediate drinking and regional BRS distribution. We therefore employed simultaneous measurement of ANG IIR and BRS with drinking in an effort to unmask a precursor-product-receptor correspondence.

Results in the literature indicate that rats in estrus have decreased spontaneous and ANG II-induced drinking (14) concomitant with increased sensitivity to ANG II-induced uterine contractility (15). We sought to determine whether brain ANG IIR and BRS are regulated during the estrous cycle and to correlate variations with reported changes in drinking and uterine ANG II responsiveness. Variations in uterine ANG IIR during estrus were recently reported (16).

We found that spontaneous drinking decreased in a highly significant ($p < 0.001$) manner on the day of estrus from 32 ml/day in proestrus or diestrus to 26 ml/day on estrus. Rats were staged by both smears and examination of uterine size at sacrifice. On estrus there was a significant increase in specific ANG II binding by the uterus (Table 2).

Table 2. Variation in ANG II receptors during estrous cycle

Tissue	ANG II specific binding fmol/g protein+		
	Proestrus	Estrus	Diestrus
Uterus	707 ± 87 (5)	2532 ± 382 (4)	645 ± 56 (7)++
Pituitary	1207 ± 235 (9)	1051 ± 105 (8)	2290 ± 242 (9)++
Olfactory bulb (ant.)	1120 ± 131 (9)	997 ± 133 (8)	929 ± 120 (9)
POA	516 ± 111 (9)	367 ± 57 (8)	577 ± 88 (9)
Septum	2115 ± 385 (9)	1656 ± 230 (8)	2436 ± 321 (9)
AH	461 ± 108 (9)	450 ± 102 (8)	342 ± 113 (9)
Thalamus	1265 ± 226 (9)	935 ± 196 (8)	938 ± 169 (9)
MBH	340 ± 108 (5)	287 ± 47 (4)	329 ± 100 (7)
MDH	790 ± 118 (5)	582 ± 84 (4)	672 ± 91 (6)
LH	1099 ± 336 (5)	653 ± 148 (4)	1197 ± 274 (7)
VTA	685 ± 221 (3)	717 (1)	380 ± 31 (4)
PAG	909 ± 85 (5)	724 ± 64 (4)	760 ± 47 (7)

+ Mean ± S.E.; numbers in parenthesis denote separate animals and separate analyses, one analysis/animal. Abbreviations: POA-preoptic area; AH-anterior hypothalamus; MBH-medial basal hypothalamus; MDH-medial dorsal hypothalamus; LH-lateral hypothalamus; VTA-ventral tegmental area; PAG-periacqueductal grey. ++ Significant at $p < 0.01$ (two tails) when comparisons were made between estrus and diestrus

In contrast to uterine ANG IIR, several brain regions exhibited decreases in specific ANG II binding on estrus compared to diestrus or proestrus. At this stage of our studies, only the pituitary shows a significant decrease while ANG IIR in the preoptic area, lateral hypothalamus and septum tend to be decreased; however, further replication is necessary to evaluate statistical significance of the results.

Changes in regional BRS (Table 3) indicate that a highly significant decrease in BRS content was found for the olfactory bulb ($p < 0.005$) at estrus compared to diestrus, while the area postrema also showed a significant ($p < 0.05$) decrease. No significant changes in BRS were found in other areas although the plasma angiotensinogen was significantly elevated ($p < 0.005$) on estrus. Since the BRS measurements for the olfactory bulb and area postrema are not yet corrected for plasma contamination, the decreases observed in these two areas should become even more significant when corrections are made. Other areas may also become significantly decreased.

These results show that the brain ANG II system is clearly regulated differently from the peripheral system. Uterine ANG IIR and PRS are up-regulated at estrus while brain ANG IIR and BRS are either down-regulated (in selected areas) or unchanged. It is important to note that BRS levels change in a selective fashion. The finding that BRS and ANG IIR change in a corresponding manner (decreasing at estrus) is interesting and warrants further replication.

Table 3. Variation in regional BRS during estrous cycle

Region/Tissue	Brain angiotensinogen µg ANG I/mg/protein+		
	Proestrus	Estrus	Diestrus
Pituitary	1.47 ± 0.20 (4)	1.22 ± 0.11 (7)	1.30 ± 0.14 (13)
Olfactory bulb	1.37 ± 0.13 (13)	0.96 ± 0.09 (12)+++	1.44 ± 0.10 (21)
POA	2.93 ± 0.38 (14)	3.19 ± 0.29 (13)	2.88 ± 0.27 (21)
Septum	2.25 ± 0.26 (14)	2.23 ± 0.26 (13)	1.92 ± 0.17 (21)
AH	4.59 ± 0.82 (14)	3.96 ± 0.46 (13)	4.21 ± 0.34 (21)
Thalamus	4.12 ± 0.31 (14)	4.00 ± 0.48 (13)	3.97 ± 0.31 (20)
MBH	3.06 ± 0.41 (14)	3.11 ± 0.30 (13)	4.27 ± 0.48 (21)
MDH	4.76 ± 1.04 (4)	4.01 ± 0.71 (7)	5.04 ± 0.79 (12)
LH	3.37 ± 0.50 (14)	3.68 ± 0.47 (13)	3.78 ± 0.37 (20)
VTA	3.59 ± 0.41 (12)	3.17 ± 0.48 (12)	3.41 ± 0.35 (21)
PAG	3.53 ± 0.39 (13)	3.62 ± 0.40 (13)	4.36 ± 0.38 (20)
AP	2.86 ± 0.57 (12)	2.37 ± 0.37 (11)++	4.21 ± 0.60 (19)
SFO	2.03 ± 0.48 (12)	3.45 ± 0.81 (13)	2.32 ± 0.51 (11)
Plasma (ng ANG I/ml)	604.00 ± 42.00 (14)	622.00 ± 26.00 (13)+++	504.00 ± 25.00 (21)

+ Mean ± S.E. Numbers in parenthesis denote separate animals and assays. Significance tests compared with diestrus: ++ $p < 0.05$. +++ $p < 0.005$. Abbreviations see Table 2

CORRELATION BETWEEN FLUID INTAKE AND REGIONAL BRS

The data summarized above together with other findings of this laboratory indicate that changes in drinking behavior are reflected by selective changes in BRS. It was of interest to examine whether water deprivation affected regional BRS. Two

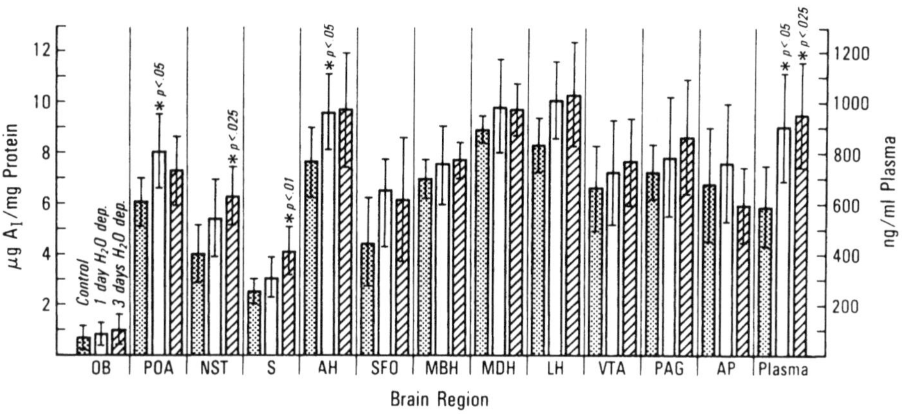

Fig. 3. Histogram of variation in regional BRS of rats, corrected for plasma contamination with 24 and 72 hours water deprivation. Mean $\pm$ S.E.

experimental approaches were utilized - water restriction and specific lesions to interrupt neural pathways regulating drinking. In the first, rats were water-deprived for 24 hours or 72 hours prior to sacrifice. There was a significant elevation of PRS at both 24 and 72 hours (Fig. 3). After correction for plasma contamination using the inulin space method (10), we find significant elevations in BRS at 24 hours in the anterior hypothalamus and the preoptic area and at 72 hours in the nucleus of the stria terminalis and the septum. Lesions of the subfornical organ or medial forebrain bundle, sites which mediate drinking behavior, also led to significant elevations of BRS in the anterior hypothalamus, septum and nucleus of stria terminalis (data not shown). In the latter experiment there was no significant elevation of PRS at 24 hours although marked elevations of regional BRS were found.

CONCLUSION

The results summarized in this paper point to a relationship between drinking, brain ANG II receptors and brain angiotensinogen in selective areas of the rat central nervous system. The central issue that is being addressed is whether an intrinsic brain angiotensin system exists and is functionally important in normal animals. The approach to this question is, of necessity, complicated by the expected redundancy of systems controlling drinking since fluid intake is a physiological behavior essential to survival of the organism. Although it is clear that the putative brain ANG II system is not the only regulatory mechanism for fluid intake, it does seem to be of major importance, since in experiments conducted in this laboratory we found highly selective changes in regional brain angiotensinogen.

Our present approach attempts to relate precursor with receptor and behavior or physiological response. Only through this multilevel analysis can a central ANG II system be established rigorously. All the data from this laboratory would indicate that a central ANG II system exists - starting with precursor and leading to angiotensin II generation and interaction with receptors. Furthermore, elements of this system appear to be regulated in a direction opposite to that of the peripheral ANG II system in an apparent counterbalancing phenomenon.These data also indicate that steroid hormones are among the messengers which regulate and couple the central and peripheral ANG II systems. In addition to our previous proposal of a linkage of the adrenal gland to the central ANG II system, we now must generalize this hypothesis to include sex-steroid hormones and propose the following scheme:

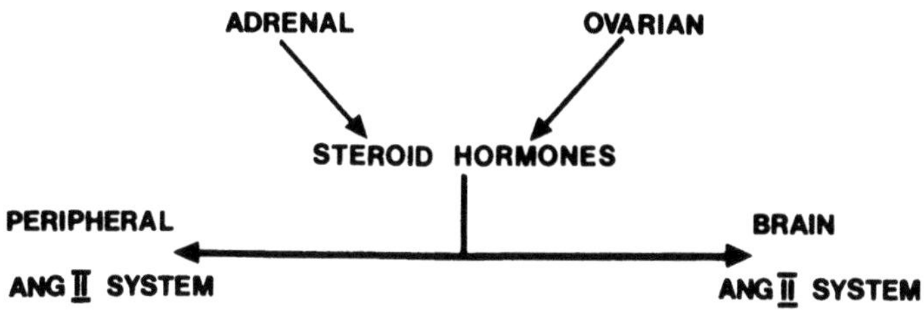

We therefore propose that steroid hormones exert a selective and localized response and are one important mechanism through which the peripheral volume state is sensed by a functional central angiotensin system.

ACKNOWLEDGEMENT

The studies summarized above were supported by a grant from the National Heart Lung Blood Institute for a Hypertension Specialized Center of Research, HL 25457. We acknowledge the valuable assistance of Ms. Jana Printz, Ms. Cindy Wead, Ms. Jennifer Sigler and John Spear in various aspects of these studies.

REFERENCES

1. BICKERTON RK, BUCKLEY JP (1961) Evidence for a central mechanism in angiotensin induced hypertension. Proc Soc Exp Biol Med 106:834-836

2. PRINTZ MP, PRINTZ JM, GREGORY TJ (1978) Identification of angiotensinogen in animal brain homogenates. Circ Res Suppl I 43:21-27

3. GANTEN D, GRANGER P, GANTEN U, BOUCHER R, GENEST J (1972) An intrinsic renin-angiotensin system in the brain. In: GENEST J and KOIW E (eds) Hypertension'72, Springer-Verlag, Berlin, p 432

4. REID IA, RAMSAY DJ (1975) The effects of intracerebroventricular administration of renin on drinking and blood pressure. Endocrinology 97:536-542

5. PRINTZ MP, LEWICKI JA (1977) Renin substrate in the central nervous system: Potential significance to central regulatory mechanisms. In: BUCKLEY JP and FERRARIO CM (eds) Central Actions of Angiotensin and Related Hormones. Pergamon Press, New York, p 57

6. PRINTZ MP,LEWICKI JA, WALLIS CJ (1980) Brain angiotensinogen: Origin and evidence for an influence by adrenal corticosteroids. In: GROSS F, VOGEL HG (eds) Enzymatic Release of Vasoactive Peptides. Raven Press, New York, p 193

7. LEWICKI JA (1979) Angiotensinogen in central compartments: Characterization, Distribution and Relationship to Plasma Prohormone. Doctoral Dissertation, University of California, San Diego

8. LEWICKI JA, FALLON JH, PRINTZ MP (1978) Regional distribution of angiotensinogen in rat brain. Brain Res 158:359-371

9. WALLIS CJ, PRINTZ MP (1980) Adrenal regulation of regional brain angiotensinogen content. Endocrinology 106:337-342

10. GREGORY TJ, WALLIS CJ, PRINTZ MP (1980) Regional changes in rat brain angiotensinogen following bilateral nephrectomy. Submitted for publication

11. CHEN FM, PRINTZ MP. A reexamination of the brain angiotensin II receptor and regional distribution in rat brain (submitted for publication)

12. SIRETT NE, McLEAN AS, BRAY JJ, HUBBARD JI (1977) Distribution of angiotensin II receptors in rat brain. Brain Res 122:299-312

13. BENNETT JP Jr, SNYDER SH (1976) Angiotensin II binding to mammalian brain membranes. J Biol Chem 251:7423-7430

14. FINDLAY ALR, FITZSIMONS JT, KUCHARCZYK J (1979) Dependence of spontaneous and angiotensin-induced drinking in the rat upon the oestrous cycle and ovarian hormones. J Endocrinol 82:215-225

15. SCHWARZ H, MASSON GMC, PAGE IH (1955) A method of assay of the oxytocic activity of angiotonin preparations on the rat's uterus. J Pharmacol Exp Ther 114:418-429

16. SCHIRAR A, CAPPONI A, CATT KJ (1980) Regulation of uterine angiotensin II receptors by estrogen and progesterone. Endocrinology 106:5-12

A Comparison of Human Renin Substrate in Plasma and Cerebrospinal Fluid

P. Eggena, T. Ito, J. D. Barrett, H. Villarreal, and M. P. Sambhi

INTRODUCTION

The administration of angiotensin II into the central nervous system (CNS) has been reported to elicit peripheral responses including an elevation in blood pressure (1,2). The presence of endogenous angiotensin-like peptides in cerebrospinal fluid (CSF) has also been documented by several investigators (3,4), but whether this vasoactive peptide is of peripheral origin or locally synthesized is not clear. Enzymes with renin-like activity (5) and angiotensin I converting enzyme (6) have both been localized in brain tissue and renin substrate has been observed both in brain tissue and in CSF (7,8). Whether the enzymes and proteins of the renin-angiotensin cascade within the CNS are identical to those of the peripheral circulation is at present not known. By comparing the physical-chemical and immunological properties of renin substrate isolated from plasma and CSF of normotensive and hypertensive human subjects we have attempted to further clarify the nature of the renin system in CNS and its possible involvement in hypertension. Studies with anephric subjects were included to establish whether the lack of peripheral renin influenced the renin system of the CNS.

METHODS AND MATERIALS

Clinical Material: Venous blood and CSF samples were obtained from male patients receiving spinal anesthesia. All subjects with hypertension, ranging from mild to severe, were on anti-hypertensive agents.

Renin Substrate Determinations: Renin substrate was quantitated both by the addition of excess homologous renin followed by radioimmunoassay of generated angiotensin I (9) and by a direct radioimmunoassay specific for the normal form of renin substrate (10).

Polyacrylamide Gel Electrophoresis: Polyacrylamide gel (7%) electrophoresis was performed at pH 8.6. Gels were sliced at 2mm intervals with a Gilson gelslicer into

Experimental Brain Research, Suppl. 4

500 μl of 100 mM sodium phosphate buffer at pH 7.4 and incubated with homologous kidney renin at 37°C to generate angiotensin I (11).

Molecular Weight Determination: Gel exclusion chromatography was performed using a 1.6 X 100 cm column with Ultrogel AcA 44 (LKB) as the support medium equilibrated at pH 7.4 in 50 mM Tris-HCl (11). Molecular weights were estimated by the method of Ackers (12).

Renin Determinations: Cerebrospinal fluid buffered at pH 7.4 with 100 mM sodium phosphate was incubated for 24 hours in the presence of angiotensinase inhibitors (13). Generated angiotensin I was assayed by radioimmunoassay (9). Cold storage (14) and trypsin activation (15) of possible inactive CSF renins were attempted in CSF alone and in a 1:1 mixture of CSF and plasma from a pool of anephric subjects. Plasma was added to overcome the possibility that CSF lacked a factor, essential for activation of inactive renin.

Isoelectric Focusing: Plasma and CSF samples were applied to isoelectric focusing gels (pH 4-6) according to the method of Allen and Mauer (16). Gels were sliced at 2 mm intervals into 500 μl water.Following pH measurement, each fraction was adjusted to pH 7.4 with a 1/10 volume 1 M sodium phosphate buffer. Cleavage of sialic acid groups by neuraminidase prior to electrofocusing was performed according to the method of Printz et al. (17).

When samples were not dialyzed against distilled deonized water prior to isoelectric focusing the term isoionic point is employed instead of isoelectric point which is reserved for samples which were dialyzed prior to focusing.

Data Analysis: All data are given with standard error of the mean. Linear least square regression coefficients were performed by previously described methods (11).

RESULTS

A comparison of the molecular weights of plasma and CSF derived renin substrate by gel exclusion chromatography indicated a molecular weight of 65-70,000 daltons for both substrates (Fig. 1). On polyacrylamide gel electrophoresis at pH 8.6 plasma and CSF renin substrate demonstrated the same electrophoretic mobility with respect to serum albumin, Rf = 0.67 $\pm$ 0.003 (Fig. 2). The vasoactive end product, generated from an incubation of plasma and CSF with exogenous homologous renal renin, indicated the same isoionic point (pH 6.6) following isoelectric focusing as synthetic angiotensin I. The pressor response of the generated angiotensin I from both plasma and CSF was indistinguishable from synthetic angiotensin I in the pentolinium ganglionic

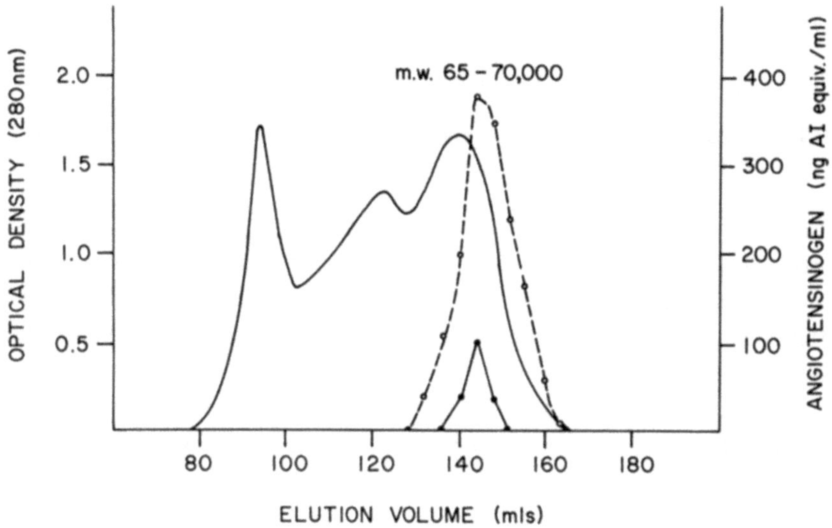

Fig. 1. Gel exclusion chromatography with Ultrogel AcA 44. o--o represents plasma, ●—● cerebral spinal fluid angiotensinogen concentrations; the solid line —— represents optical density measurements of plasma. (Courtesy of Hypertension)

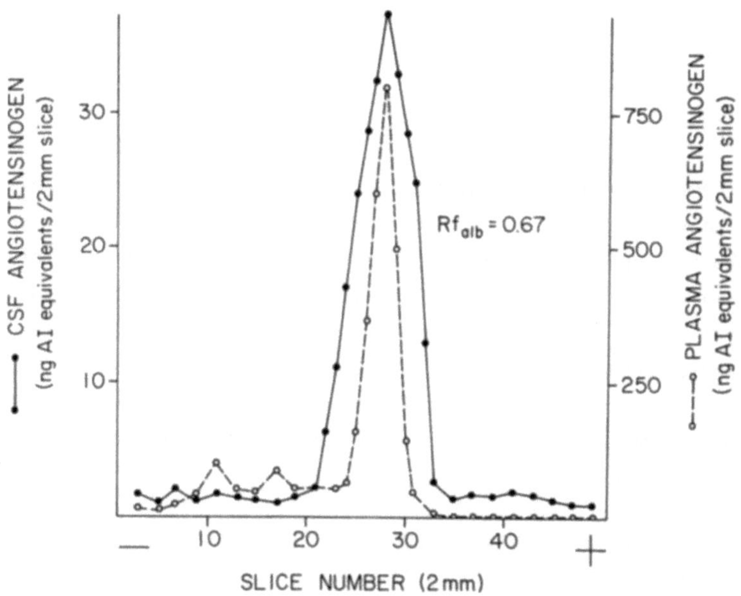

Fig. 2. Polyacrylamide gel electrophoresis (pH 8.6) of plasma and CSF angiotensinogen. (Courtesy of Hypertension)

blocked rat (18). Renin activity of CSF, if present at all was below the detection level of our radioimmunoassay (10 pg ANG I/ml). Inactive renin was not detected either by trypsin or coldstorage. Despite the absence of CSF renin activity, the circulating CSF angiotensin I level from a pool of normal CSF was 121.4 pg/ml. Circulating angiotensin II levels (n = 6) of CSF were 47.9 ± 9.6 pg/ml.

Despite the similarities of plasma and CSF renin substrate with respect to electrophoretic mobility, molecular weight and generated end product, clear differences were observed in the isoionic points of these proteins. As shown in Table 1 and Fig. 3, plasma renin substrate has a single isoionic point at pH 4.4 whereas CSF renin substrate demonstrates isoelectric heterogeneity with isoionic points at pH 4.8, 5.2 and 5.8. The contribution of sialic acid groups to this heterogeneity was investigated by incubation of CSF with neuraminidase. Following exposure of CSF renin substrate to neuraminidase, only a single component was evident on isoelectric focusing at pH 6.4, with no significant loss in total renin substrate concentration.

Table 1. Isoionic points of CSF and plasma renin substrate

	Normotensive	Hypertensive
CSF	4.80 ± 0.03 (n=6)	4.76 ± 0.02 (n=5)
	5.23 ± 0.03	5.16 ± 0.04
	5.75 ± 0.03	5.76 ± 0.04
Plasma	4.40 ± 0.04 (n=4)	4.43 ± 0.03 (n=4)

Immunologic differences between plasma and CSF renin substrate were also observed. An antibody which was prepared against the major form of plasma renin substrate was employed for the development of a direct radioimmunoassay for renin substrate. This antibody to normal plasma renin substrate did not crossreact with the renin substrate from CSF. Removal of sialic acid groups by neuraminidase from CSF substrate did not improve immunologic recognition and did not alter the antibody crossreactivity with neuraminidase-treated plasma renin substrate indicating that differences in sialic acid residues of plasma and CSF renin substrate were not the cause of the immunological distinction between these substrates.

A comparison of the physico-chemical parameters of CSF renin substrate in normal and hypertensive patients indicated no statistically significant differences with respect to molecular weight, electrophoretic mobility or isoelectric points. An exception was one subject with malignant hypertension where an increase in isoelectric heterogeneity was apparent (Fig. 3, lower panel).

Fig. 3. Upper part: Typical isoionic focusing pattern of CSF angiotensinogen of either a normotensive or hypertensive subject. Lower part: Pattern of a subject with malignant hypertension and cerebrovascular accident (Courtesy of Hypertension)

As seen in Table 2, there was a significant difference in the CSF renin substrate concentration of normal and hypertensive subjects. This was evident despite equivalent plasma substrate levels. Although the mean CSF renin substrate level of long term anephric subjects was generally below normal, this difference did not gain statistical significance (Table 2). Linear correlation coefficients between plasma and CSF renin substrate were not evident in any of the patient groups (Table 2) and no correlation was observed between CSF renin substrate concentration and either mean or systolic blood pressure.

Table 2. Comparison of plasma and CSF renin substrate levels (ng ANG I equivalents/ml) and linear least squares regression coefficients

	Normal (n=10)		Hypertensive (n=18)		Anephric (n=5)
Plasma	3589 ± 305		3518 ± 152		3542 ± 622
CSF[+]	37.6 ± 6.8		131.7 ± 19		23.5 ± 3.8
		$p < 0.005$		$p < 0.010$	
Plasma-CSF correlation coefficient[++]			0.15		0.11

[+]Measured by radioimmunoassay of angiotensin I following incubation of CSF with renin. [++]In a previous study (8) no linear correlation was observed in this subject group r=0.32. No statistically significant difference was observed between the plasma values or the CSF renin substrate of normal and anephric subjects

DISCUSSION

The similarity of human plasma and CSF renin substrate with respect to molecular weight and electrophoretic mobility is in agreement with reports of other investigators using dogs (19) and sheep (17). This equivalence of CSF and plasma renin substrate is further supported by the observation that the generated vasoactive end product in both plasma and CSF substrate was not distinguishable from synthetic angiotensin I with respect to isoelectric point or pressor response. Despite these similarities, clear differences in CSF and plasma renin substrate are apparent. Whereas CSF renin substrate was heterogenous with respect to isoionic points, only one component was evident in plasma. Incubation with neuraminidase prior to isoelectric focusing indicated the prominant influence of sialic acid groups on isoelectric heterogeneity, in agreement with the studies of Printz et al. (17).

Evidence in support of structural differences between CSF and plasma substrate was also obtained from immunological studies. We have previously reported that an antibody raised to normal plasma renin substrate does not crossreact with several forms of renin substrate induced by steroid therapy (10,11). On polyacrylamide gel electrophoresis at pH 8.6 these noncrossreacting forms have Rf values of 0.16 and 0.34; the major form with an Rf = 0.67, although elevated in these subjects, crossreacts completely with the antibody. These results are in contrast to the present data from CSF, where both plasma and CSF substrates migrate with an Rf = 0.67, but the CSF substrate is not recognized by the antiserum. This data indicates yet another immunologically distinct form of renin substrate. Immunological crossreactivity was not improved by preincubation of CSF with neuraminidase demonstrating that differences in sialic acid groups of plasma and CSF substrate do not contribute significantly to the immunological differences observed.

The lack of a correlation between the plasma and CSF renin substrate concentrations support a hypothesis that the major plasma renin substrate does not penetrate into the CSF. This data is in agreement with that of Reid and Day (19) demonstrating that stimulation of plasma renin substrate does not cause an increase in CSF substrate. Further evidence in support of a CNS source for renin substrate is the selective enrichment of CSF with substrate when compared with plasma on a mg protein per ml basis, a finding which supports similar observations by Printz et al. (17). Whereas plasma renin substrate levels were found to be equivalent in normotensive, long term anephric and hypertensive subjects, significantly elevated levels of CSF substrate were evident in hypertensive patients. Although in general no significant differences in the characteristics of CSF renin substrate of normal and hypertensive subjects were observed, an increased isoelectric heterogeneity was very apparent in one patient with malignant hypertension. The lower levels in anephric subjects may possibly reflect a higher utilization of this protein to compensate for the suppressed renin system in the plasma of these subjects. Whether increased levels of renin substrate in the CSF of hypertensive subjects contribute to the maintenance of hypertension is at present not clear. An argument against a physiologic role for CSF substrate is the absence of CSF renin. Infusion of renin into the CNS though has been shown to elicit an increase in blood pressure (20). Although we found neither circulating active or inactive renin in the CSF, renin and/or renin-like enzymes are present in brain tissue (5). The observation that the levels of angiotensin I and II in the CSF are equivalent to those of plasma may indicate that CSF renin substrate is cleaved by renin containing tissues in the CNS to produce angiotensin locally.

In conclusion, this study has shown sufficient differences in the physico - chemical and immunological character of plasma and CSF renin substrate to postulate a non-peripheral origin for CSF substrate. Whether the significantly elevated levels of

CSF renin substrate in hypertensive subjects and the suppressed levels in anephric patients are of physiological significance in the control and regulation of blood pressure remains to be determined.

ACKNOWLEDGEMENT

This work was supported by research funds from the Veterans Administration, NIH Grant HL 22773 and a grant from the American Heart Association, Los Angeles Chapter.

REFERENCES

1. SEVERS WB and DANIELS-SEVERS AE (1973) Effects of angiotensin on the central nervous system. Pharmacol Rev 25:415-449

2. KEIL LC, SUMMY-LONG J and SEVERS WB (1975) Release of vasopressin by angiotensin II. Endocrinology 96:1063-1065

3. CHANGARIS DG, DEMERS LM, KEIL LC and SEVERS WB (1977) Immunopharmacology of angiotensin I in brain. In: BUCKLEY J, FERRARIO CM and LOKHANDWALA MF (eds) Central actions of angiotensin and related hormones. Pergamon Press, New York, Frankfurt, p 233

4. SEVERS WB, CHANGARIS DG, KAPSHA JM, KEIL LC, PETRO DJ, REID IA and SUMMY-LONG JY (1977) Presence and significance of angiotensin in cerebrospinal fluid. In: BUCKLEY J, FERRARIO CM and LOKHANDWALA MF (eds) Central actions of angiotensin and related hormones. Pergamon Press, New York, Frankfurt, p 225

5. GANTEN D, FUXE K, GANTEN U, HÖKFELT T and BOLME P (1977) The brain isorenin-angiotensin system: Localization and biological function. In: DE JONG W, PROVOOST AP, SHAPIRO AP (eds)Hypertension and brain mechanisms. Elsevier Scientific Publishing, Amsterdam, Vol. 47, p. 155

6. IGIC RP, ROBINSON CJG and ERDOS EG (1977) Angiotensin I converting enzyme activity in the choroid plexus and in the retina. In: BUCKLEY J, FERRARIO CM and LOKHANDWALA MF (eds) Central actions of angiotensin and related hormones. Pergamon Press, New York, Frankfurt, p 23

7. PRINTZ MP, LEWICKI JA (1977) Renin substrate in the CNS: Potential significance to central regulatory mechanism. In: BUCKLEY J, FERRARIO CM and LOKHANDWALA MF (eds) Central actions of angiotensin and related hormones. Pergamon Press, New York, Frankfurt, p 57

8. ITO T, EGGENA P, BARRETT JD, KATZ D, METTER J, SAMBHI MP (1980) Studies on angiotensinogen of plasma and cerebrospinal fluid in normal and hypertensive human subjects. Hypertension 2:432-436

9. EGGENA P, BARRETT JD, SAMBHI MP and WIEDEMAN CE (1974) The validity of comparing the measurements of angiotensin I generated in human plasma by radioimmunoassay and bioassay. J Clin Endocrinol Metab 39:865-870

10. EGGENA P, BARRETT JD, HIDAKA H, CHU CL, THANANOPAVERN C, GOLUB M and SAMBHI MP (1977) A direct radioimmunoassay for human renin substrate and identification of multiple substrate types in plasma. Circ Res 41:(Suppl.II), 34-37

11. EGGENA P, HIDAKA H, BARRETT JD and SAMBHI MP (1978) Multiple forms of human plasma renin substrate. J Clin Invest 62:367-372

12. ACHERS GK (1967) A new calibration procedure for gel filtration columns. J Biol Chem 242:3237-3238

13. BARRETT JD, EGGENA P and SAMBHI MP (1976) Influence of angiotensinase inhibitors on the enzymatic activity of renin. Biochem Med 16:157-168

14. OSMOND DH, ROSS LJ and SCIAFF KD (1973) Increased renin activity after cold storage of human plasma. Canad J Physiol Pharmacol 51:705-708

15. EGGENA P, BARRETT JD, WIEDEMAN CE and SAMBHI MP (1979) Studies on renin activation in normal human plasma. Hypertension 1:523-528

16. ALLEN RC and MAUER HP (1974) Electrophoresis and isoelectric focusing in polyacrylamide gel. WALTER DE GRUYTER, 1974

17. PRINTZ MP, PRINTZ JM and GREGORY TJ (1978) Identification of angiotensinogen in animal brain homogenates. Circ Res 43:(Suppl. I) 21-27

18. SAMBHI MP and BARRETT JD (1967) Development of tachyphylaxis to aspartyl-1 and not to asparginyl-1 angiotensin II in the rat. Circ Res 21:327-332

19. REID IA and DAY RP (1977) Interactions and properties of some components of the renin-angiotensin system in brain. In:BUCKLEY J, FERRARIO CM and LOKHANDWALA MF (eds) Central actions of angiotensin and related hormones. Pergamon Press, New York, Frankfurt, p 267

20. REID IA and RAMSAY DJ (1975) The effects of intracerebroventricular administration of renin on drinking and blood pressure. Endocrinol 97:536-542

Regulation of Angiotensinogen in Cerebrospinal Fluid and Plasma

P. Schelling, D. Felix, and J. F. Liard

INTRODUCTION

The injection of renin into the brain tissue or into the intraventricular space (i.v.t.) initiates central actions similar to those obtained after the i.v.t. application of angiotensin II (ANG II), namely drinking (5,6,25,30,36), blood pressure increase (25,29,30,34,36), release of vasopressin (20,29) and ACTH (29,34) and stimulation of sympathetic tone (34,36). Angiotensin II-receptor blockade and inhibition of ANG II synthesis abolishes these effects (5,25,30,34,36), thus pointing to an ANG II-generating pathway within the brain. Indeed, angiotensinogen, the renin substrate, and other components of a renin-angiotensin system (RAS) could be identified in brain tissue (for review see:11,33).

Angiotensinogen was reported to be present at high levels in the brain and in the cerebrospinal fluid (CSF) of animals and men (9,15,17,20,26,27,29,30,31,32). It was shown to be a glycoprotein similar to plasma angiotensinogen (26) but was subsequently proven to differ from the latter in some physicochemical and biochemical aspects (27). More recently, angiotensinogen was definitively demonstrated to be of central origin by Sernia and Reid (35), since it could be synthesized in vitro from brain tissue slices.

The prohormone levels in CSF are much higher than those in plasma when expressed in terms of protein (27,29,31,32). They appear to originate from tissue rather than to be blood-borne, since stimulation of plasma renin substrate in dogs following nephrectomy (NX) or treatment with dexamethasone had no effect on the angiotensinogen levels in CSF (29). In a more recent study of Wallis and Printz (38), however, adrenals were shown to influence the regional distribution of angiotensinogen within the brain of rats in a similar way as known for plasma angiotensinogen (3,23,24), since adrenalectomy (ADX) decreased the prohormone content in some brain regions, while subcutaneous (s.c.) glucocorticoid treatment restored the angiotensinogen deficit in those areas. Thus adrenals represent probably a common regulator of angiotensinogen synthesis in the liver and in the brain.

Experimental Brain Research, Suppl. 4

It is not known as yet, whether angiotensinogen has a regulatory function in CSF or whether it represents a spill-over from the brain only. Nevertheless, fluctuations of the prohormone levels in CSF might be of importance in view of the role of periventricular structures as target organs mediating ANG II-related water intake and blood pressure increase (2,7,25). Thus we examined the angiotensinogen levels in CSF under conditions, which are known to either increase or decrease the plasma renin substrate concentrations.

MATERIALS AND METHODS

Subjects

All experiments were done in male Sprague Dawley rats (250 - 350 g body weight) purchased from Ivanovas, Kisslegg, Germany. The animals were housed individually and were adjusted to the colony room for 2 weeks. The environmental conditions were: lighting, from 0700 h to 1900 h; temperature, 23°C; relative humidity, 55 - 65%.

Experimental Procedure

Operated and/or treated rats were always investigated in parallel with appropriate controls. All surgery was carried out under ether anesthesia.

Rats were nephrectomized 16 h before CSF and blood collection; controls were sham nephrectomized. Adrenalectomy was performed 5 days prior to study and one group of animals was nephrectomized in addition on the 4th day and 16 h before study. After adrenalectomy, and sham adrenalectomy as well, 1% NaCl was given to drink.

Rats were treated with the converting enzyme inhibitor Captopril^{R} (Squibb, Princeton, New Jersey) by addition to the tap water in a concentration which guaranteed a daily intake of 50 mg/kg body weight. The glucocorticoid treatment was performed as follows: a crystalline suspension of hydrocortisone (Streuli, Uznach, Switzerland) was injected subcutaneously (s.c.) twice a day at 0900 h and 1800 h during 2 days (15 mg/kg/day). Blood and CSF were collected on the next day. In another experiment, previously adrenalectomized rats (4 days; 1% NaCl for drinking) received 3 s.c. injections of prednisolone (Graueb, Bern, Switzerland) at intervals of 10 h. The animals were treated either with 15 mg/kg, with 30 mg/kg or with 60 mg/kg body weight of prednisolone. Blood and CSF were withdrawn from the rats 4 h after the last treatment.

Collection of CSF and Blood

Rats were anesthetized with sodium thiopental (40 mg/kg body weight, intraperitoneally). The femoral artery was catheterized for blood collection and a 20-gauge cannula was inserted into the fourth brain ventricle for CSF sampling; the methodological details have been described elsewhere (32). Both, CSF and plasma were collected on ice into plastic vials containing 5% (v/v) of angiotensinase inhibitors (25 mM o-phenantroline and 71 mM ethylenediaminetetraacetic acid). Blood and CSF were centrifuged in the cold at 5000 x g for 10 min. Plasma samples of 50 μl were pipetted into incubation vials and stored frozen at -30°C until assayed. The possible contamination of CSF with blood was checked after the centrifugation by visual examination and by measuring the protein concentration (19). Blood free CSF was distributed in aliquots of 50 μl into incubation vials and kept frozen at -30°C.

Measurements of Angiotensinogen and Renin

For the angiotensinogen estimation, 50 μl of either CSF or plasma were incubated at 37°C with 200 mM tris-maleate buffer, pH 5.5, containing 0.05 U hog kidney renin (NBC, Cleveland, Ohio) and an angiotensinase inhibitor cocktail (10) in a final volume of 750 μl. Aliquots were taken from the incubation after 1, 2 or more hours and were diluted 1:2 with ice-cold 100 mM tris-acetate buffer, pH 7.4 (radioimmunoassay (RIA) buffer), and boiled at once for 5 min. The generated ANG I was measured after centrifugation in the enzyme free supernatant by a specific RIA for ANG I (12). The results are expressed as ng ANG I generating capacity per ml fluid or per mg protein.

Virtually the same disposition of the incubation was used for renin measurement, except that purified angiotensinogen from rat plasma (12) was added in excess, instead of hog kidney renin. The data are given as ng ANG I generated per h and per ml plasma.

The whole procedure of incubation and of RIA was performed several times after the addition of ANG I (8 ng). The recovery of the added ANG I served as a validity test.

Other Measurements

Sodium and potassium were measured by flame photometry in plasma which was diluted 1 : 200. The method of Lowry et al. (19) was used for protein measurement. Plasma and CSF were diluted 1 : 200 and 1 : 2 respectively before assay.

Statistics

Results are expressed as means ± standard error of the mean (SEM). Statistical significance of differences was calculated by Student's unpaired t-test.

RESULTS

The indirect estimation of angiotensinogen by exhausting the substrate and measuring the generated ANG I fulfilled validity tests. 1) The amount of ANG I formed during the incubation with 0.05 U hog kidney renin for 1, 2 or more hours was the same, thus the substrate was entirely exhausted within one hour of incubation. 2) Angiotensin I added to the incubation medium was completely recovered. 3) The generated product was measured with a specific RIA for ANG I in the enzyme free supernatant of the incubation mixture.

A typical pattern of angiotensinogen in plasma and CSF of untreated male rats is shown on Fig. 1. The angiotensinogen concentration is about 10-times higher in plasma than in CSF as expressed per ml of fluid, while just the reverse is true if the results are given as specific concentration per mg protein. This is due to the low protein content in CSF (Table 1).

Table 1. Protein content in plasma and CSF of rats after adrenalectomy and subsequent treatment with prednisolone

	plasma (mg prot./ml)	CSF (mg prot./ml)
CO	64.73 ± 0.93 (10)	0.457 ± 0.019 (10)
ADX	59.95 ± 0.86 (9)$^{+++}$	0.495 ± 0.018 (10)
ADX 5 mg pred.	74.67 ± 1.12 (10)$^{+++}$	0.436 ± 0.030 (9)
ADX 10 mg pred.	71.87 ± 0.75 (10)$^{+++}$	0.375 ± 0.022 (10)$^{+}$
ADX 20 mg pred.	73.55 ± 1.68 (10)$^{+++}$	0.371 ± 0.018 (8)$^{++}$

Controls (CO) and adrenalectomized rats (ADX) received 1% Na Cl for drinking. Adrenalectomized rats were treated 3-times every 10 hours either with 5 mg, 10 mg or 20 mg prednisolone (pred.). Plasma and CSF were obtained 5 hours after the last treatment. Results are given as means ± SEM and number of measurements are indicated in parentheses.

$^{+++}$ $p < 0.005$, $^{++}$ $p < 0.01$, $^{+}$ $p < 0.05$ versus controls

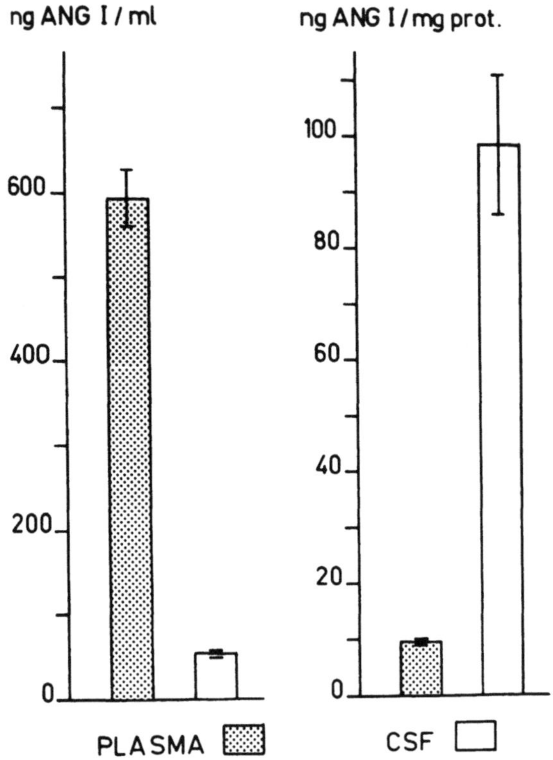

Fig. 1. Angiotensinogen levels in plasma (shadowed columns) and in CSF (open columns) of male Sprague Dawley rats are shown (means ± SEM; n : 9). On the left panel results are expressed as ng ANG I/ml; on the right panel as ng ANG I/mg prot

The means of angiotensinogen levels measured in CSF under control conditions varied from one experiment to another from 39.71 ± 3.27 ng ANG I/ml to 58.31 ± 1.56 ng ANG I/ml. Furthermore they changed during day time, being lowest at noon ($p < 0.05$): 56.09 ± 3.17 ng ANG I/ml, n = 14 (0800 h - 1030 h); 47.88 ± 2.15 ng ANG I/ml, n = 12 (1130 h - 1330 h); 54.74 ± 2.47 ng ANG I/ml, n = 11 (1430 h - 1630 h). No such variation of angiotensinogen existed in plasma.

The influence of several treatments on the prohormone levels in plasma and in CSF are summarized on Fig. 2. Nephrectomy increased angiotensinogen levels in both plasma and CSF. No correlation existed between the individual values from either sources (r = -0.128; n = 15). Adrenalectomy decreased peripheral and central prohormone levels. The combination of both, adrenalectomy and nephrectomy induced a dissociation of the renin substrate in the two compartments investigated. Angiotensinogen rose in plasma and fell in CSF. The treatment with hydrocortisone elevated peripheral and central prohormone levels in parallel. There was a significant linear correlation between the individual values from plasma and CSF (r = 0.65; n = 10; $p < 0.05$). The converting enzyme inhibitor, captopril, administered during 4 weeks had a strong inhibitory effect on plasma renin substrate. The angiotensinogen concentration in CSF was diminished too; however, without being significant versus controls, as expressed in absolute terms. All other results featured on Fig. 2 were significantly different from the appropriate controls given either per ml of fluid or per mg of protein.

The parallel regulatory effect of glucocorticoids on angiotensinogen in plasma and CSF was further studied by the application of prednisolone at different doses in adrenalectomized rats (Fig. 3). Prednisolone more than restored the renin substrate deficit in plasma caused by adrenalectomy already at the lowest dose, and induced a dose related increase of plasma angiotensinogen levels. The plasma renin concentrations were strongly elevated after adrenalectomy ($p < 0.001$) and were restored to normal after subsequent application of prednisolone. The following values were measured: 14.97 ± 1.68 ng ANG I/ml/h, n = 9 (adrenalectomy); 3.26 ± 0.39 ng ANG I/ml/h, n = 10 (untreated controls) and 3.18 ± 0.42, n = 9 (adrenalectomy + prednisolone). Only the highest dose of prednisolone restored electrolytes after adrenalectomy to normal.

The effects of adrenalectomy and of steroid treatment on the CSF prohormone were moderate compared to changes in plasma renin substrate but they were significant (Fig. 3). A clear dose-related increase of the CSF angiotensiogen levels was obtained if data were calculated as specific concentrations per mg protein. The expression of these results per ml of CSF showed minor differences and revealed an increase of CSF angiotensinogen after the highest steroid dose only. This discrepancy is explained by the effect of the glucocorticoids on the protein content

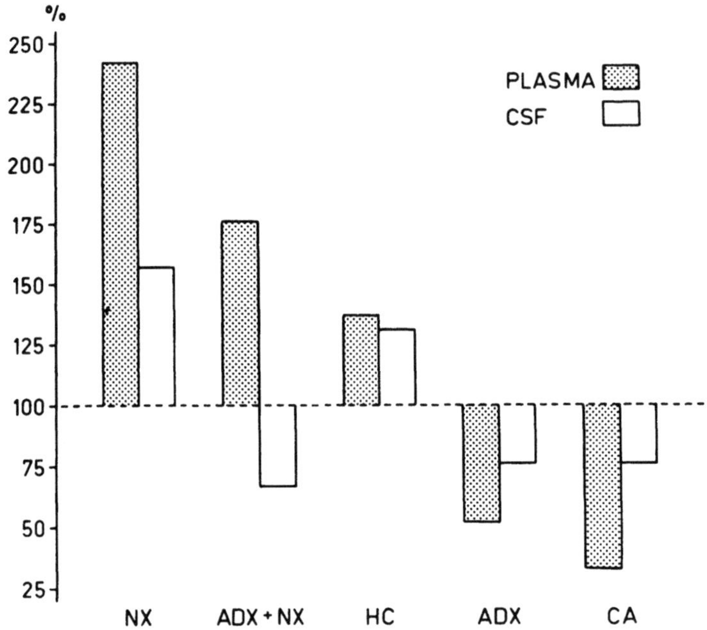

Fig. 2. Angiotensinogen levels in plasma (shadowed column) and in CSF (open column) are given as percentage of controls after the following treatments: NX, nephrectomy (16 h); ADX, adrenalectomy (5 days + 1% Na Cl for drinking); ADX + NX, adrenalectomy (5 days + 1% NaCl for drinking) with subsequent nephrectomy (16 h). HC, hydrocortisone treatment (15 mg/kg/day for 2 days); CA, captopril treatment (50 mg/kg/day during 4 weeks)

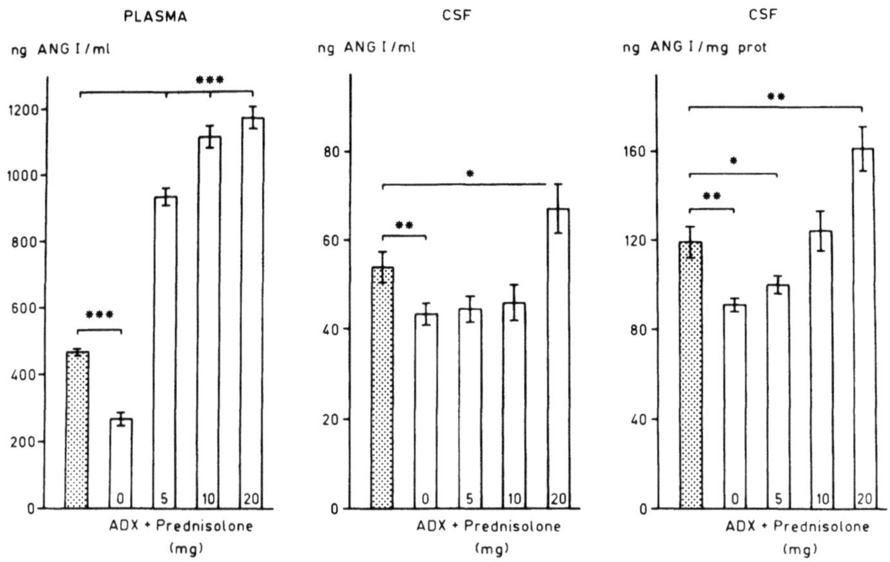

Fig. 3. Angiotensinogen was measured in untreated adrenalectomized rats (ADX) or in ADX rats substituted with 5, 10 or 20 mg prednisolone/rat and compared with controls (shadowed columns). Values in plasma are given as ng ANG I/ml on the left panel (means $\pm$ SEM; n = 10); the corresponding CSF levels are expressed as ng ANG I/ml on the middle panel and as ng ANG I/mg prot. on the right panel, $^{+}$ $p < 0.05$ $^{++}$ $p < 0.02$ $^{+++}$ $p < 0.001$

in CSF. The application of prednisolone decreased the protein concentration in CSF and, on the other hand, increased it in plasma as shown on Table 1. Adrenalectomy had the reverse effect on plasma protein but did not affect the protein content in CSF.

DISCUSSION

Confirming previous studies (3,13,23,24), we found a prominent increase of the plasma angiotensinogen levels after nephrectomy. The implication of the adrenal glands on regulation of plasma renin substrate was also confirmed (3,23,24). This was shown either by adrenalectomy inducing a decrease of angiotensinogen in plasma or by glucocorticoid treatment having the reverse effect. Furthermore, the low renin substrate values in plasma of adrenalectomized rats were restored after the treatment with prednisolone and rose to high values in a dose related manner. The increase of plasma renin substrate after nephrectomy appears to be independent of the glucocorticoids, since it was not abolished by previous adrenalectomy. These results probably point to a factor in plasma of nephrectomized rats which might stimulate the hepatic synthesis of renin substrate (13). However, our data could also be explained by the eliminated substrate consumption after nephrectomy (3,24); all the more as the renin concentrations were clearly increased in adrenalectomized rats. The converting enzyme inhibitor captopril was reported to inhibit the renin substrate levels in plasma of men (8,28). This is in accordance with our results in rats, where captopril strongly suppressed plasma angiotensinogen. An increased consumption of the substrate due to an elevated renin concentration (1) and the missing stimulatory effects of ANG II (16,24) and steroids (23,24) on angiotensinogen synthesis under captopril treatment (1) may probably contribute to the decrease of plasma angiotensinogen.

The variations of the angiotensinogen values within an experimental group as well as between different groups of untreated rats were greater in CSF than in plasma. This could be due to regulatory mechanisms not yet known or to methodological problems such as the flow rate of CSF which was changing from one animal to another. We observed a fluctuation of the angiotensinogen levels in CSF during daytime being low at noon, pointing to an endogenous circadian rhythm of angiotensinogen in CSF. The plasma cortisone levels in rats were also reported to follow a circadian rhythm being highest at 1900 h and lowest at 0700 h (21). Thus, there is probably no relationship between the daily fluctuation of glucocorticoids in plasma and those of angiotensinogen in CSF. Our plasma renin substrate measurements did not reveal a circadian rhythm so far.

Nephrectomy induced a rise of angiotensinogen levels in both, plasma and CSF. This synchronism between CSF and plasma was also apparent after adrenalectomy and after the hydrocortisone treatment attenuating the prohormone levels in the former case and stimulating them in the latter. These results might be explained by the following mechanisms: first a diffusion of blood-borne angiotensinogen into the CSF, second an active uptake of the molecule into the intraventricular space and third a regulation of peripheral and central angiotensinogen over a common link.

The first proposal is not consistent with our findings, since (1) the individual CSF values did not correlate with their plasma counterparts after nephrectomy although plasma renin substrate varied considerably from 1000 ng ANG I/ml to 3000 ng ANG I/ml and (2) since the protein content in CSF of either nephrectomized or steroid treated rats were not increased. In view of the low protein content and the high specific angiotensinogen concentrations in CSF it had to be expected that a diffusion of plasma proteins into the intraventricular space would have a pronounced effect on CSF protein content and little weight on angiotensinogen concentration. Finally the high molecular weight of rat plasma renin substrate reported to be 56000 daltons (14) does not favour a passive entry of the circulating molecule. The second proposal of a selective transport of blood-borne angiotensinogen into the intraventricular space is not very likely at all since adrenalectomy followed by nephrectomy induced a clear dissociation of the peripheral and central angiotensinogen. Plasma renin substrate increased and CSF angiotensinogen remained at the low level found after adrenalectomy alone. Furthermore, the view of a stimulated active uptake of blood-borne angiotensinogen due to adrenal steroids is not consistent with the fact that prednisolone treatment diminished the protein content in CSF as shown on Table 1. Glucocorticoids were reported to be useful in the treatment of brain edemas and to reduce the extravasation of proteins into the brain (18). Thus, they might inhibit the uptake of proteins into the central nervous system rather than stimulate it. In the third proposal a common regulation of the peripheral and central angiotensinogen is suggested. Our studies favour this view and point to the adrenal steroids as the link, all the more as these lipid soluble molecules are known to pass the blood brain barrier (4). Thus, the changes of the CSF angiotensinogen levels under the experimental conditions investigated can be explained best by the action of steroid hormones. The excision of the adrenals reduced the prohormone levels and the subsequent treatment with glucocorticoids restored them in a dose related manner. The adrenal glands also mediated the central stimulation of angiotensinogen after nephrectomy since the previous adrenalectomy abolished it completely. The long-term treatment with the converting enzyme inhibitor, captopril, induced a clear suppression of plasma renin substrate as already discussed. The CSF angiotensinogen levels tended also to decrease. Whether glucocorticoids were inhibited under these conditions is unknown. Aldosterone was reported to be suppressed (1) and renin to be stimulated (1) under captopril treatment. The pronounced increase of brain renin concentration in captopril treated rats as reported by Unger et al. (37) may be of special interest for the central angiotensinogen regulation. It raises the question whether the attenuated prohormone level in CSF reflects an increased consumption of angiotensinogen due to the elevated amount of renin present in the brain under captopril treatment.

The implication of adrenal steroids on the angiotensinogen content in brain tissue of the rat was convincingly shown by Wallis and Printz (38). In support of this we report now that a similar regulation of angiotensinogen may occur in the CSF of rats. Reid and Day (29) investigating the relationship of peripheral and central prohormone in dogs, however, did not find any change in CSF angiotensinogen levels after either nephrectomy or dexamethasone treatment, though the plasma renin substrate levels behaved as expected. Thus from their studies as well as from ours it is clear that the blood-CSF barrier is impermeable for circulating angiotensinogen. As to the diversity between both studies a species difference has to be considered. Furthermore, the collection of CSF in rats might have shifted brain interstitial fluid into CSF by a decrease in intraventricular pressure.

Angiotensinogen synthetized intracellularly by brain tissue (35) may attain the interstitial fluid (22) of the brain and probably be drained into the intraventricular space. A specific secretion of angiotensinogen should nevertheless also be considered in view of the high specific concentration of the molecule in CSF; all the more as the most striking central effects were elicited by the effector peptide of the RAS if it is injected i.v.t. (2,7,25). Thus, it cannot be decided as yet whether angiotensinogen in CSF reflects the prohormone content in brain and whether it is of physiological significance. However, it has been shown that the angiotensinogen levels in CSF are regulated and, they were reported to be elevated in hypertensive human subjects (15).

ACKNOWLEDGEMENTS

The research presented here was supported by grant 3.948.80 of the Swiss National Science Foundation and by the Emil Barell Foundation. We wish to thank S. Müller and M. Donzallaz for excellent technical assistance and F. Liard for typing the manuscript.

REFERENCES

1. ATKINSON AB, BROWN JJ, FRASER R, LECKIE B, LEVER AF, MORTON JJ, ROBERTSON JIS (1979) Captopril in hypertension with renal artery stenosis and in intractable hypertension; acute and chronic changes in circulating concentrations of renin, angiotensin I and II and aldosterone, and in body composition. Clin Sci 57: 139s-143s

2. BRODY MJ, JOHNSON AK (1980) Role of the anteroventral third ventricle region in fluid and electrolyte balance, arterial pressure regulation, and hypertension. In: MARTINI L, GANONG WF (eds) Frontier in Neuroendocrinology,vol 6. Raven Press, New York, p 249

3. CARRETERO O, GROSS F (1967) Evidence for humoral factors participating in the renin-substrate reaction. Circ Res 20/21 (Suppl II):115-126

4. CHRISTY NP, FISHMAN RA (1961) Studies of the blood-cerebrospinal fluid barrier to cortisol in the dog. J Clin Invest 40:1997-2006

5. COOLING MJ, DAY MD (1974) Inhibition of renin-angiotensin induced drinking in the cat by enzyme inhibitors and by analogue antagonists of angiotensin II. Clin Exp Pharm Physiol 1:389-396

6. EPSTEIN AN, FITZSIMONS JT, ROLLS BJ (1970) Drinking induced by injection of angiotensin into the brain of the rat. J Physiol 210:457-474

7. FELIX D, SCHELLING P (1981) Angiotensin and single neurons. This book

8. FYHRQUIST F (1981) In: HOROVITZ ZP (ed) Angiotensin Converting Enzyme Inhibitors, Mechanisms of Action and Clinical Implications. Urban & Schwarzenberg, Baltimore Munich, p 100

9. GANTEN D, MARQUEZ-JULIO A, GRANGER P, HAYDUK K, KARSUNKY KP, BOUCHER R, GENEST J (1971) Renin in dog brain. Am J Physiol 221:1733-1737

10. GANTEN D, SCHELLING P, HOFFMAN WE, PHILLIPS MI, GANTEN U (1978) The measurement of extrarenal isorenins. In: KRAUSE DK et al. (eds) Radioimmunoassay-Renin-Angiotensin. Thieme Verlag, Stuttgart, p 144

11. GANTEN D, SPECK G (1978) The brain renin-angiotensin system: a model for the synthesis of peptides in the brain. Biochem Pharmacol 27:2379-2389

12. GANTEN D, SPECK G, MEYER D, LOOS HE, SCHELLING P, RETTIG R, UNGER TH (1980) Biochemical and functional aspects of brain renin: In: GROSS F, VOGEL G (eds) Enzymatic Release of Vasoactive Peptides. Raven Press, New York, p 31

13. HASEGAWA H, TATEISHI H, MASSON GMC (1973) Evidence for an angiotensinogen-stimulating factor after nephrectomy. Can J Physiol 51:563-566

14. HILGENFELDT U, HACKENTHAL E (1979) Purification and characterization of rat angiotensinogen. Biochim Biophys Acta 579:375-385

15. ITO T, EGGENA P, BARRETT JD, KATZ D, METTER J, SAMBHI MP (1980) Studies on angiotensinogen of plasma and cerebrospinal fluid in normal and hypertensive human subjects. Hypertension 2:432-436

16. KHAYYAL M, MACGREGOR J, BROWN JJ, LEVER AF, ROBERTSON JIS (1973) Increase of plasma renin substrate concentration after infusion of angiotensin in the rat. Clin Sci 44:87-90

17. LEWICKI JA, FALLON JH, PRINTZ MP (1978) Regional distribution of angiotensinogen in rat brain. Brain Res 158:359-371

18. LONG DM, MAXWELL RE, FRENCH LA (1972) The effects of glucosteroids upon experimental brain edema. In: REULEN JH, SCHÜRMANN K (eds) Steroids and Brain Edema. Springer, Berlin, p 65

19. LOWRY DM, ROSEBROUGH NJ, FARR AI, RANDALL RJ (1951) Protein measurement with the phenol reagent. J Biol Chem 193:265-275

20. MALAYAN SA, REID IA (1976) Antidiuresis produced by injection of renin into the third cerebral ventricle of the dog. Endocrinology 98:329-335

21. MOORE RY (1981) The suprachiasmatic nucleus, circadian rhythms, and regulation of brain peptides. In: MARTIN JB et al. (eds) Neurosecretion and Brain Peptides. Raven Press, New York, p 449

22. MORRIS BJ, REID IA (1978) The distribution of angiotensinogen in dog brain studied by cell fractionation. Endocrinology 103:492-500

23. NASJLETTI A, MASSON GMS (1972) Studies on angiotensinogen formation in a liver perfusion system. Circ Res 30/31 (Suppl II):187-198

24. PEACH MJ (1977) Renin-angiotensin system: biochemistry and mechanisms of action. Physiol Rev 57:313-370

25. PHILLIPS MI, FELIX D, HOFFMAN WE, GANTEN D (1977) Angiotensin-sensitive sites in the brain ventricular system. In: COWAN WM, FERRENDELLI JA (eds) Approach to the Cell Biology of Neurons (Society for Neuroscience Symposia, Vol II). Society for Neuroscience, Bethesda, p 308

26. PRINTZ MP, LEWICKI JA (1977) Renin substrate in the CNS: potential significance to central regulatory mechanisms. In: BUCKLEY JP, FERRARIO CM (eds) Central Actions of Angiotensin and Related Hormones. Pergamon Press, New York, p 57

27. PRINTZ MP, PRINTZ JM, GREGORY TJ (1978) Identification of angiotensinogen in animal brain homogenates. Circ Res 43 (Suppl I):21-27

28. RASMUSSEN S, DAMKJAER NIELSEN M, GIESE J (1981) Captopril combined with thiazide lowers renin substrate concentration: implications for methodology in renin assays. Clin Sci 60:591-593

29. REID IA, DAY RP (1977) Interactions and properties of some components of the renin-angiotensin system in brain. In: BUCKLEY JP, FERRARIO CM (eds) Actions of Angiotensin and Related Hormones. Pergamon Press, New York, p 267

30. REID IA, RAMSAY DJ (1975) The effects of intracerebroventricular administration of renin on drinking and blood pressure. Endocrinology 97:536-542

31. SCHELLING P, GANTEN D, HECKL R, HAYDUK K, HUTCHINSON JS, SPONER G, GANTEN U (1977) On the origin of angiotensin-like peptides in cerebrospinal fluid. In: BUCKLEY JP, FERRARIO CM (eds) Central Actions of Angiotensin and Related Hormones. Pergamon Press, New York, p 519

32. SCHELLING P, GANTEN U, SPONER G, UNGER T, GANTEN D (1980) Components of the renin-angiotensin system in the cerebrospinal fluid of rats and dogs with special consideration of the origin and the fate of angiotensin II. Neuroendocrinology 31:297-308

33. SCHELLING P, SPECK G, UNGER T, GANTEN D (1980) The brain renin-angiotensin system: biochemistry, localization and functional aspects. In: PARVEZ H, PARVEZ S (eds) Advances in Experimental Medicine. Elsevier/North-Holland, Biomedical Press, Amsterdam, p 243

34. SCHÖLKENS BA, STEINBACH R, JUNG W, GANTEN D (1980) Induction of angiotensin formation in brain: biological effects and pharmacological interferences. In: GROSS F, VOGEL HG (eds) Enzymatic Release of Vasoactive Peptides. Raven Press, New York, p 365

35. SERNIA C, REID IA (1980) Release of angiotensinogen by rat brain in vitro. Brain Res 192:217-225

36. SEVERS WB, DANIELS-SEVERS AE (1973) Effects of angiotensin on the central nervous system. Pharmacol Rev 25:415-449

37. UNGER T, ROCKHOLD RW, KAUFMANN-BÜHLER I, HÜBNER D, SCHÜLL B, SPECK G, GANTEN D (1981) Effects of angiotensin converting enzyme inhibitors on the brain. In: HOROVITZ ZP (ed) Angiotensin Converting Enzyme Inhibitors, Mechanisms of Action and Clinical Implications. Urban & Schwarzenberg, Baltimore, Munich, p 55

38. WALLIS CJ, PRINTZ MP (1980) Adrenal regulation of regional brain angiotensinogen content. Endocrinology 106:337-342

Definite Evidence for the Presence of [Ile^5]-Angiotensin I and [Ile^5]-Angiotensin II in the Brain of Rats

K. Hermann, D. Ganten, C. Bayer, T. Unger, R.E. Lang, and W. Rascher

INTRODUCTION

The mere presence of renin, angiotensinogen and converting enzyme in brain tissue does not prove that these form part of a biologically active brain renin-angiotensin system (RAS) and that they interact to generate the effector peptide of the RAS namely angiotensin II (ANG II) in vivo. A step in this direction is to demonstrate that ANG II is present in brain tissue or cerebrospinal fluid (CSF). There are several reports concerning the presence of ANG II in cerebrospinal fluid (CSF) and brain tissue (for ref. see 11) but there have been objections to some of these experiments. In addition, other investigators have attempted but failed to demonstrate ANG II in the brain and CSF (6,12,15). Immunocytochemical studies revealed a specific distribution of ANG II-like peptides in the brain (1,3,5,9) but others, using different antibodies, reported either a different distribution (1) or negative results (10). The present experiments were carried out to clarify the controversy whether ANG II was present in brain and CSF and to identify by direct biochemical methods brain angiotensin.

MATERIALS AND METHODS

Male Wistar rats weighing between 200 to 300 g bodyweight were used in all experiments.

For the collection of CSF rats were anesthetized with ether and mounted in a stereotaxic apparatus. The membrana atlanto-occipitalis was exposed and a 20 gauge cannula connected to polyethylene tubing was inserted into the fourth brain ventricle. CSF flowed freely by gravity and was collected in polyethylene tubes containing angiotensinase inhibitors (11). Immediately after collection the sample was boiled for 5 minutes in a water bath and centrifuged at 10.000 x g. Blood contamination of CSF was verified as described (11).

Experimental Brain Research, Suppl. 4

For blood collection a laparatomy was performed on ether anesthetized rats and blood collected from the abdominal aorta into syringes containing 5 vol% of 12.5 mM EDTA and 2.6 mM 1-10 orthophenanthroline as angiotensinase inhibitors. The blood samples were immediately transferred into distilled polyethylene tubes, centrifuged at 10.000 x g for 10 minutes at 4°C and the plasma stored at -20°C (8).

The brains were perfused in situ with warm 0.9% saline solution containing 25 U Heparin per 1 ml at 37°C. Transcardiac perfusion was performed by puncture of the still beating left heart ventricle, ligature of the abdominal aorta and allowing free flow of the perfusate through the disrupted vena jugularis. Perfusion was continued until the perfusate was clear. The brains were then quickly removed, frozen on dry ice and stored at -20°C. For the extraction of peptides from the brain, tissue was homogenized in 0.1 N HCl in a ratio of 1:10 (w/v) and centrifuged 30.000 x g for 30 minutes. The clear supernatant was collected and the pellet was reextracted with 0.1 N HCl;the supernatants were pooled and lyophylized.

Plasma and the brain extracts were redissolved in 10 ml of 0.001 N HCl, concentrated and purified on octadecasilyl-silica (ODS-silica) (Sep-Pak C 18) cartridges. In this procedure, the cartridges were first moistened with 3 ml of methanol and then washed with 10 ml of a 1% trifluoracetic acid (TFA) solution in water. In order to minimize nonspecific absorption, the cartridges were next coated with 1 ml of a 1% polypeptide solution and washed again with a mixture of methanol/water/TFA = 80/19/1 (v/v) and with a 1% TFA solution. The sample was applied and the gel washed twice with 5 ml of a mixture of 1% TFA/1% NaCl in a ratio of 1:1 (v/v). The retained peptides were eluted from the cartridge with 2 x 1 ml methanol/water/TFA = 80/19/1 (v/v) (Table 1). The organic phase of the eluent was removed under a flow of nitrogen and then taken to dryness in a vacuum box. The dry residue was dissolved in 4 x 250 µl of 0.001 N HCl and transferred into polyethylene tubes. This solution was used for direct radioimmunoassay and application to the high pressure liquid chromatography (HPLC) column.

Table 1. Peptide purification on SEP-PAK C_{18} cartridges (ODS-SILICA)

1)	3 ml methanol
2)	10 ml 1% TFA
3)	1 ml of 1% polypeptide solution
4)	10 ml methanol/water/TFA = 80/19/1 (v/v)
5)	10 ml 1% TFA
6)	Sample application (2x)
7)	2 x 5 ml 1% TFA/1% NaCl = 1:1 (v/v) (wash)
8)	2 x 1 ml methanol/water/TFA = 80/19/1 (v/v) (elution)
9)	dry under a flow of nitrogen and subsequent vacuum

The HPLC system consisted of two pumps (Model 6000A), a model 660 solvent programmer, an U6K injection system coupled to a variable wavelength UV-detector (Model 450) (Waters Assoc.). All separations were done on ODS-silica reversed phase columns (Bondapak C18, 300 x 4 mm, Waters Assoc.) with a particle size of 10 microns using two different gradient elution systems. 1) For the methanol gradient, 10 mM ammonium acetate buffer was adjusted to pH 4.5 or pH 5.4 with acetic acid, and methanol was increased linearly from 30 to 80 vol% within 35 minutes. 2) For the isopropanol gradient, 10 mM triethylammoniumphosphate (TEAP) was adjusted to pH 3.0 with concentrated phosphoric acid. In a linear gradient, isopropanol concentration was increased from 5 to 40 vol% within 45 minutes.

Sample injection volumes varied between 200 and 400 μl; flow rate was 1.0 ml/minute and fractions were collected for 30 or 60 seconds into polyethylene tubes and subjected directly to RIA. Blank controls for Sep-Pak extraction and HPLC comprised the complete procedure as described but without sample application. Recovery was tested by the addition of [Ile^5]-ANG I and [Ile^5]-ANG II to CSF or tissue homogenate.

Radioimmunoassays (RIA) were performed as described previously (8). Briefly, iodination of ANG I or ANG II was performed by the chloramine T method and ^{125}I labeled ANG was purified on a DEAE Sephadex A 25 column. Antibodies were obtained from white New Zealand rabbits immunized with ANG coupled to bovine serum albumine by the carbodiimide method. ANG I and ANG II antibodies were used in dilution of 7.6 x 10^{-4} and 2.4 x 10^{-6} respectively. Interferences with other peptides are listed in Table 2 and Table 3.

CSF was collected from untreated rats and pooled from 5 animals for one experiment. In another series of experiments rats were injected with 100 mU hog kidney renin into the brain ventricles 15 minutes prior to CSF sampling. For in vitro ANG I generation a pool of 1 ml of CSF was incubated with 200 mU of hog kidney renin at 37°C for 4 hours in tris-maleate buffer, pH 5.5, containing angiotensinase inhibitors (11). All experiments were run in triplicate or more. Human CSF was obtained by lumbar puncture.

Brain tissue was collected from untreated rats and from rats which had been nephrectomized 24 h previously. In both instances, brains were perfused with saline, as described.

Substances

All chemicals used were of analytical grade. Ethylenediaminetetraacetic acid (EDTA), 1-10 phenanthroline monohydrate, trishydroxymethylaminomethane (TRIS), maleic acid, BAL, trifluoracetic acid (TFA), isopropanol (Uvasol^R), were from Merck, Darmstadt, Germany. Ammonium acetate and triethylamine (TEA) were purchased from Fluka AG,

Switzerland. Baker chemicals, Holland, provided 1 **N** hydrochloric acid, methanol and water, spectrophotometric grade for use in HPLC; polypeptide solution was from Sigma chemicals, St. Louis, USA; [Ile^5]-ANG I, [Ile^5]-ANG II, [Ile^5]-ANG III and tetradecapeptide (TDP) were purchased from Bachem, Geneva, Switzerland, [Val^5]-ANG II from Ciba Geigy, Wehr, Germany, Saralasin was a gift from Norwich Pharmacal, USA. Hog kidney renin was purchased from NBC, Cleveland, USA. Rat plasma angiotensinogen was kindly provided by Dr. Hilgenfeldt, Heidelberg.

Table 2. Crossreactivity of angiotensin I antibody (K6) with various peptides

Peptide	% Crossreactivity
ANG I	100
ANG II	0.05
[Ala^8]-ANG (2-8) heptapeptide	0.01
ANG (2-8) heptapeptide	0.01
ANG (3-8) hexapeptide	0.05
ANG (4-8) pentapeptide	0.05
[Sar^1, Ala^8]-ANG (1-8) octapeptide (Saralasin)	0.05
Thyrotropin releasing factor (TRF)	0.005
Substance P	0.005
Gonadotropin releasing hormone (LHRH)	0.005

Table 3. Crossreactivity of angiotensin II antibody (CE) with various peptides

Peptide	% Crossreactivity
[Ile^5]-ANG II	100
[Ile^5]-ANG I	1
[Ile^5]-ANG (2-8) heptapeptide	100
[Ile^5]-ANG (3-8) hexapeptide	100
[Ala^8]-ANG (2-8) heptapeptide	0.02
[Sar^1, Val^5, Ala^8]-ANG (1-8) octapeptide (Saralasin)	0.1
Rat plasma angiotensinogen 95% pure	0.001
Renin substrate (1-14) tetradecapeptide	0.001
Thyrotropin releasing factor (TRF)	0.001
Desaminovasotocin	0.001
Bradykinin	0.001
Lys-bradykinin	0.001
Met-Lys-bradykinin	0.001
Substance P	0.001
Desaminooxytocin	0.001
Vasopressin (AVP, ADH)	0.001
Leu-enkephalin	0.001
Physalaemin	0.001
Gonadotropin releasing hormone (LHRH)	0.001

RESULTS AND DISCUSSION

Separation of Peptides on HPLC

Angiotensin and its fragments as well as a number of other neuropeptides can be separated from each other in a reproducible way on the HPLC systems described here. Peptides with similar retention times on the isopropanol gradient system (Fig. 1) could either be further isolated on the methanol gradient (Fig. 2) or did not interfere in the radioimmunoassays for [Ile^5]-ANG I and [Ile^5]-ANG II (Tables 2 and 3). This provided confidence that peptides extracted from the brain and CSF and eluted with retention times identical to synthetic [Ile^5]-ANG I, [Ile^5]-ANG II and [Ile^5]-ANG III on both chromatographic systems corresponded to the standard peptides and probably have the same amino acid sequence.

For all angiotensin peptides baseline separation was achieved on both chromatography systems used in this study (Fig. 1,2,3). It is noteworthy that [Ile^5]-ANG II was separated from [Val^5]-ANG II and from [Ile^5]-ANG III, the latter having a longer retention time than [Ile^5]-ANG II on the methanol gradient and a shorter retention time on the isopropanol gradient system.

Analysis of Angiotensin Peptides in CSF

The levels of angiotensin in CSF in rats have been reported to vary between undetectable levels to 200 fmol per ml or higher (11). The variability of angiotensin concentration in native CSF has been confirmed in this study and frequently undetectable levels of angiotensin have been observed in some but not all samples of native CSF of rats (Fig. 4, top panel).

In those cases where angiotensin-like immunoreactivity was detected, this could be identified as true angiotensin on HPLC. In a series of CSF angiotensin measurements, mean values of 17.6 $\pm$ 1.5 fmol/ml [Ile^5]-ANG I and 29.5 $\pm$ 5.9 fmol/ml [Ile^5]-ANG II were reported (11). There was no relationship between the CSF levels and plasma angiotensin concentration or with blood pressure.

If renin was injected into the brain ventricles and CSF was collected 15 minutes later, significant amounts of [Ile^5]-ANG I and [Ile^5]-ANG II were detected in CSF and the peptides eluted identically with the synthetic standard peptides previously added to the CSF (Fig. 4, middle panel).Small amounts of [Ile^5]-ANG III were also formed under these conditions (Fig. 5 top, middle panels). When CSF was incubated under in vitro conditions with the addition of inhibitors of converting enzyme and angiotensinases (11) only [Ile^5]-ANG I was generated from CSF angiotensinogen (Fig. 5,bottom panel). [Ile^5]-ANG I and [Ile^5]-ANG II extracted from both rat plasma

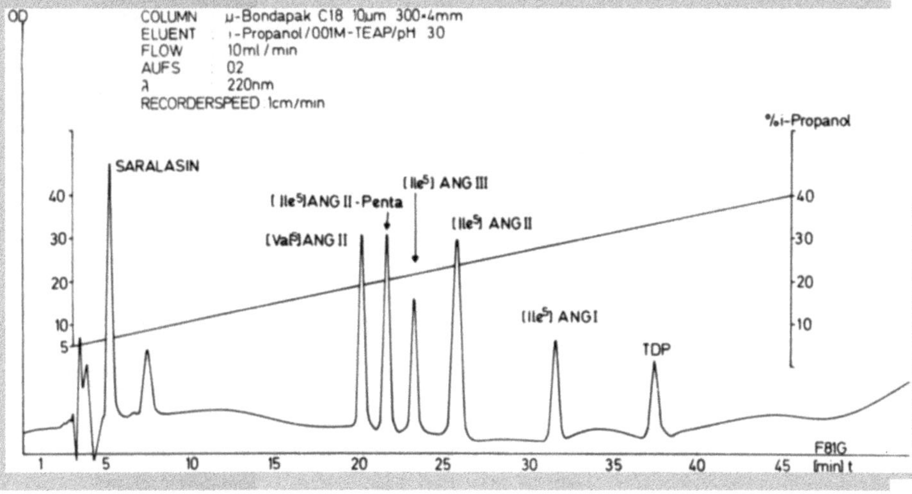

Fig. 1. Characterization of angiotensin peptides on HPLC; column characteristics are indicated on the left, TEAP: Triethylammonium phosphate. Note the baseline separation of all angiotensin peptides, especially of [Val^{5}]-angiotensin II and [Ile^{5}]-angiotensin II. On this isopropanol HPLC gradient system [Ile^{5}]-angiotensin II has a longer retention time than [Ile^{5}]-angiotensin III (5)

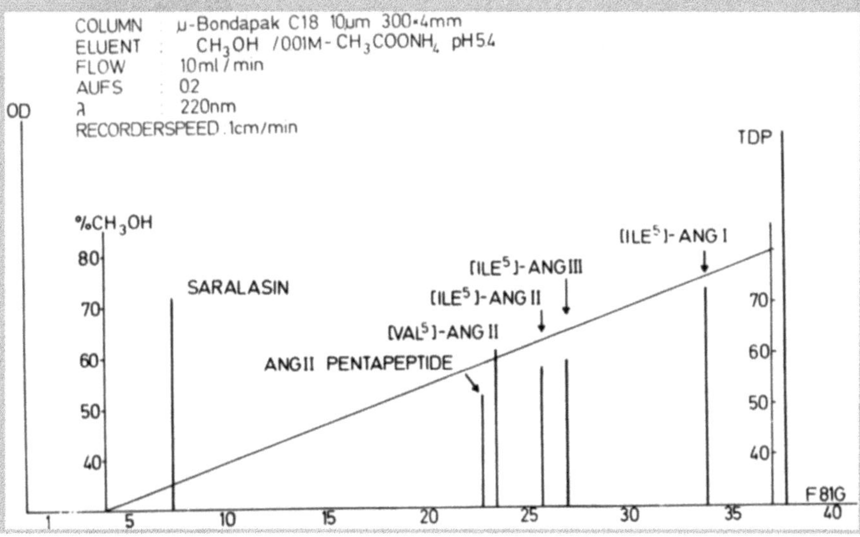

Fig. 2. Characterization of angiotensin peptides on a methanol HPLC gradient system; methanol was linearly increased from 30 to 80%. Note again baseline separation of [Val^{5}]-angiotensin II from [Ile^{5}]-angiotensin II; [Ile^{5}]-angiotensin III has a longer retention time than [Ile^{5}]-angiotensin II

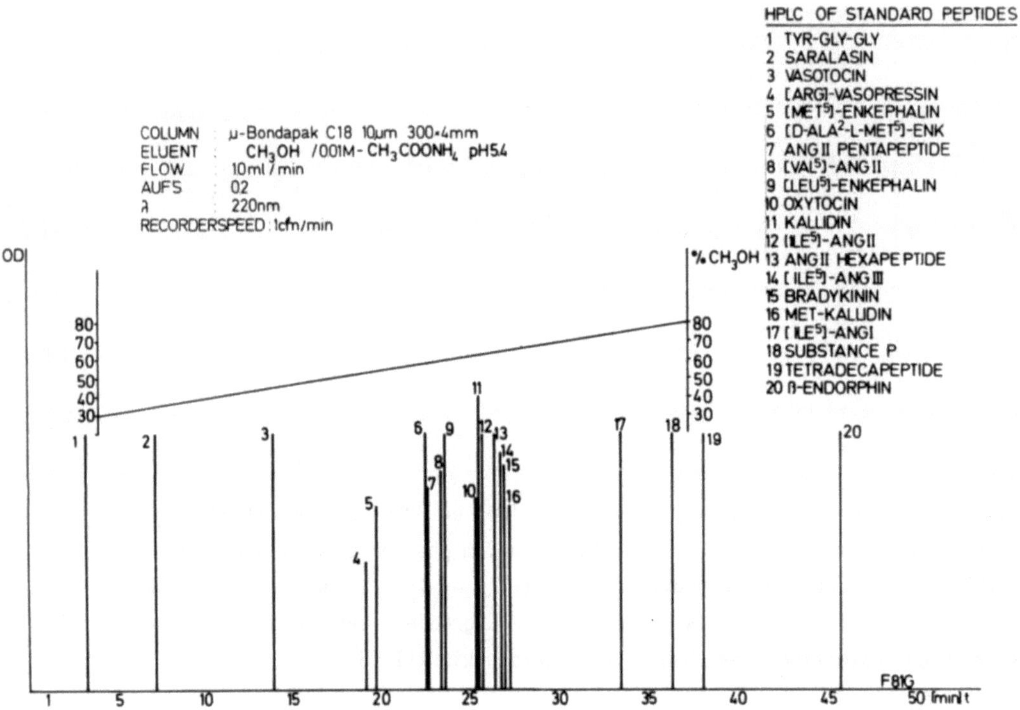

Fig. 3. Characterization of brain peptides on HPLC: Note baseline separation of a number of brain peptides. Further separation of angiotensin peptides between 20 and 30 min retention time is illustrated in Fig. 1 and Fig. 2. Cross reactivity of these peptides with angiotensin antibodies is documented in Table 1 and Table 2

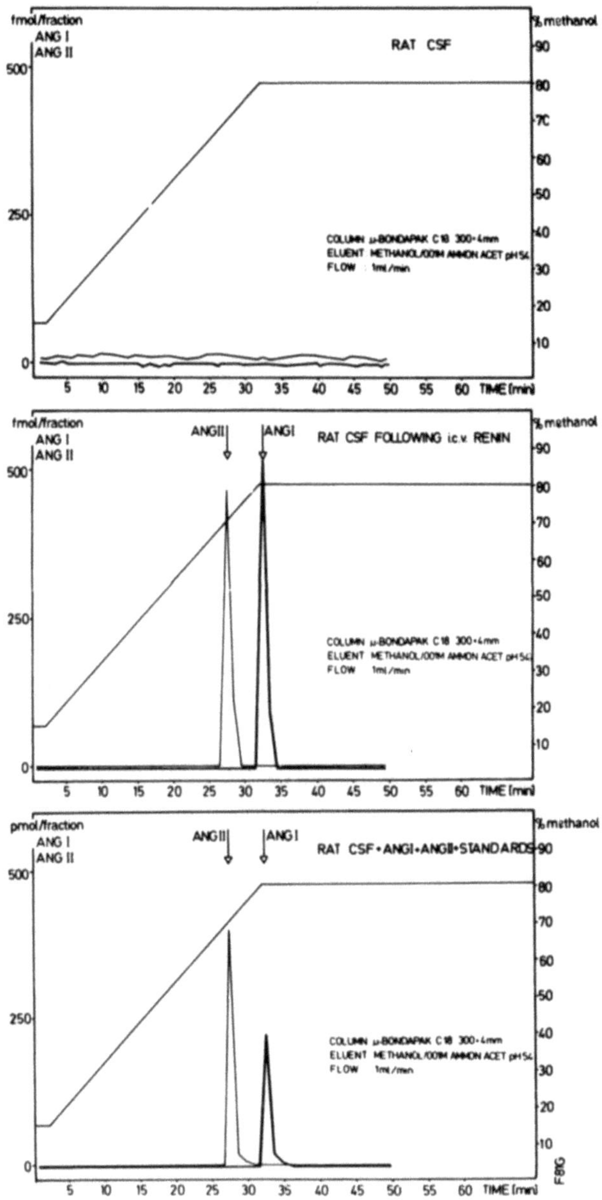

Fig. 4. Characterization of angiotensin peptides from cerebrospinal fluid (CSF) on HPLC. Note that this sample of native CSF from rats does not contain angiotensin peptides (upper panel, see also text). Following injection of hog kidney renin into the brain ventricles and withdrawal of CSF 15 minutes later, angiotensin I and angiotensin II are being formed from CSF angiotensinogen (middle panel). The angiotensin I and angiotensin II generated from CSF angiotensinogen correspond to synthetic [Ile^5]-angiotensin I and [Ile^5]-angiotensin II standards which have been added to the native CSF (lower panel)

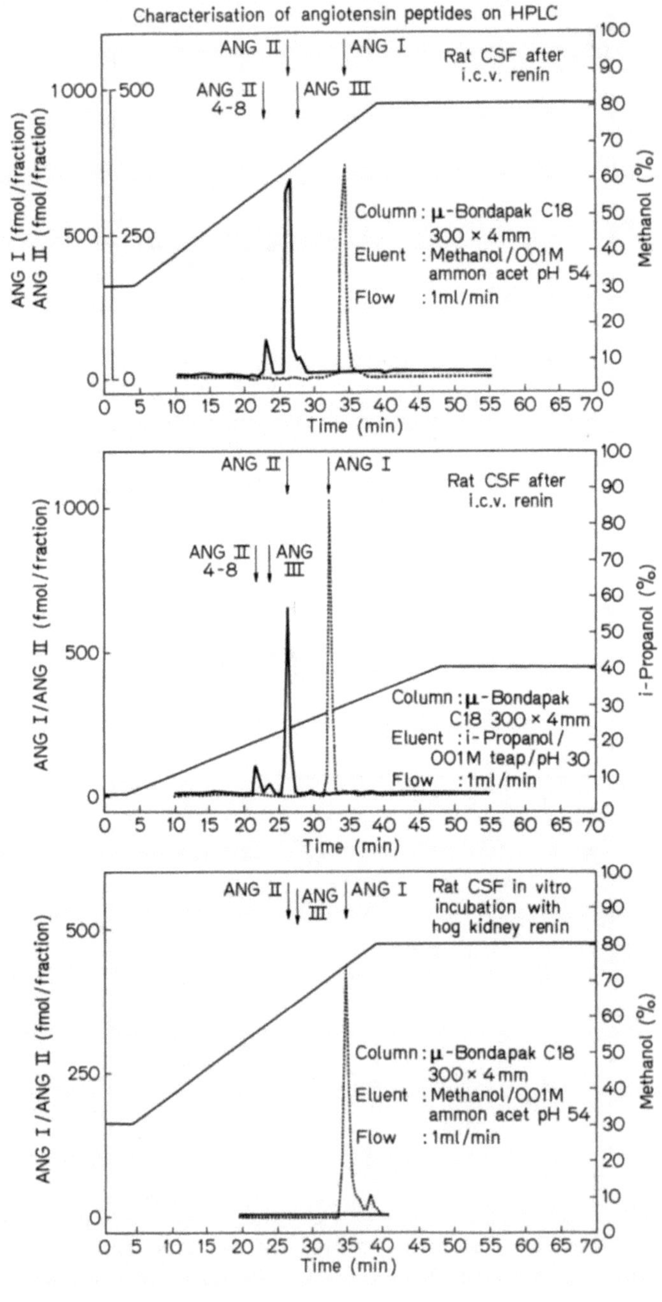

Fig. 5. Characterization of angiotensin peptides in CSF following in vivo generation of angiotensin by intracerebroventricular (i.c.v.) injection of renin. Note the identity of the cleavage products from CSF angiotensinogen with synthetic $[Ile^5]$-ANG I and $[Ile^5]$-ANG II on both chromatography systems (arrows, upper and middle panel); small amounts of $[Ile^5]$-ANG III are equally present. Following in vitro incubation only $[Ile^5]$-ANG I is formed (lower panel)

and CSF had identical retention times (Fig. 6). It can therefore be concluded that their physicochemical properties, i.e. their amino acid sequence, are similar or identical.

The identification of [Ile^5]-ANG I as the cleavage product of renin from CSF angiotensinogen and from plasma angiotensinogen could indicate that the high molecular weight precursors of angiotensin in plasma and in brain are identical and that they originate from the plasma. This is unlikely, however, as has been discussed elsewhere in detail (5,9,11,15) and as will be shown below.

Angiotensin in Human CSF

Similar to the rat, human CSF also contains variable amounts of angiotensin. In those samples where angiotensin levels were sufficiently high, the CSF peptide could be shown to be identical with synthetic [Ile^5]-ANG I and [Ile^5]-ANG II standards (Fig. 7). This is in contrast to other reports in which it was claimed that the peptides crossreacting with the angiotensin antibody in the radioimmunoassay were neither [Ile^5]-ANG I, [Ile^5]-ANG II nor [Ile^5]-ANG III (12).

Extraction of Angiotensin from Brain

Rat brain tissue itself contained measurable amounts of angiotensin which was identified as being [Ile^5]-ANG I, [Ile^5]-ANG II and to a lesser extent [Ile^5]-ANG III upon rechromatography on two different HPLC systems. Preliminary results indicate that rat brain stem (Fig. 8) and the median eminence (Fig.9) have high angiotensin content. Despite satisfactory recovery of 70 to 80% of synthetic angiotensin added to the brain extract, no quantitative values can be given at the present time for angiotensin concentrations in brain tissue, the values varying between 12 to 1 700 fmol [Ile^5]-ANG I or [Ile^5]-ANG II per g wet tissue. The relatively constant ratio between [Ile^5]-ANG I [Ile^5]-ANG II of less than one (Table 4) indicates, however, that the extraction rate is similar for both peptides. It also supports previously reported data, namely that brain angiotensin does not originate from the plasma since the ratio [Ile^5]-ANG I [Ile^5]-ANG II in the plasma is above 1. The dissimilarity of the [Ile^5]-ANG I [Ile^5]-ANG II ratio in plasma and in brain apart from the intracellular localization of the peptide and its persistance after nephrectomy is an additional argument speaking against the uptake of angiotensin from plasma into the brain and speaking in favour of local synthesis of angiotensin.

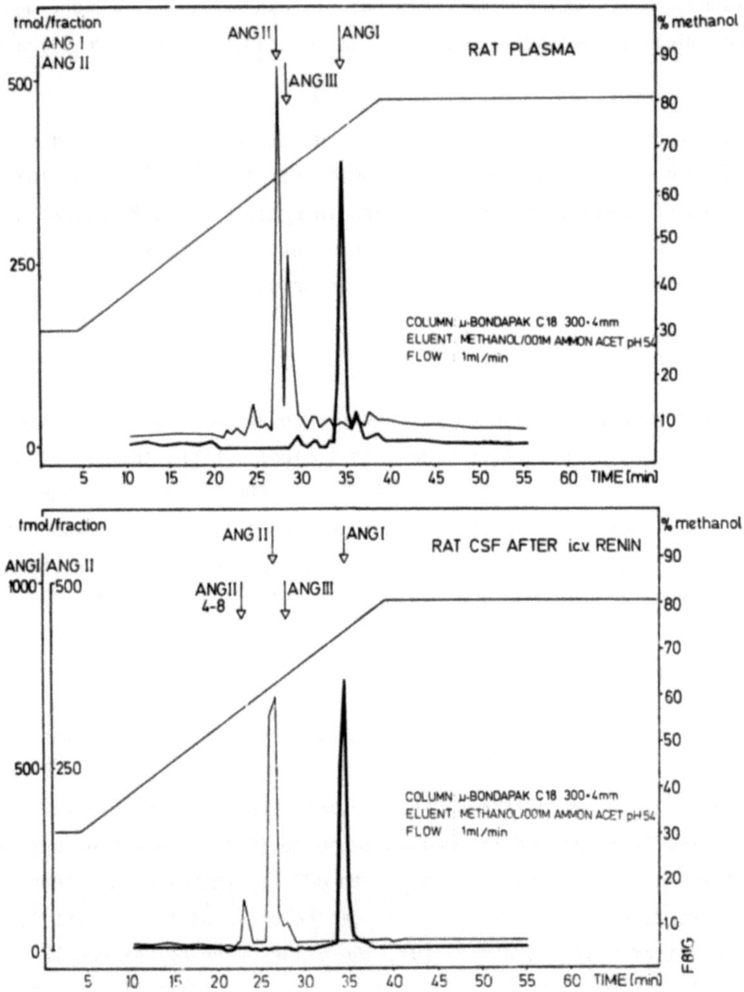

Fig. 6. Comparison and identity of angiotensin excracted from rat plasma and from CSF

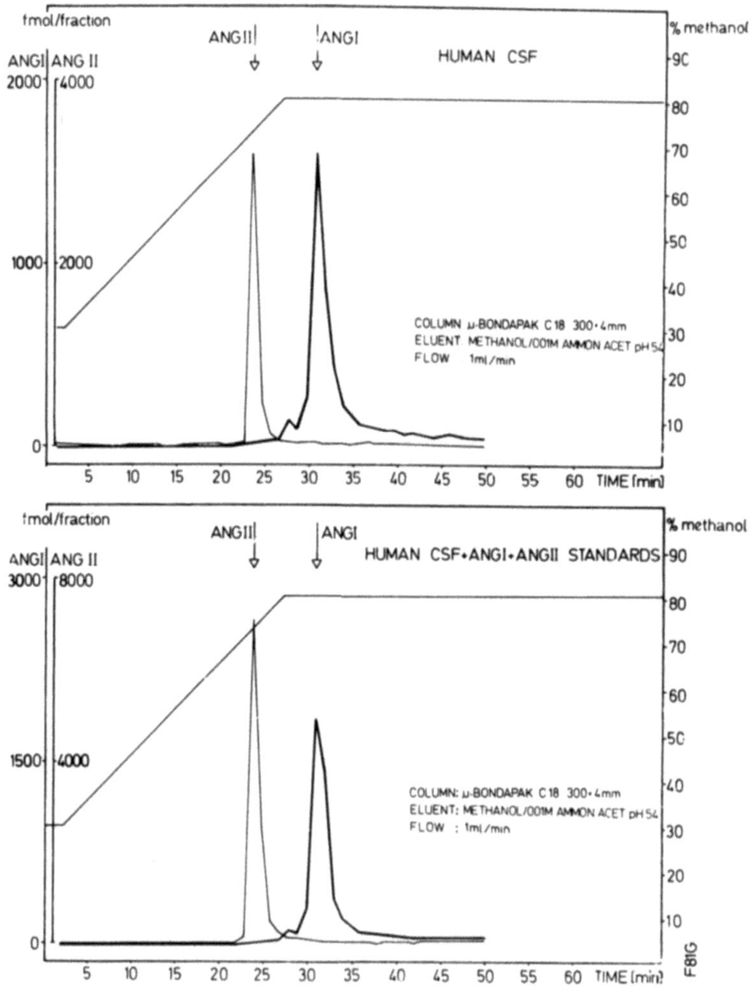

Fig. 7. Comparison and identity of CSF angiotensins from man with synthetic [Ile^{5}]-ANG I and [Ile^{5}]-ANG II standards added to the CSF prior to sample application

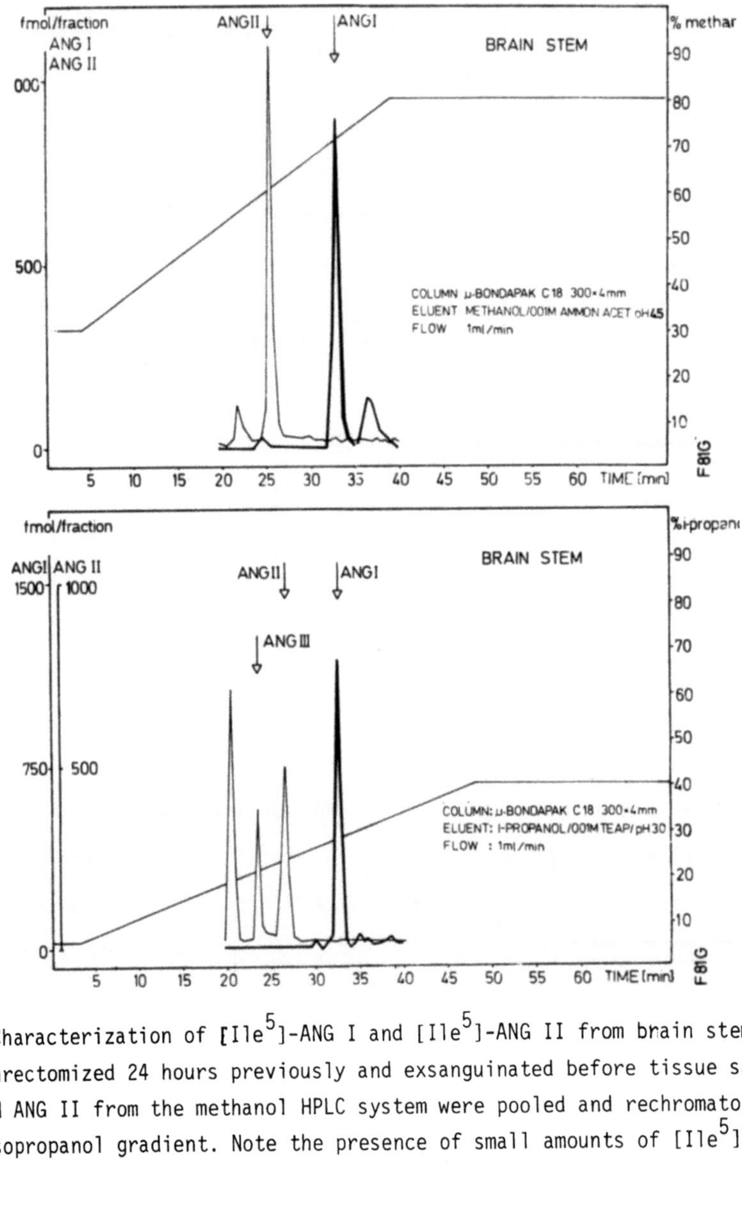

Fig. 8. Characterization of [Ile^{5}]-ANG I and [Ile^{5}]-ANG II from brain stem. Rats were nephrectomized 24 hours previously and exsanguinated before tissue sampling. ANG I and ANG II from the methanol HPLC system were pooled and rechromatographed on the isopropanol gradient. Note the presence of small amounts of [Ile^{5}]-ANG III

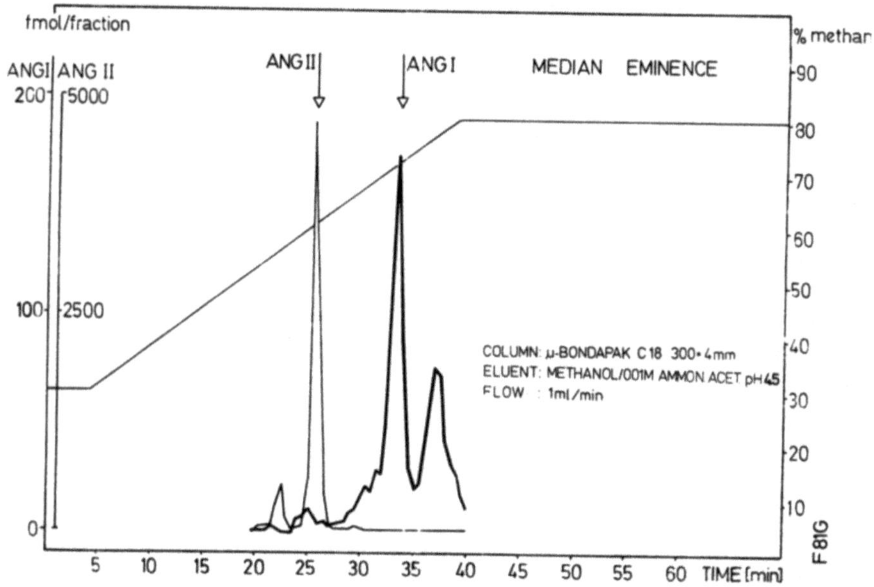

Fig. 9. Characterization of [Ile^5]-ANG I and [Ile^5]-ANG II extracted from median eminence of brain. Rats were nephrectomized 24 hours previously and exsanguinated before killing

Table 4. ANG I and ANG II concentration (fmol/g wet tissue) in brain extracts of rats

Experiment	ANG I	ANG II	ANG I/ANG II
1. No perfusion No nephrectomy	337	1603	0.21
2. a) Perfusion	175	709	0.26
b) Perfusion + Nephrectomy	1047	1677	0.62
3. a) Perfusion	12	31	0.38
b) Perfusion + Nephrectomy	23	77	0.29
4. a) Perfusion	29	114	0.25
b) Perfusion + Nephrectomy	12	46	0.26

In summary [Ile^5]-ANG I, [Ile^5]-ANG II and [Ile^5]-ANG III have been identified in the brain and CSF of rats and in human CSF. The brain peptides are identical with plasma angiotensin on two different HPLC systems indicating that the amino acid sequences of brain and plasma angiotensin are probably the same. This solves the controversy as to whether angiotensin is present in the brain and supports the idea that the brain renin angiotensin system is active in vivo.

REFERENCES

1. CHANGARIS DG, DEMERS LM, KEIL LC, SEVERS WB (1978) Angiotensin II immunohistochemistry of the rat brain. Neuroendocrinology 25:257-264

2. FISCHER-FERRARO C, NAHMOD VI, GOLDSTEIN DJ, FINK S (1971) Angiotensin and renin in rat and dog brain. J exp Med 133:353-361

3. FUXE K, GANTEN D, HÖKFELT T, BOLME P (1976) Immunohistochemical evidence for the existence of angiotensin II-containing nerve terminals in the brain and spinal cord in the rat. Neurosci Lett 2:229-234

4. GANTEN D, MARQUEZ-JULIO A, GRANGER P, HAYDUK K, KARSUNKY KP, BOUCHER R, GENEST J (1971) Renin in dog brain. Am J Physiol 221:1733-1737

5. GANTEN D, FUXE K, PHILLIPS MI, MANN JFE, GANTEN U (1978) The brain isorenin-angiotensin system; biochemistry, localization and possible role in drinking and blood pressure regulation. In: GANONG WF and MARTIN L (eds) Frontiers in Neuroendocrinology. Vol 5, Raven Press, New York, pp 61-99

6. HORVATH JS, BAXTER C, FURBY F, TILLER D (1977) Endogenous angiotensin in the brain. In: De JONG W, PROVOOST AP and SHAPIRO AP (eds) Progress in brain research. Vol 47, pp 161-165

7. HUTCHISON JS, CSICSMANN J, KORNER PI, JOHNSTON CI (1978) Characterization of immunoreactive angiotensin in canine cerebrospinal fluid as Des-Asp[1]-angiotensin II. ClinSci Mol Med 54:147-151

8. MANN JFE, JOHNSON AK, GANTEN D (1980) Plasma angiotensin II: dipsogenic levels and angiotensin generating capacity of renin. Am J Physiol 238:R 372-377

9. PHILLIPS MI (1978) Angiotensin in the brain. Neuroendocrinology 25:354-377

10. REID IA (1979) The brain renin-angiotensin system: a critical analysis. Fed. Proc 38:2255-2259

11. SCHELLING P, GANTEN U, SPONER G, UNGER TH and GANTEN D (1980) Components of the renin-angiotensin system in the cerebrospinal fluid of rats and dogs with special consideration of the origin and the fate of angiotensin II. Neuroendocrinology 31:297-308

12. SEMPLE ES, MACRAE WA, MORTON JJ (1980) Angiotensin II in human cerebrospinal fluid may be an immunoassay artifact. Clin Sci 59:615-645

13. SIRETT NE, BRAY JJ, HUBBARD JI (1981) Localization of immunoreactive angiotensin II in the hippocampus and striatum of rat brains. Brain Res 217:405-411

14. SPECK G, POULSEN K, UNGER TH, RETTIG R, BAYER C, SCHÖLKENS B, GANTEN D (1981) In vivo activity of purified mouse brain renin. Brain Res.219:371-384

15. WALLIS CJ, PRINTZ MP (1980) Adrenal regulation of the regional brain angiotensinogen content. Endocrinology 106:337-342

Immunocytochemical Demonstration of Angiotensin II and Renin-Like Immunoreactive Nerve Cells in the Hypothalamus. Angiotensin Peptides as Comodulators in Vasopressin and Oxytocin Neurons and Their Regulation of Various Types of Central Catecholamine Nerve Terminal Systems

K. Fuxe, D. Ganten, K. Anderson, L. Calza, L. F. Agnati, R. E. Lang, K. Poulsen, T. Hökfelt, and P. Bernardi

INTRODUCTION

There exist indications for the existence of angiotensin II-like immunoreactivity in nerve cell bodies and nerve terminals in the brain and the spinal cord of mammals (3,7,8,11,14,20). These groups have described a similar distribution of angiotensin II immunoreactive nerve terminals in the central nervous system. Angiotensin II-like immunoreactivity in the nerve cell bodies is only seen following pretreatment of rats with colchicine intraventricularly. These cell bodies mainly belong to the magnocellular component of the paraventricular hypothalamic nucleus and of the supraoptic nucleus (3). There exists evidence that angiotensin II-like immunorecativity is costored with vasopressin in the same neuron (14). It has recently also been possible to localize renin-like immunoreactivity in the magnocellular component of the paraventricular hypothalamic nucleus and the supraoptic nucleus in the rat and mouse brain (10). In order to establish if renin-like material is mainly costored in the vasopressin or oxytocin neuron systems the distribution of vasopressin, oxytocin, angiotensin, neurophysin I and neurophysin II-like immunoreactivity has been compared with the distribution of renin-like and angiotensin II-like immunoreactivity found in adjacent sections of the rat hypothalamus containing the magnocellular hypothalamic nuclei.

Another aim of the present study has also been to continue the analysis of the interactions of the central renin-angiotensin systems with the central catecholamine pathways (8). In the previous analyses it could be demonstrated that intraventricularly injected renin increases dopamine (DA) turnover in the medial and lateral palisade zone (MPZ, LPZ) of the median eminence and selectively reduces noradrenaline (NA) turnover in the posterior magnocellular component of the paraventricular hypothalamic nucleus containing mainly vasopressin but also oxytocin immunoreactive nerve cell bodies. Furthermore, a weak increase in NA turnover was induced by i.v.t. renin in the dorsomedial hypothalamic nucleus (DM). In the forebrain a highly selective increase of DA turnover was produced by i.v.t. renin within the dorsal and posterior part of the nucleus accumbens which seems to be

Experimental Brain Research, Suppl. 4

rich in cholecystokinin (CCK) positive DA nerve terminals (13). All the effects of renin were counteracted by an angiotensin I converting enzyme inhibitor. Furthermore, biochemical studies indicated that the overall change in the hypothalamus was an increase of adrenaline (A) and NA turnover following intraventricular renin injections, while the NA and DA levels and turnover were unaffected in the anteromedial frontal cortex (8,9). Finally, a selective increase of A turnover could be demonstrated within the dorsal midline area of the caudal medulla oblongata (DCMO) containing e.g. nucleus tractus solitarii and nucleus motorius dorsalis nervi vagi (9). In these experiments intraventricular renin injections reduced prolactin secretion and growth hormone secretion (8). These studies have now been continued in the present paper comparing the effects of intraventricular and intravenous injections of renin on discrete DA and NA nerve terminal systems of the hypothalamus.

IMMUNOCYTOCHEMICAL METHODOLOGY

The problem on how angiotensin II-like the angiotensin immunoreactivity demonstrated in central neuron systems is, has recently been discussed by Brownfield, Reid, Ganten and Ganong (3) using the same antibodies as used in our own work of the mapping of the angiotensin II-like immunoreactive central and peripheral neuron systems. Early in our work it was noted that only some of the angiotensin II antibodies (DE and CE) could be used for the immunocytochemical analysis of immunoreactive angiotensin in the rat peripheral and central neuron system. The angiotensin II antibodies were obtained by coupling 5-Ile-angiotensin II to bovine serum albumin using glutaraldehyde as a coupling agent and injecting this complex into rabbits. The DE and CE angiotensin II antisera crossreact to 100% with angiotensin II and angiotensin III. The crossreactivity with angiotensin I is 1-4% and with the tetradecapeptide synthetic renin substrate (TDP) 0.1-0.94% (3). It has recently been found, however, that not only angiotensin II and angiotensin III but also angiotensin I and TDP, although in higher concentrations, can prevent the staining obtained with these antibodies (3). Therefore, in the following we will only refer to the immunoreactive material as angiotensin-like immunoreactivity although it may still represent the presence of angiotensin II. To further characterize the immunoreactive angiotensin, a chemical identification is necessary. It must be emphasized, however, that the radioimmunological characterization of the CE and DE angiotensin II antibodies show hardly any crossreactivity with peptides outside the angiotensin family, such as vasopressin and oxytocin, enkephalins, releasing factors etc. In agreement with this it has not been possible to prevent staining with these antibodies by preadsorption of the antibodies with vasopressin and oxytocin in a concentration of 50 μg/ml.

Brain renin crossreacts with monospecific antibodies raised against the mouse submaxillary renin (12,17), and therefore it has become possible to demonstrate renin-like material in mouse and rat brain. The renin-like immunoreactivity was mainly found within some magnocellular components of the nucleus paraventricularis hypothalami and supraopticus. Renin-like immunoreactivity could also be demonstrated in the Purkinje cell bodies and dendrites of the cerebellar cortex (10). In the present study the distribution of renin-like immunoreactivity within the nucleus paraventricularis and nucleus supraopticus have been compared to that of neurophysin I, neurophysin II, oxytocin, vasopressin and angiotensin immunoreactivity demonstrated in adjacent sections of the hypothalamus. These antibodies have been characterized in radioimmunoassay procedures (10,11,15); crossreactivity of neurophysin I with neurophysin II antibodies was 5%, and neurophysin II with neurophysin I 10%.

The indirect immunofluorescence procedure of Coons was used (5). The immunohistochemical procedure (7) involved perfusion fixation with 4% icecold formaline followed by cryostat sectioning. The dilution of the DE angiotensin II antiserum ranged from 1:40 to 1:60, the renin antiserum from 1:20 to 1:60, the oxytocin antiserum from 1:300 to 1:1000, the vasopressin antiserum from 1:100 to 1:300, the neurophysin II antiserum from 1:300 to 1:1000 and the neurophysin I antiserum from 1:200 to 1:500.

DISTRIBUTION OF ANGIOTENSIN IMMUNOREACTIVE NERVE CELL BODIES IN THE CENTRAL AND PERIPHERAL NERVOUS SYSTEM

Most angiotensin immunoreactive cell bodies are found in the nucleus supraopticus and in the nucleus paraventricularis hypothalami. The angiotensin II cell bodies are found both in the magno- and parvo-cellular part of the paraventricular hypothalamic nucleus. However, the angiotensin positive cell bodies in the parvocellular part still belong to the magnocellular type of nerve cell bodies, and Swanson and Sawchenko (24) have described a medial magnocellular component of the paraventricular hypothalamic nucleus. Therefore, all the angiotensin immunoreactive nerve cell bodies appear to belong to the classical magnocellular neurophysin positive neurons of the supraoptic and paraventricular nucleus. These nerve cell bodies are seen to send angiotensin immunoreactive fibres into the median eminence and the posterior pituitary gland after turning around the fornix on its lateral side (11,12). Scattered angiotensin immunoreactive nerve cell bodies have also been observed in the perifornical area and in the dorsal and ventral part of the lateral hypothalamic area and in the medial amygdaloid cortex (Fig. 1). These angiotensin positive nerve cell bodies could mainly be observed following colchicine pretreatment (75 μg/30 μl, given intraventricularly 24 h before perfusion). The magnocellular nerve cell bodies noticed in the lateral hypothalamus seem to belong to the accessory components of the supraoptic and paraventricular nuclei. In all the other regions of the brain and the spinal cord no angiotensin immunoreactive nerve cell bodies have been observed. Therefore, it seems possible that most of the angiotensin immunoreactive nerve terminals demonstrated in various parts of the brain and the spinal cord belong to the magnocellular neurons of the hypothalamus, especially those within the paraventricular hypothalamic nucleus (24), since the supraoptic nerve cells mainly project to the posterior pituitary gland. Some cell bodies of the spinal ganglia are also angiotensin immunoreactive and large numbers of angiotensin immunoreactive nerve cell bodies can be demonstrated in the plexus myentericus of the duodenum, stomach and colon (Fig. 2).

DISTRIBUTION OF ANGIOTENSIN IMMUNOREACTIVE NERVE TERMINALS IN THE CENTRAL AND PERIPHERAL NERVOUS SYSTEM

The distribution of these systems has been summarized in previous publications (8,11). In the present paper we will for the first time illustrate the distribution of angiotensin immunoreactive nerve terminals within the nucleus dorsomedialis hypothalami (weakly to moderately immunoreactive terminals of low to moderate density), within the ventrolateral part of the caudal medulla oblongata, where NA cell bodies are located (moderately to strongly immunofluorescent varicose terminals of a low density), within the sympathetic lateral column of the upper part of the thoracic spinal cord (moderately immunoreactive fine nerve terminals of a high density) (Fig. 3) and within layer I and II of the spinal cord (substantia gelatinosa) (Fig. 4). Layers I and II represent the substantia gelatinosa of the spinal cord. It is seen in Fig. 4 that two types of angiotensin innervation exist in this area. Thus, within the marginal zone (layer I) one type is found forming patches of angiotensin immunoreactive terminals (see arrow). Within the inner part of layer II another innervation field is found forming a high density of angiotensin immunoreactive nerve terminals especially in the lateral part, which appear as dots in the transverse section shown. It appears from preliminary studies on the effects of transection of the dorsal root that at least the angiotensin immunoreactive terminals in layer II belong to the central branches of primary sensory neurons. Instead the angiotensin terminals in the sympathetic lateral column and possibly also in the marginal zone belong to descending angiotensin pathways as shown in transection experiments of the spinal cord. It now seems likely that these descending pathways originate from the magnocellular part of the paraventricular hypothalamic nucleus. This nucleus rich in angiotensin immunoreactive cell bodies probably also give rise to the dense networks of angiotensin nerve terminals

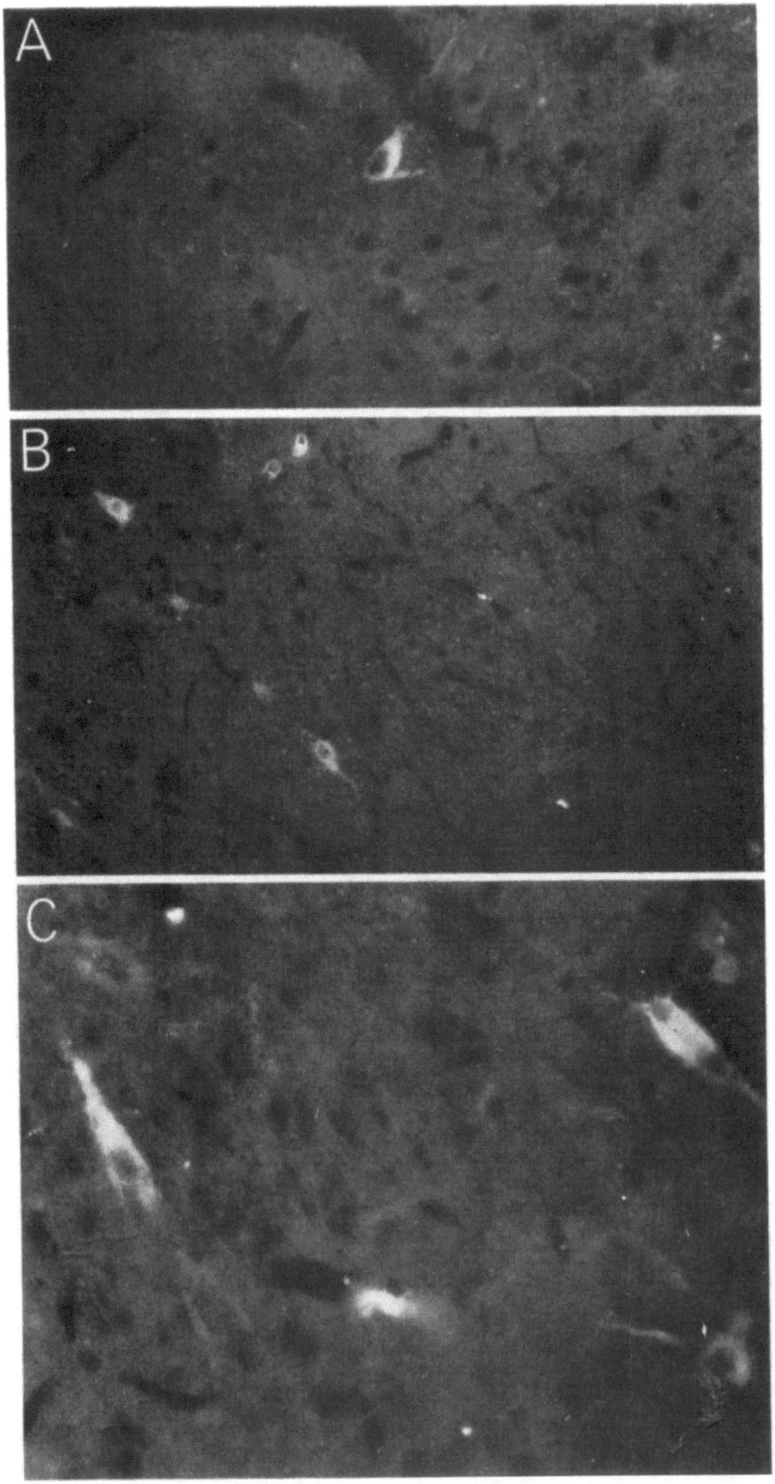
A
B
C

found in the external layer of the median eminence mainly within the medial palisade zone (14).

In Fig. 2 the angiotensin immunoreactive cell bodies within the plexus myentericus are seen. They probably give rise to the angiotensin immunoreactive fibres innervating the circular muscular layer, the outer longitudinal muscle layer as well as the plexus myentericus itself. Only a few positive fibres are found within the submucous plexus. Some angiotensin fibres also appear to innervate some blood vessels.

EXISTENCE OF RENIN-LIKE IMMUNOREACTIVITY IN OXYTOCIN NEURONS OF THE PARAVENTRICULAR AND SUPRAOPTIC NUCLEUS

Renin-like immunoreactivity has been demonstrated in nerve cell bodies of the nucleus supraopticus and paraventricularis hypothalami (Fig. 5) (10) and also in nerve terminals of the posterior pituitary (10) (Fig. 6). This analysis was performed by comparing the distribution pattern of renin (Fig. 5) and oxytocin immunoreactive nerve cell bodies in thin adjacent sections of the paraventricular and supraoptic nucleus (Fig. 7 and 8). As seen, at least some renin positive cells are oxytocin immunoreactive (arrow heads). In addition, it was possible to quantitate the degree of coexistence by the use of a method of Agnati and coworkers (1). According to this procedure three thin adjacent sections are incubated with renin antiserum, oxytocin antiserum and renin plus oxytocin antiserum. The degree of coexistence can then be obtained by subtracting the number of cells obtained in the section incubated with both antibodies from the sum of the number of immunoreactive cells obtained in the sections incubated with renin antiserum alone and oxytocin antiserum alone. Thus, if coexistence exists for oxytocin and renin,

◁ Fig. 1. Angiotensin-like immunoreactivity in normal male Sprague-Dawley rat. Indirect immunofluorescence technique. Denise angiotensin II antibodies 1:40. 24 h before killing the animals were treated with colchicine intraventricularly in a dose of 75 µg/20 µl. In this way the antigen is accumulated in the nerve cell bodies and angiotensin immunoreactivity can be demonstrated e.g. in the medial amygdaloid cortex (A;300 x), within the lateral hypothalmic area, often in the perifornical region (B;120 x, C;300 x). The angiotensin positive nerve cell bodies shown are of the magnocellular type like the cell bodies in the supraoptic and paraventricular nuclei

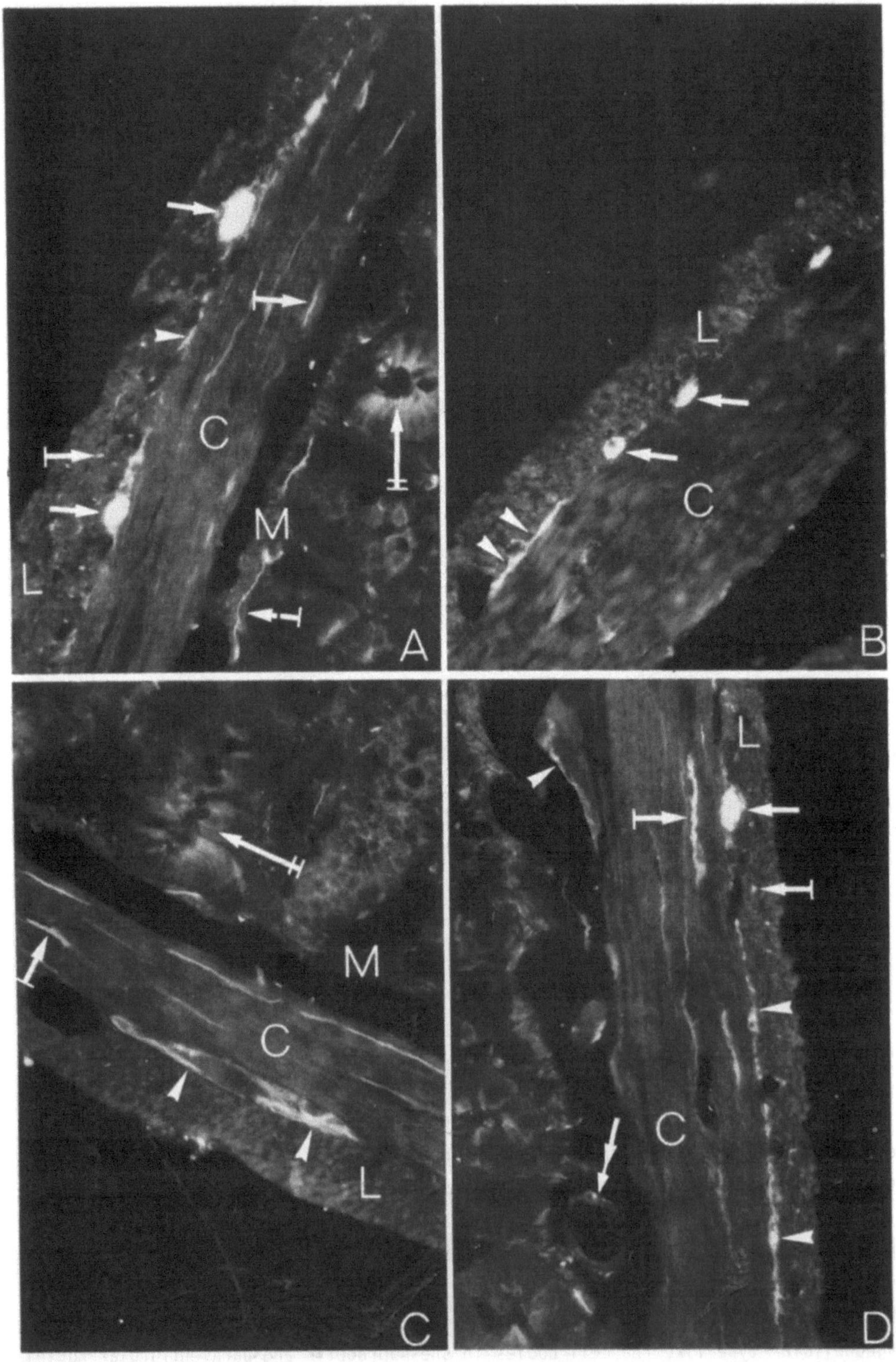
C
M
L
A
L
C
B
M
C
L
C
L
C
D

◁ Fig. 2. Angiotensin-like immunoreactivity in the duodenum of the normal male Sprague-Dawley rat. Indirect immunofluorescence technique. Denise angiotensin II antibodies 1:40. Strongly immunoreactive nerve cell bodies are seen in the plexus myentericus (arrows). Angiotensin immunoreactive nerve fibres are seen especially in the circular muscular layer (↦) and within the plexus myentericus itself where they may establish contact with non-angiotensin immunoreactive nerve cell bodies (see arrow heads). A few transversely cut angiotensin immunoreactive fibres are also seen in the longitudinal muscle layer (L). A few immunoreactive fibres are also present in the plexus submucosus (M) where some of them may establish contacts with blood vessels (double arrows). Unspecific fluorescence is seen in the apical parts of the epithelial cells of the duodenum (↦). C=circular layer. Magnification in A, C and is 300 x and in B 120 x

Fig. 3. Angiotensin-like immunoreactivity in nerve terminals. Indirect immunofluorescence technique. Denise angiotensin II antibodies 1:40. A:Nuc. dorsomedialis hypothalami (DM) of a 1 day nephrectomized male Sprague-Dawley rat. A low to moderate density of fine nerve terminals is seen, 300 x. B:The ventrolateral part of the caudal medulla oblongata of normal male Sprague-Dawley rat. This area is rich in NA cell bodies. A low density of strongly fluorescent varicose fibres is seen, 300 x. C, D:Sympathetic lateral column of the male Sprague-Dawley rat cranial to a spinal cord transection performed 7 days before killing. A dense plexus seen mainly as densely packed dots is observed surrounding the nerve cells of this nucleus. L.F. = lateral funiculus. C = magnification 120 x. D = magnification 300 x (Fig. 3 see p. 216)

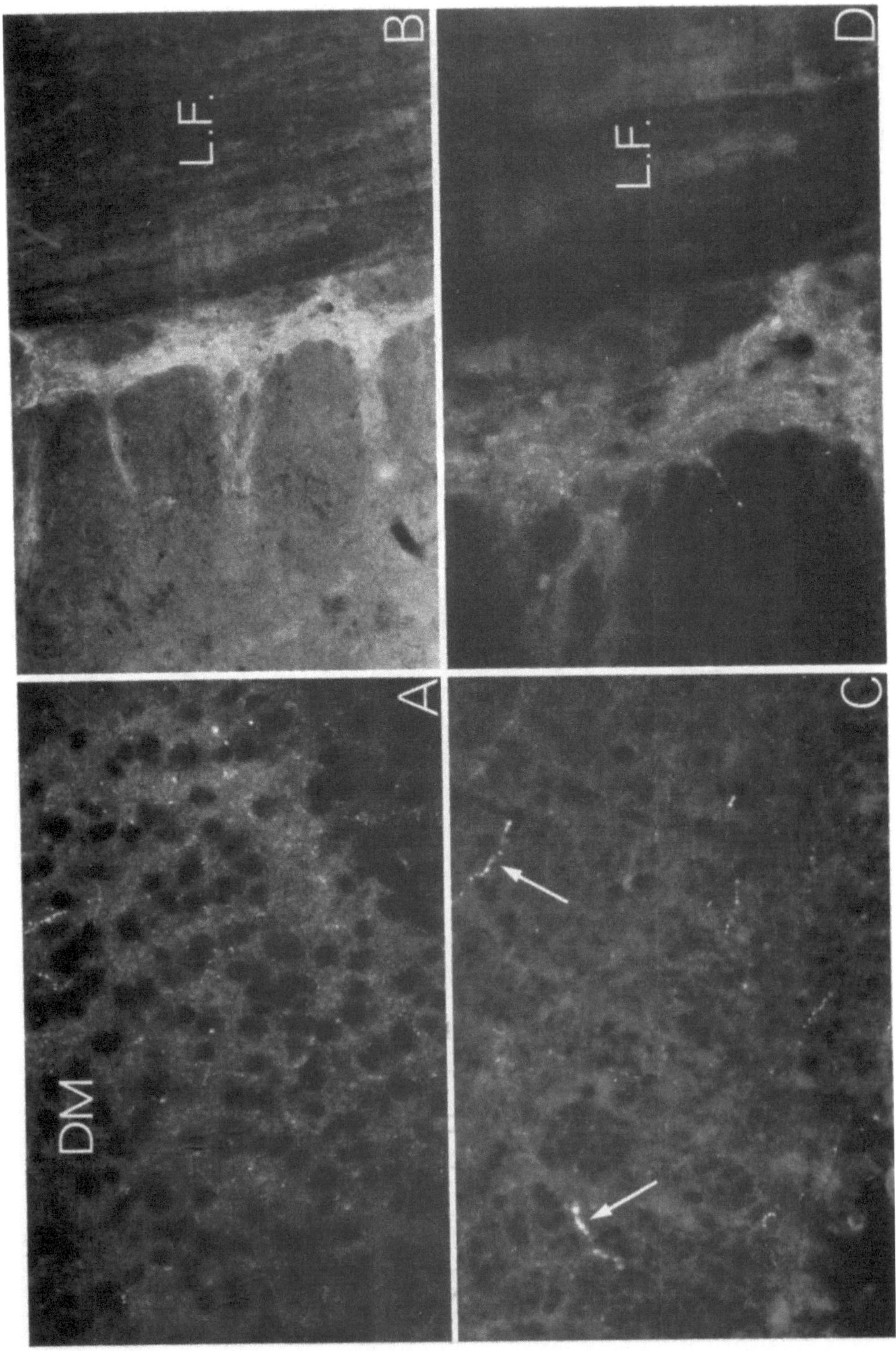
DM
A
L.F.
B
C
L.F.
D

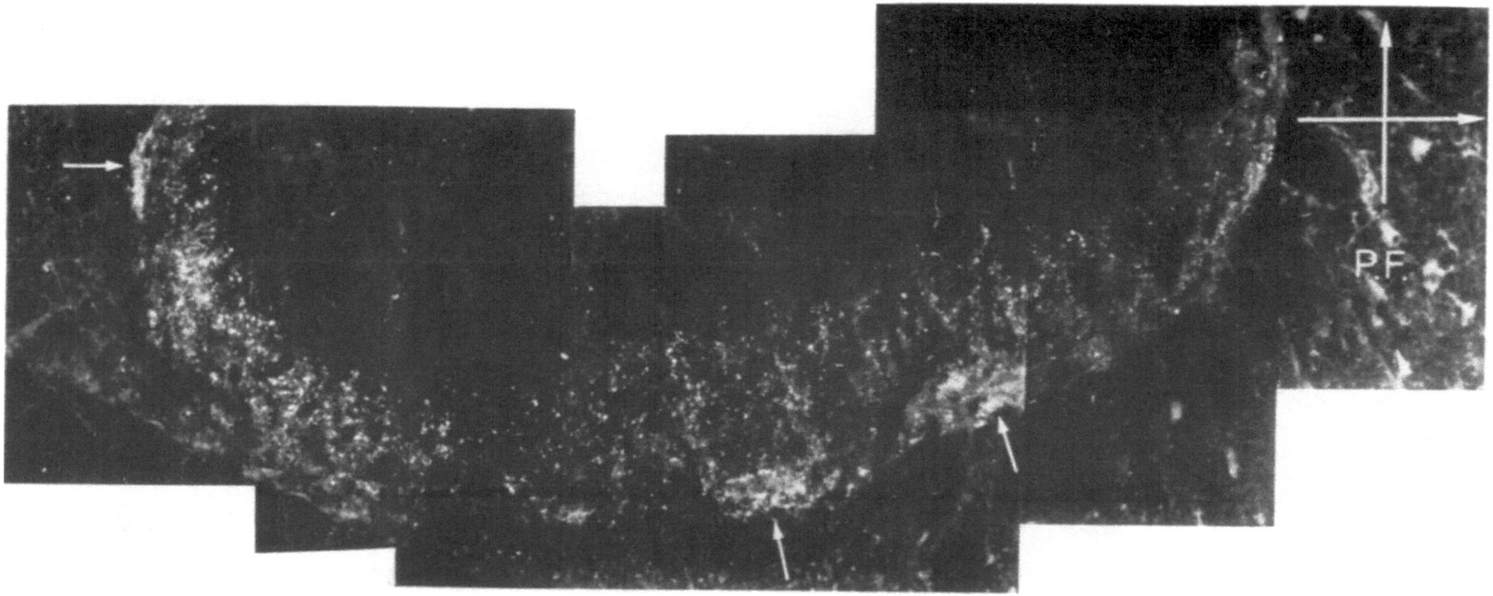

Fig. 4. Angiotensin-like immunoreactivity in nerve terminals.The cervical spinal cord of male Sprague-Dawley rat. A dense plexus of ANG II-positive nerve terminals is found within layer I and layer II of the dorsal horn. They have been transversely cut and therefore have the appearance of dots. The topography of the picture is indicated by the two large arrows present in the far right part of the picture. The horizontal arrow points to the medial part of the spinal cord. The vertical arrow points to the anterior part of the spinal cord. The small arrows point to probable terminal axons innervating the layer I of the gray matter. P.F.= posterior funiculus. Note that the terminals are present as a continuous band from the medial to the lateral part of the substantia gelatinosa. Magnification 200 x

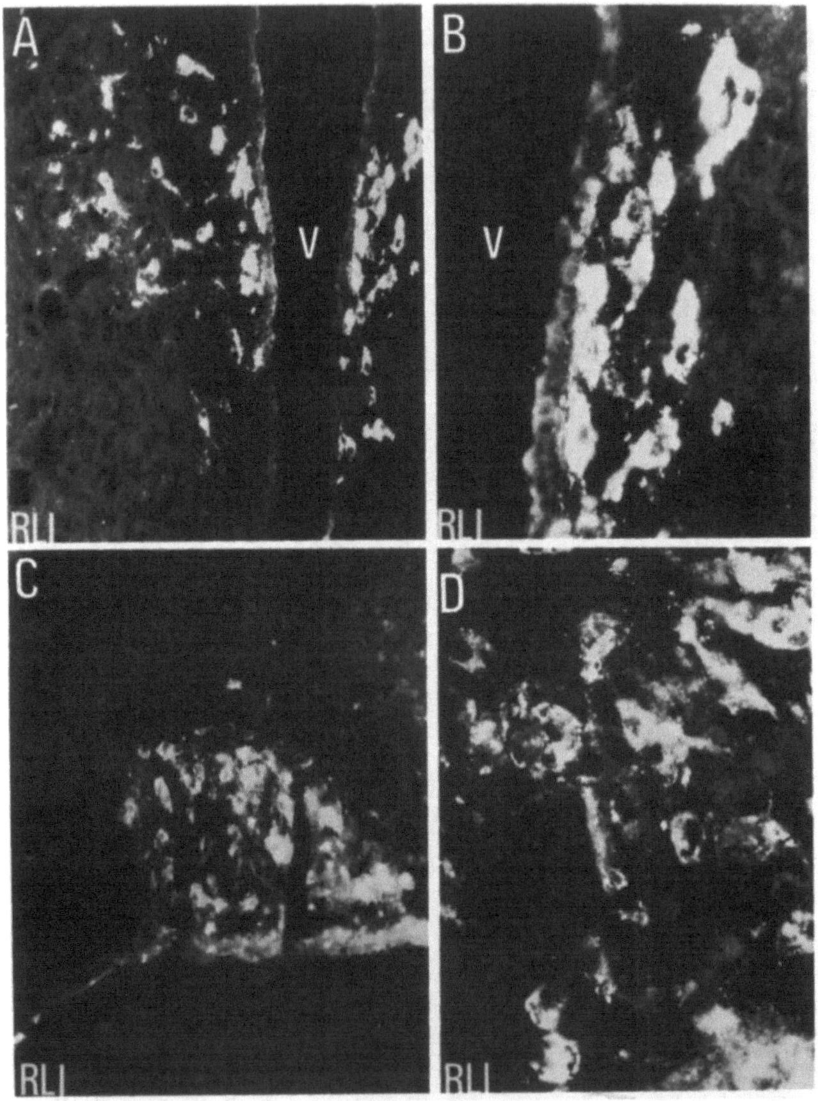

Fig. 5. Renin-like immunoreactivity in the normal male Sprague-Dawley rat, injected with colchicine intraventricularly in a dose of 75 µg 24 h before killing. Indirect immunofluorescence technique. Antibody dilution 1:80. In A and B the immunoreactivity is seen within the anterior and medial part of the magnocellular component of the nucleus paraventricularis hypothalami. A = 120 x ; B = 300 x. In C and D renin-like immunoreactivity is found in some of the magnocellular cells of the nucleus supraopticus. C = 120 x; D = 300 x. The distribution is similar to that of the oxytocin nerve cell bodies. V = third ventricle

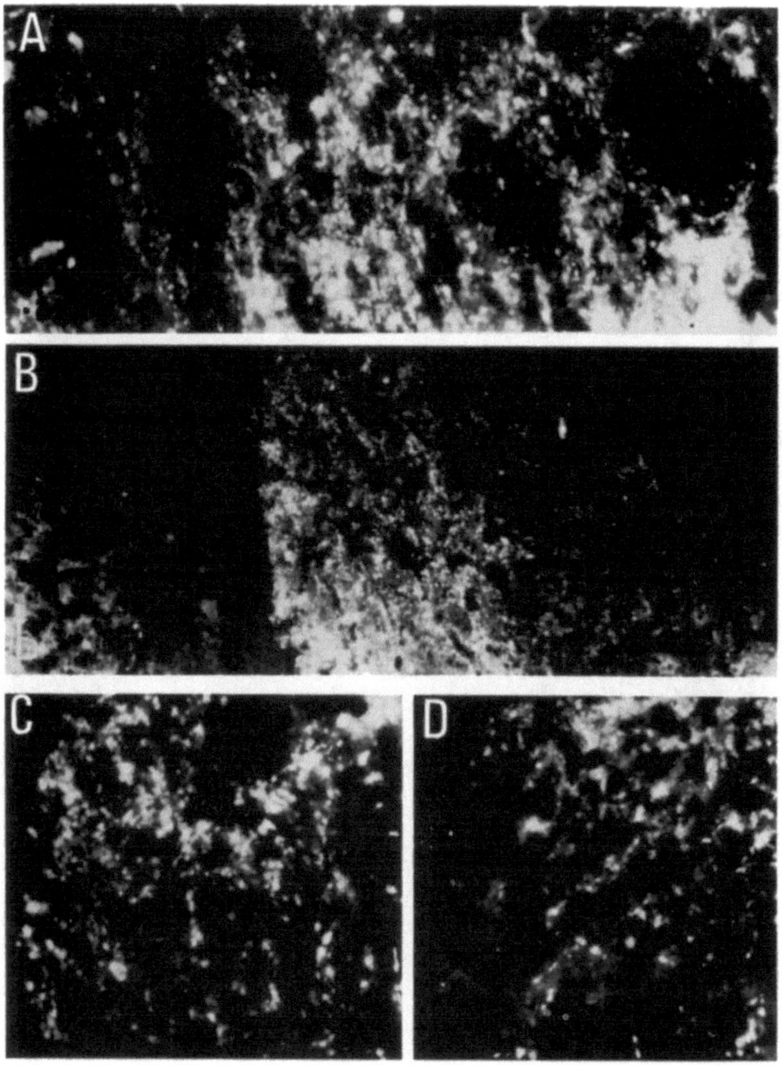

Fig. 6. Renin-like immunoreactivity in the normal male Sprague-Dawley rat. Indirect immunofluorescence technique. Antibody dilution 1:80. A is taken from the infundibular stalk and from the posterior pituitary gland. The pars intermedia is present to the left in the picture. Higher magnification of the strongly fluorescence zone in B is found in C and D. Magnification 300 x except in B where it is 120 x. In both A, B, C and D, strong immunoreactivity is seen in rounded structures probably representing nerve terminals belonging to the magnocellular hypothalamic systems of the nucleus supraopticus and nucleus paraventricularis hypothalami. Renin-like immunoreactive structures are mainly found in the infundibular stalk and in the dorsal part of the posterior pituitary gland usually close to the pars intermedia. This distribution pattern resembles that of oxytocin immunoreactive fibres

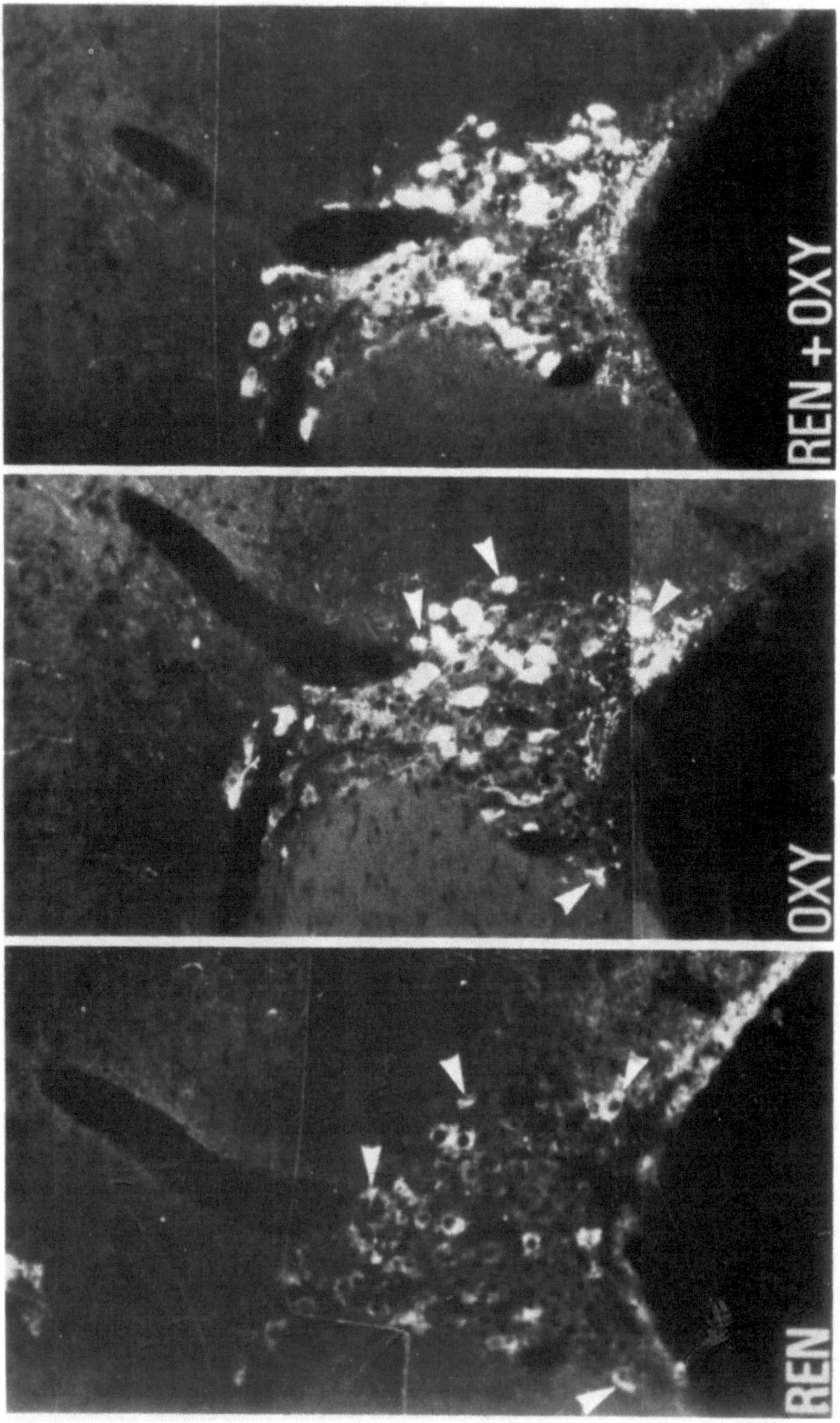
REN + OXY
OXY
REN

coexisting cells should be demonstrated both with renin antiserum alone and with oxytocin antiserum alone as well as with the combined use of the renin and oxytocin antiserum. The overall similarity in distribution pattern of the renin and oxycotin immunoreactive cell bodies is also apparent from the figures (Fig. 7 and 8). By means of a similar procedure we could also demonstrate that neurophysin II immunoreactivity has an identical distribution pattern to that of the oxytocin immunoreactive cells (24). Thus, the present analysis indicates, that renin-like immunoreactivity is almost exclusively located within the oxytocin and neurophysin II immunoreactive nerve cell bodies of the supraoptic and paraventricular nuclei. The distribution pattern of the vasopressin immunoreactive neurons is completely different from that of the renin immunoreactive nerve cell bodies, clearly indicating that at least the majority of the vasopressin immunoreactive cells do not contain renin-like material. Instead it was found that neurophysin I immunoreactivity was linked to the vasopressin immunoreactive neurons (24). All studies were performed in the rat brain. There is evidence that angiotensin immunoreactivity exists in vasopressin immunoreactive cells (14).

Taken together the present and previous results suggest that the renin- and angiotensin-like immunoreactivity within the paraventricular hypothalamic nucleus and within the supraoptic nucleus are almost exclusively present in neurophysin II and oxytocin immunoreactive and neurophysin I and vasopressin immunoreactive nerve cell bodies, respectively. One explanation for these findings would be that the levels of renin in the vasopressin cell bodies are too low to be detected with the present antibodies and/or that the renin in the vasopressin systems has a different type of immunoreactivity so that it cannot react with the present antibodies. Furthermore, renin in the oxytocin cell bodies may be in the inactive form and therefore no angiotensin immunoreactivity can be seen in the cell bodies.

In view of the present discovery it is of substantial interest that oxytocin immunoreactive fibres in the rat (4,19,23) have been shown to innervate the brain stem and to descend as far as the filum terminale located in the dorsolateral funiculus

◁ Fig. 7. Renin-like and oxytocin-like immunoreactivity in thin adjacent sections (10 µ thick). The third section (to the right) has been incubated with both renin antiserum and oxytocin antiserum. For further details, see text. Indirect immunofluorescence technique. Antibody dilution for renin 1:40. Antibody dilution for oxytocin 1:1000. Colchicine treatment (75 µg/30 µl, intraventicularly) 24 h before killing. Some of the magnocellular nerve cells of the nucleus supraopticus which show both renin-like and oxytocin-like immunoreactivity are indicated with arrows, 300 x

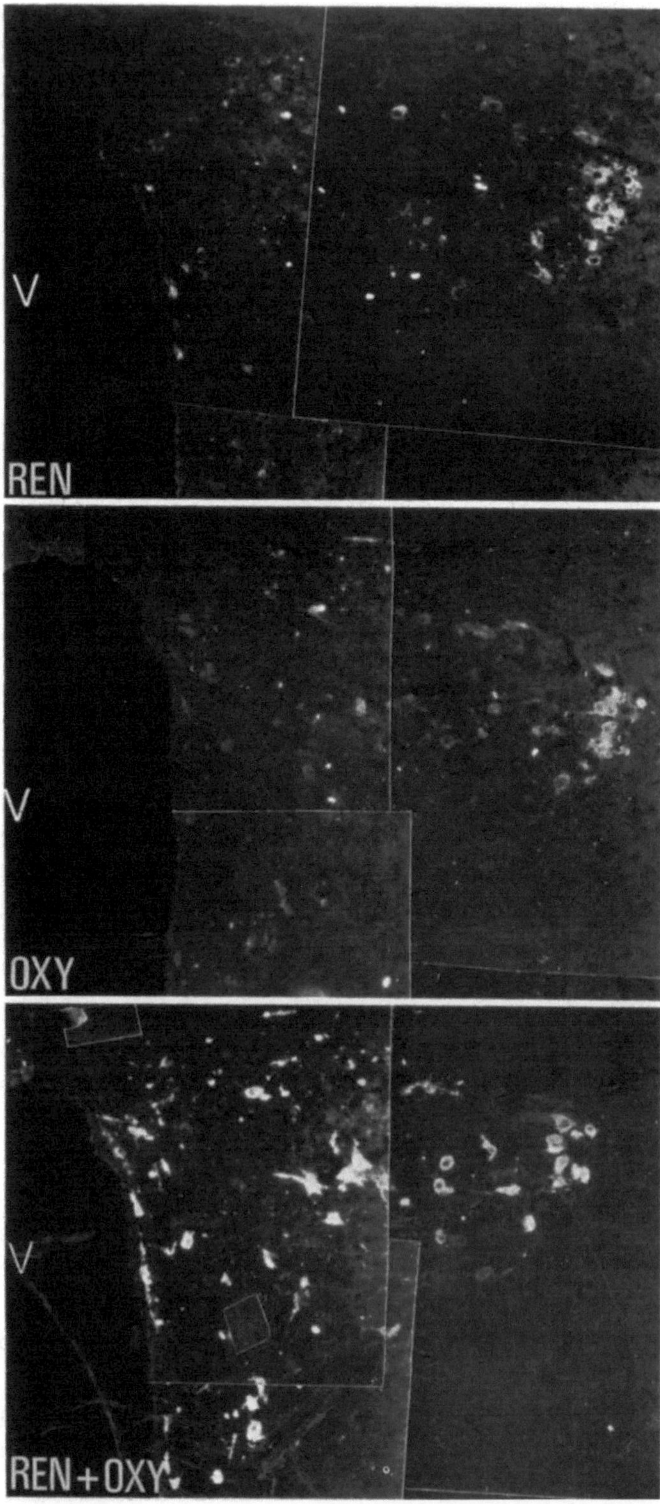
V
REN
V
OXY
V
REN+OXY

and to innervate the marginal zone and the central grey at all levels of the spinal cord. Furthermore, oxycotinergic fibres innervate the sympathetic lateral column, although with varying density depending on the level analyzed which is also true for the angiotensin immunoreactive terminals. Oxytocin immunoreactive cells in the paraventricular hypothalamic nucleus also appear to project into the dorsal vagal complex (24) where also angiotensin immunoreactive nerve terminals have been demonstrated. In view of the similarity in the distribution of some oxytocin and angiotensin immunoreactive terminals it may be speculated that angiotensin peptides could still be formed in the oxytocin neurons by renin but exclusively in the fibres and nerve terminals where active renin may exist. If this is so angiotensin peptides could be important comodulators not only in vasopressinergic systems but also in oxytocinergic systems. In agreement with this speculation is the fact that intraventricular injections of angiotensin II result in marked activation of oxytocinergic nerve cells and release of oxytocin, actions which are antagonized by a competitive angiotensin II antagonist (2,15,16).

It should also be mentioned that angiotensin II can produce vasopressin release (19) not only in vivo but also in organ cultures of the hypothalamo-neurophypophyseal system by generating action potentials (22). Thus, it seems possible that the comodulator role played by angiotensin related peptides in the oxytocin and vasopressin systems may also involve the production of recurrent facilitation of the entire oxytocin and vasopressin systems possibly via recurrent collaterals. Angiotensin related peptides formed by renin in the oxytocin neurons and by renin or other angiotensin forming enzymes in the vasopressin neurons may then be released by branches of the oxytocin and vasopressin neurons close to the nerve cell bodies. The angiotensin related peptides released can subsequently activate practically all oxytocin and vasopressin neurons by binding to specific angiotensin receptors located on the cell bodies and dendrites of these neurons. Thus, angiotensin related peptides seem to have the capacity to synchronize activity in the oxytocin-vasopressin systems via a mechnism which may be called "recurrent facilitation".

◁ Fig. 8. Renin-like and oxytocin-like immunoreactivity are demonstrated in adjacent sections of the nucleus paraventricularis hypothalami. For further details, see text to Fig. 7. A shows renin-like immunoreactivity, B oxytocin-like immunoreactivity and in C the sections have been incubated with both renin and oxytocin antiserum, 300 x

EFFECTS OF CENTRAL AND PERIPHERAL ADMINISTRATION OF RENIN ON THE AMINE LEVELS AND TURNOVER IN DISCRETE DOPAMINE AND NORADRENALINE NERVE TERMINAL SYSTEMS OF THE HYPOTHALAMUS

As in series of experiments (8) normal specific pathogen free Sprague-Dawley rats (bodyweight 150 g) were used in the present studies on the effects of renin on the DA and NA levels and turnover. Changes in the CA turnover were evaluated by studying the decline of the CA stores following the injection of the tyrosine hydroxylase inhibitor, α-methyltyrosine methyl ester (H 44/68, 250 mg/kg, 1 or 2 h before killing). The CA levels in various types of DA and NA nerve terminal systems were determined by the quantitative use of the Falck-Hillarp methodology involving quantitative microfluorimetrical measurements of the formaldehyde induced CA fluorescence present in paraffine sections of the hypothalamus. The specificity of the renin action (0.01 - 1.0 units per rat, dissolved in saline and given into the lateral brain ventricle; National Biochemicals Corp., Cleveland, Ohio, U.S.A.) was tested by giving the angiotensin I converting enzyme inhibitor, SQ 14225 (3-mercapto-2-d-methylpropanoyl-1-proline, Captopril$^{(R)}$) intraventricularly in a dose of 300 μg 5 min before the intraventricular administration of the various doses of renin. This dose of the inhibitor can clearly block the marked increase in arterial blood pressure induced by 0.1 units of renin. H 44/68 was administered at the same time as SQ 14225. In the intravenous infusion experiments which also were performed on intact awake and unstressed animals with brain cannulas implanted 2 days earlier H 44/68 was given at the onset of the intravenous infusion of renin (0.03 to 0.3 units per rat). The total dose administered was contained within 1 ml of saline. The infusion lasted for 1 h, after which the animals were killed by rapid decapitation. For details on the quantitative microfluorimetrical procedures see (6).

In agreement with previous results it was found that the angiotensin I converting enzyme inhibitor alone did not influence the DA or NA turnover in the hypothalamic CA nerve terminal systems analyzed (Fig. 9). With the highest i.v.t. dose of renin (1 unit) SQ 14225 could not block the increase of DA turnover found within the MPZ and LPZ of the external layer of the median eminence nor could the inhibitor block the renin induced reduction of NA turnover in the magnocellular part of the paraventricular hypothalamic nucleus (PA FM). Principally the data are in good agreement with our previous results (8), indicating angiotensin peptide regulation mainly of tubero-infundibular DA systems and of the NA nerve terminals innervating the vasopressin and oxytocin rich posterior part of the magnocellular paraventricular hypothalamic nucleus. In view of the present and previous observations that angiotensin peptides may be a comodulator of vasopressin neurons it is suggested that the vasopressin neurons innervating the median eminence may regulate dopamine turnover and release in this area. The oxytocin innervation of this area is small compared with

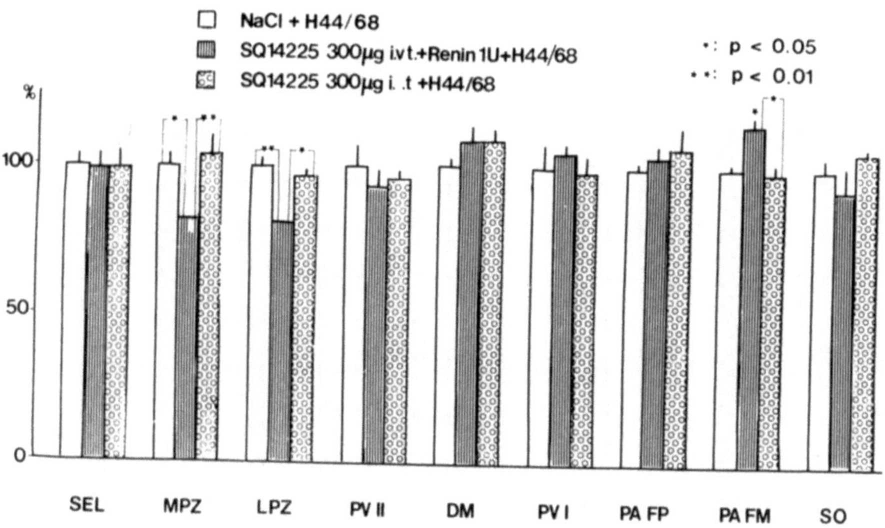

Fig. 9. The effects of renin on the H 44/68 induced catecholamine fluorescence disappearance in various hypothalamic regions of normal male rat. Saline or renin (1 unit) was injected intraventricularly 5 minutes after α-methyltyrosine methyl ester (H 44/68, 250 mg/kg i.p., 2 h before killing) in the presence of an angiotensin I converting enzyme inhibitor SQ 14225 (300 μg/30 μl, given intraventricularly immediately after H 44/68). Means ± S.E.M. in per cent of saline + H 44/68 treated group mean value; n = 4-5. Statistical analysis according to Wilcoxon: one-way classification comparing all possible pairs of treatments. *: $p < 0.05$; ** : $p < 0.01$. SEL = subependymal layer of the median eminence (100% = 127 ± 5 pmol/g); MPZ = medial palisade zone of median eminence (100% = 115 ± 4); LPZ = lateral palisade zone of median eminence (100% = 306 ± 11); PV I = anterior periventricular hypothalamic region (100% = 107 ± 8); PV II = posterior periventricular hypothalamic region (100% = 95 ± 7); DM = dorsomedial hypothalamic nucleus (100% = 163 ± 6); PA FP = parvocellular part of nucleus paraventricularis hypothalami (100% = 146 ± 2); PA FM = magnocellular part of nucleus paraventricularis hypothalami (100% = 187 ± 3); SO = nuc. supraopticus (100% = 57 ± 2)

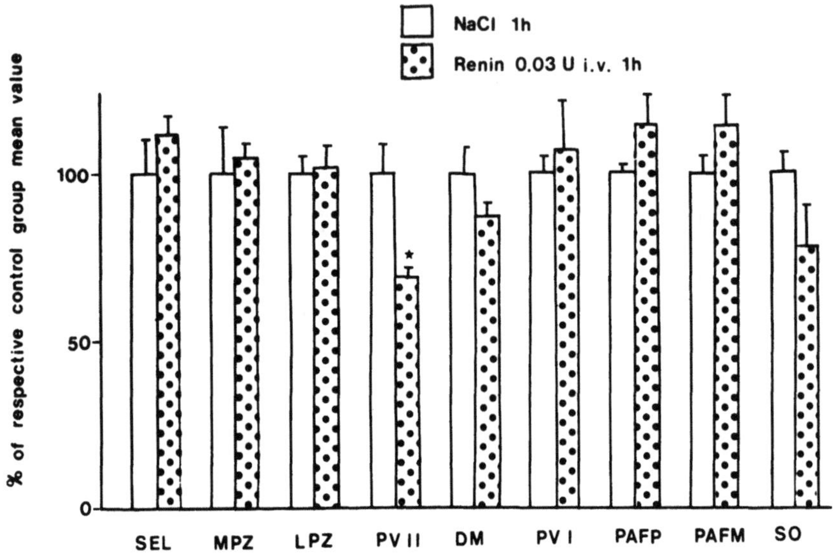

Fig. 10. The effects of an intravenous infusion of renin (0.03 units) on catecholamine fluorescence in various types of dopamine and noradrenaline nerve terminals in the hypothalamus of the normal male rat. The infusion time was 1 h and the volume 1 ml. Means ± S.E.M. are shown in per cent of saline treated group mean value; n = 4-5. Mann-Whitney U-test. For abbreviation, see text to previous figure. SEL: 100% = 155 ± 15 pmol/g; MPZ: 100% = 140 ± 20; LPZ: 100% = 500 ± 25; PV II: 100% = 105 ± 10; DM: 1C0% = 163 ± 6; PV I: 100% = 107 ± 8; PA FP: 100% = 146 ± 2; PA FM: 100% = 187 ± 3; SO: 100% = 75 ± 5

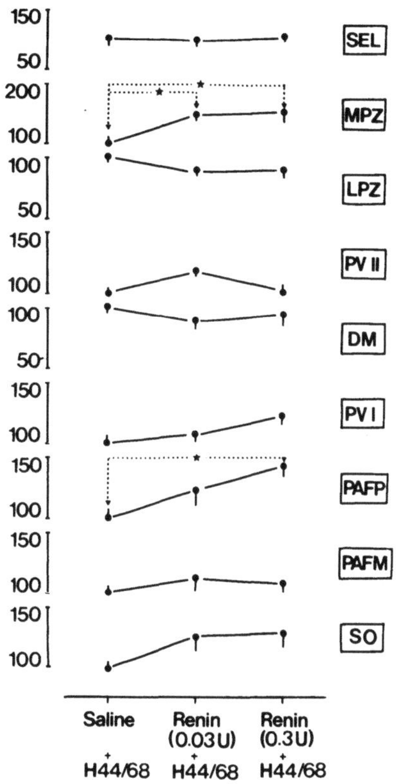

Fig. 11. The effects of an intravenous infusion of renin on the H 44/68 induced catecholamine fluorescence disappearance in various types of hypothalamic dopamine and noradrenaline nerve terminal systems of the normal male rat. Renin was infused in a dose of 0.03 or 0.3 units and H 44/68 (250 mg/kg, i.p.) was given at the onset of the renin infusion. The rats were killed after an infusion time of 1 h. For abbreviations and statistical procedures, see text to Fig. 9. * : $p < 0.05$; The significance refers to the saline + H 44/68 group mean value. The absolute values for the untreated control group are shown in text to Fig. 10

that of vasopressin. The reduction of NA turnover in the paraventricular hypothalamic nucleus may simply be a response secondary to the marked increase in oxytocin and vasopressin secretion produced by renin, assuming that these noradrenergic terminals have an excitatory influence on the vasopressin and oxytocin immunoreactive cell bodies.

As seen in Figs. 10 and 11, intravenous infusion of renin both in the low and high dose used produced an opposite effect on DA turnover in the MPZ compared to that obtained with intraventricular injections of renin. Thus, a clearcut reduction of DA turnover was observed in the MPZ, while the LPZ DA turnover was unaffected in contrast to the case after i.v.t. injection. Another difference between intraventricular and intravenous experiments was that renin in a dose of 0.3 units following intravenous infusion produced a selective reduction of NA turnover in the parvocellular part of the paraventricular hypothalamic nucleus (PA FP) (Fig. 11). In contrast, the NA turnover in the magnocellular portion was not significantly reduced unlike the results obtained in the intraventricular experiments. Furthermore, the low dose of renin in the intravenous experiment produced a significant and selective reduction of the NA stores in the posterior periventricular hypothalamic region (PV II) (Fig. 10).
These results are of substantial interest and will be discussed in detail. With regard to the changes in the MPZ it can be assumed that angiotensin II formed from renin in the i.v. experiments via the systemic circulation can reach the external layer of the median eminence, since this part of the brain lacks the blood brain barrier. Thus, one possible mechanism can be that angiotensin II receptors located on the DA terminals in this region can inhibit the release of DA. The failure to see this action following intraventricular injection can be explained on the basis that in this case renin may not directly reach the external layer and can instead induce marked increases of firing in probably all the oxytocin and vasopressin immunoreactive neurons innervating the external layer of the median eminence (mainly a vasopressinergic innervation). Vasopressin peptides via their receptors on the DA terminals may have excitatory influences on the DA release process in the median eminence. The vasopressin receptors may then be in a balance with the angiotensin receptors, which have an opposite influence. This hypothesis can be tested by an intravenous infusion of vasopressin and/or oxytocin. If this view is correct angiotensin peptides formed in the vasopressin terminals of the median eminence may act as endogenous antagonists inhibiting the effects of vasopressin receptor activation on DA turnover. Another explanation for the reduction of DA turnover in the MPZ after i.v. renin could, however, also be the ability of peripheral angiotensin II to reduce vasopressin and oxytocin secretion in the median eminence via a central feedback action.

When discussing the selective reduction of NA turnover in the parvocellular part of paraventricular hypothalamic nucleus upon renin infusion it should be remembered that it probably contains cells projecting into the median eminence. Other types of nerve cells in this region may also give rise to large descending projections innervating the dorsal vagal complex, such as the nucleus tractus solitarii, the parabrachial nucleus as well as the sympathetic lateral column of the spinal cord (24). Thus, this area may represent a site of integration of neuroendocrine and autonomic mechanisms. It may be speculated that angiotensin II formed in the periphery e.g. by activation of volume receptors and/or baroreceptors due to the increased arterial blood pressure and/or aldosterone secretion can produce a reduction of activity in a medullo-paraventricular NA pathway, facilitating activity in various paraventriculo-infundibular systems with neuroendocrine functions and in descending pathways with autonomic nervous system cardiovascular functions. Also the dendrites of the magnocellular paraventricular vasopressin and oxytocin neurons may be regulated by the NA terminals in the parvocellular part of the nucleus. Thus, it seems possible that the peripheral renin-angiotensin signal via reducing activity in an ascending facilitory noradrenergic pathway can inhibit the central renin-angiotensin signal originating within the oxytocinergic and vasopressinergic neurons which may have e.g. important cardiovascular function by their descending projections (mainly oxytocinergic).

The selective reduction of NA levels in the posterior periventricular hypothalamic region (PV II), by the low dose of renin (0.03 units) is probably due to a reduction mainly of NA synthesis, since the NA turnover was significantly reduced.

The functional significance of this change is unkown. It seems possible that renin via activation of certain types of visceral afferent pathways which finally end in the hypothalamus can change the activity in the local circuits of PV II so that NA synthesis becomes exclusively inhibited in this area. Such a selective inhibition of NA synthesis may be brought about e.g. via activation of presynaptic receptors located on the NA terminals, which can exert inhibitory influences on NA synthesis. Such an action by a peripheral renin-angiotensin signal may indicate that this NA system may have a role in regulatory functions related to the renin-angiotensin systems such as cardiovascular functions. As a matter of fact NA in the posterior hypothalamic area may exert vasopressor actions.

SUMMARY

The present and previous papers provide evidence for the existence of oxytocin-renin and vasopressin-angiotensin immunoreactive nerve cell bodies in the nucleus supraopticus and paraventricularis. Thus, angiotensin peptides may act as comodulators in these neurons. Intraventricular renin injections increase selectively activity in the tuberoinfundibular DA systems (both MPZ + LPZ) and decrease NA turnover in nucleus paraventricularis (magnocellular part). Intravenous infusions of renin in contrast leads to a selective reduction of DA turnover in MPZ and to a reduction of NA turnover in the paraventricular hypothalamic nucleus (parvocellular part). In addition, the renin infusion reduces noradrenaline synthesis in the posterior periventricular hypothalamic area. It is therefore speculated that the peripheral renin-angiotensin system can influence central neuroendocrine mechanisms via effects on CA turnover in the median eminence and in the paraventricular hypothalamic nucleus (parvocellular part) and cardiovascular function mainly via effects on NA turnover in the latter region which may lead to a changed activity in the descending oxytocinergic and vasopressinergic pathways regulating cardiovascular functions. In these systems angiotensin related peptides themselves may act as comodulators.

ACKNOWLEDGEMENTS

This work has been supported by grants from the Swedish Medical Research Council (04X-4246), the Magnus Bergvall's Stiftelse, the Knue & Alice Wallenberg's Foundation, and the Deutsche Forschungsgemeinschaft (DFG) within the SFB90 "Cardiovasculäres System".

REFERENCES

1. AGNATI LF, FUXE K, LOCATELLI V, CALZA L, HÖKFELT T (1981 submitted) A method for the quantitative evaluation of coexistence of substances in a nerve cell group. J Neurosci Methods

2. AKAISHI T, NEGORO H, KOBAYASHI S (1980) Responses of paraventricular and supraoptic units to angiotensin II, [Sar^1, Ile^8] -angiotensin II and hypertonic NaCl administered into the cerebral ventricle. Brain Res 188:499-511

3. BROWNFIELD MS, REID IA, GANTEN D, GANONG WF (to be published) Differential distribution of immunoreactive angiotensin and converting enzyme in rat brain

4. BUIJS RM (1978) Intra- and extrahypothalamic vasopressin and oxytocin pathways in the rat. Pathways to the limbic system, medulla oblongata and spinal cord. Cell Tiss Res 192:423-435

5. COONS AH (1958) Fluorescent antibody methods. In: DANIELLI JF (ed) General Cytochemical Methods. Academic Press, New York, p 399

6. EINARSSON P, HALLMAN H, JOHANSSON G (1975) Quantitative microfluorimetry of formaldehyde-induced fluorescence of dopamine in the caudate nucleus. Med Biol 53:15-24

7. FUXE K, GANTEN D, HÖKFELT T, BOLME P (1976) Immunohistochemical evidence for the existence of angiotensin II containing nerve terminals in the brain and spinal cord of the rat. Neurosci Lett 2:229-234

8. FUXE K, ANDERSSON K, GANTEN D, HÖKFELT T, ENEROTH P (1980) Evidence for the existence of an angiotensin II-like immunoreactive central neuron system and its interactions with the central catecholamine pathways. In: GROSS F, VOGEL G (eds) Enzymatic Release of Vasoactive Peptides. Raven Press, New York, p 161

9. FUXE K, GANTEN D, BOLME P, AGNATI LF, HÖKFELT T, ANDERSSON K, GOLDSTEIN M, HÄRFSTRAND A, UNGER TH, RASCHER W (1980) The role of central catecholamine pathways in spontaneous and renal hypertension in rats. In: FUXE K, GOLDSTEIN M, HÖKFELT B, HÖKFELT T (eds) Central Adrenaline Neurons: Basic Aspects and their Role in Cardiovascular Functions. Pergamon Press, Oxford and New York, p 259

10. FUXE K, GANTEN D, HÖKFELT T, LOCATELLI V, POULSEN K, STOCK G, RIX E, TAUGNER R (1980) Renin-like immunocytochemical activity in the rat and mouse brain. Neurosci Lett 18:245-250

11. GANTEN D, FUXE K, PHILLIPS MI, MANN JFE, GANTEN U (1978) The brain isorenin-angiotensin system: histochemistry, localization and possible role in drinking and blood pressure regulation. In: GANONG WF, MARTINI B (eds) Frontiers in Neuroendocrinology, Vol 5, Raven Press, New York, p 61

12. HIROSE S, YOKOSAWA H, INAGAMI T, WORKMAN RS (1980) Renin and prorenin in dog brain: ubiquitous distribution and high concentration in the pituitary and pineal. Brain Res 191:489-499

13. HÖKFELT T, REHFELD JF, SKIRBOLL L, IVEMARK B, GOLDSTEIN M, MARKEY K (1980) Evidence for coexistence of dopamine and CCK in mesolimbic neurones. Nature 285:476-478

14. KILCOYNE MM, HOFFMAN DL, ZIMMERMAN EA (1981) Immunocytochemical localization of angiotensin II and vasopressin in rat hypothalamus: evidence for production in the same neuron. Clin Sci 59:515-605

15. LANG RE, RASCHER W, HEIL J, UNGER TH, WIEDEMANN G, GANTEN D (1981) Angiotensin stimulates oxytocin release. Life Sci 29:1425-1428

16. LANG RE, RASCHER W, HEIL J, UNGER TH, GANTEN D (1981 in press) Angiotensin stimulates oxytocin release: Impaired response in rats with genetic hypothalamic diabetes insipidus. Ann N Y Acad Sci

17. MALLING C, POULSEN K (1977) A direct radioimmunoassay for plasma renin in mice and its evaluation. Biochem Biophys Acta (Amst) 491:532-541

18. NICOLL RA, BARJER JL (1971) Excitation of supraoptic neurosecretory cells by angiotensin II. Nature New Biol 233:172-174

19. NILAVER G, ZIMMERMAN EA, WILKINS J, MICHAELS J, HOFFMAN D, SILVERMAN A-J (1980) Magnocellular hypothalamic projections to the lower brain stem and spinal cord of the rat. Neuroendocrinology 30:150-158

20. PHILLIPS MI, WEYHENMEYER J, FELIX D, GANTEN D, HOFFMAN WE (1979) Evidence for an endogenous brain renin-angiotensin system. Fed Proc 38:2260-2266

21. RIX E, GANTEN D, SCHÖLL B, UNGER TH, TAUGNER R (1981) Converting enzyme in the choroid plexus, brain, and kidney: immunocytochemical studies in rats. Neurosci Lett 22:125-130

22. SLADEK CD, JOYNT RJ (1979) Angiotensin stimulation of vasopressin release from the rat hypothalamo-neurohypophyseal system in organ culture. Endocrinology 104:148-153

23. SWANSON LW, McKELLAR S (1979) The distribution of oxytocin- and neurophysin-stained fibers in the spinal cord of the rat and monkey. J Comp Neurol 188: 87-106

24. SWANSON LW, SAWCHENKO PE (1980) Paraventricular nucleus. A site for the integration of neuroendocrine and autonomic mechanisms. Neuroendocrinology 31: 410-417

Brain Angiotensin II Receptors

N. E. Sirett, J. I. Hubbard, and J. J. Bray

INTRODUCTION

Since peptide hormones are known to act through receptors we have studied in vitro [^{125}I]-angiotensin II (ANG II) binding to rat brain membranes (11,14,16,17). The central region of the brain containing the hypothalamus, thalamus, septum and midbrain (HTSM) was used to characterize the receptors (14). Several features of the binding suggest that physiological receptors are present. Judged from progressive inhibition by increasing concentrations of ANG II agonists and antagonists the binding is specific, reversible, temperature dependent and occurs with high affinity. The equilibrium dissociation constant (Kd) of the specific binding is 0.9 nM/L and the maximum binding (B max) is 11 fmol mg^{-1} protein. Similar values have been reported by others (2,3,6,13).

LOCALIZATION OF SPECIFIC ANGIOTENSIN II BINDING IN RAT BRAIN AND ITS PHYSIOLOGICAL SIGNIFICANCE

Specific angiotensin II (ANG II) binding was localized in central structures of the brain, i.e. septum, midbrain, hypothalamus, thalamus, and the medulla (Fig. 1A and 1B). Specific binding in striatum, cerebellum, hippocampus and cortex, regions which constitute about 70% of the brain, was negligible. Localization of ANG II receptor binding to the central structures of the brain correlates well with in vivo studies of the central actions of ANG II (12) and regions adjacent to the lateral and third ventricles have been shown to be particularly ANG II sensitive.

The subfornical organ (SFO) which is included in the septum (Fig. 1A) is a small nodule of nervous tissue (diameter 500 μ and thickness 300 μ) projecting into the medial aspect of the rostral wall of the third ventricle. The ability of the SFO to respond to local changes in ANG II concentration has been demonstrated both in vivo (12) and in vitro (5). No difference in receptor binding was detected between the dorsal septum including the SFO (14) and ventral regions. A vertical

Experimental Brain Research, Suppl. 4

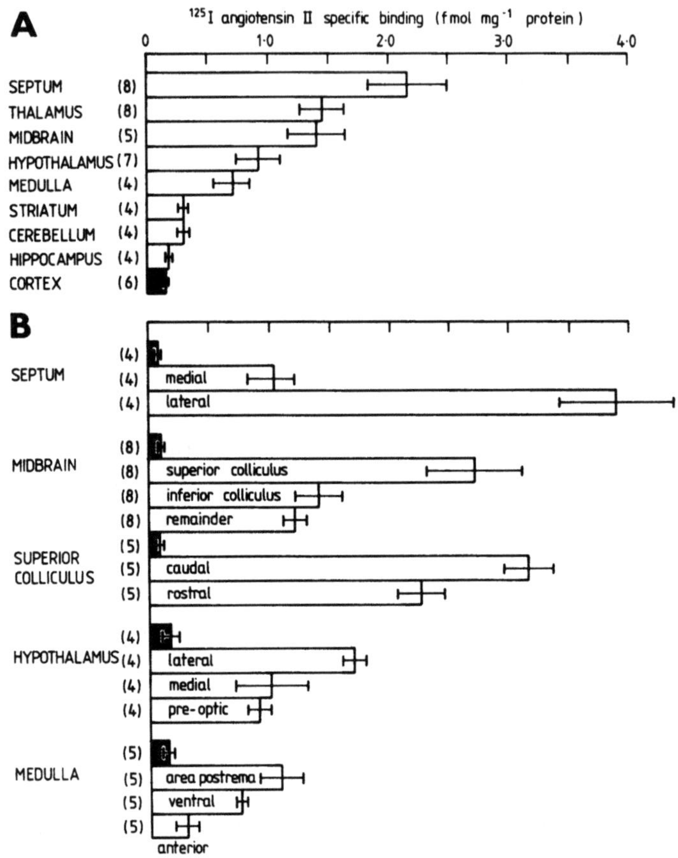

Fig. 1. Specific binding of [^{125}I]-angiotensin II in brain tissue (mean $\pm$ SEM), ligand concentration 0.15 $\pm$ 0.18 nM (14,16). Number of rats (n). Fig. 1A. Whole brain regions. Fig. 1B. Regions showing additional localization of binding. Black bars = cortex

division of septum into lateral and medial regions which localized the SFO to the medial region (Fig. 1B) disclosed a significant localization in the lateral septum. Failure to localize binding in the regions of septum containing the SFO need not to eliminate the SFO as an important receptor for ANG II because it is so small, its mass is not a significant portion of the whole.

The midbrain has been identified as a site for the central pressor action of ANG II (12). There is a significant localization of ANG II binding in the superior colliculus and division of the superior colliculus into a caudal and rostral region disclosed a further localization of receptors in the caudal region (Fig. 1B). We tested the importance of the caudal region of superior colliculus by continuously monitoring the blood pressure in pentobarbitone anesthetized rats during aspiration of the midbrain (16). ANG II applied locally to the midbrain results in a transient increase in blood pressure. The pressor response was abolished when a discrete region of the midbrain and periaqueductal grey which includes the caudal region of the superior colliculus was removed.

Hypothalamic localization was not confined to any particular region (Fig. 1B). The highest ANG II binding was in the lateral hypothalamus which contains the medial forebrain bundle and the fornix pathways connecting the midbrain, septum and hippocampus. No particular localization was seen in the preoptic region although it contains the preoptic recess of the third ventricle within which lies the organum vasculosum laminae terminalis (OVLT), reported to be the most ANG II sensitive region in the brain (12).

In the medulla, ANG II binding was localized in the area postrema and the medulla ventral to area postrema (posterior medulla) (Fig. 1B). The area postrema is claimed to be essential for the pressor effects of intra-arterially injected ANG II (12). However, rat area postrema in vitro contains no units excited by ANG II (4) and we, and others, have concluded that the area postrema is not a target for ANG II action (8). The receptors detected by us in the area postrema presumably lie in the vagal nuclei which are ventral to the area postrema and likely to be included in the dissection.

Receptor Localization in other Species

ANG II receptor distribution was studied in the brains of monkey (n=3), cat (n=3) and rabbit (n=2). The specific binding was always less than in rat HTSM; as a percentage of the total it was always less than 40% and frequently negligible, whereas in the rat it is between 65-80%. Reduced specific binding has been noted in rodents other than rats (7) and negligible binding has been reported in calf brain except for cerebellar cortex (3). In monkey and cat, specific binding representing 20-30% of the total was detected in septum, midbrain, posterior medulla, hypothalamus, striatum and cerebellum and 30-40% specific binding was detected in rabbit hypothalamus, midbrain, septum and hippocampus. Because of the inconsistencies between species, the physiological relevance of the regional distribution of brain ANG II receptors must be regarded cautiously. However, a wide range of mammals and birds give both drinking and pressor responses to intracranial injections of ANG II.

New Zealand Genetically Hypertensive Rat

It has been suggested that the hypertension in the New Zealand strain of genetically hypertensive (GH) rat is maintained neurogenically (17). Specific ANG II receptor binding was similarly distributed in the brains of normal and GH rats (Fig. 2). In superior colliculus, septum, thalamus, remainder of midbrain and posterior medulla, the ratio (GH/normal) was significantly less than unity. Reported comparisons of brain ANG II binding in the HTSM of spontaneously hypertensive (SH) and their control (WKY) rats at any age showed no difference in the Kd or B max (6). A claim that ANG II binding in the OVLT of the SHR strain is increased (19) can be discounted because of the difficulty of separating the tiny OVLT from the underlying hypothalamus, and the values reported for OVLT and SFO probably represent the binding of hypothalamus and septum.

The effects of centrally administered ANG II and its antagonist [Sar^1, Ala^8] angiotensin (saralasin) were also studied in conscious and anesthetized normal and GH rats (17). In conscious rats of both strains, intraventricular (IVT) injection of ANG II resulted in similar increases in blood pressure and drinking and IVT injection of saralasin produced a non-significant fall in blood pressure. Under pentobarbitone anesthesia, a 10-fold increase in the dose of ANG II (0.5 ng to 5 ng) was required to produce a comparable increase in blood pressure in both strains and the elevation in the GH rats was significantly less than in the normals. Intracerebroventricular (IVT) saralasin produced a greater fall in anesthetized GH than in normal rats. Further, IVT infusion of saralasin for 6 days caused a prompt and persistent fall in blood pressure in the conscious SHR but no change in WKY rats (10).

Development of ANG II Receptors with Age

In a study of brain ANG II receptor binding in relation to age (15), specific binding was determined in HTSM from rats, aged less than 1 day, 1-30 days, 60-80 days and adult rats (300 g). Concurrently, the ANG II receptor binding to adrenal membranes was also determined. As Table 1 shows, the specific ANG II binding in both HTSM and adrenal membranes changed rapidly over the first 30 days, building up to a peak at 14 days and thereafter slowly declining. The changes in binding to adrenal membranes from juvenile rats followed a similar time course to that seen in the brain. Similar changes in brain ANG II receptor binding capacity with age has been reported by others (2,6). Specific ANG II binding has been reported in cultures from foetal rat brain cells (13).

Table 1. Specific ^{125}I ANG II binding in rat brain HTSM and adrenals in relation to age (fmol mg^{-1} protein, mean ± SEM, ligand concentration 0.15 nM). P value for the difference from the value for 9-14 days. Number of rats (n)

Age Days	HTSM	P	Adrenals	P
1	2.4 ± 4.0 (5)	< 0.001		
1- 3	2.9 ± 0.4 (7)	< 0.01	31 ± 3 (6)	< 0.05
9-14	6.5 ± 0.5 (14)		46 ± 3 (13)	
20-23	4.2 ± 0.3 (7)	< 0.02	17 ± 3 (7)	< 0.001
60-80	3.1 ± 0.4 (7)	< 0.001	12 ± 1 (6)	< 0.001
Adult	1.5 ± 0.1 (4)	< 0.001	15 ± 1 (4)	< 0.001

DOES PLASMA ANG II REGULATE BRAIN ANG II RECEPTORS?

Adrenal ANG II receptor binding is under a positive regulation from circulating ANG II (1). Sodium restriction in rats increases both plasma ANG II and adrenal ANG II receptor binding and conversely the presence of high plasma sodium decreases both plasma ANG II and adrenal ANG II binding. There are conflicting reports of the effect of plasma ANG II and sodium restriction on ANG II binding to rat brain membranes, being either no change after nephrectomy (2) to no change or a small decrease after sodium restriction (6,9). We have examined brain ANG II receptor binding after treatments which either increase or decrease plasma ANG II (Table 2). Rats were fed a low salt diet or implanted subcutaneously with silastic rubber strips containing deoxycorticosterone acetate (DOCA, 10 mg/100 g bodyweight) and fed 1% NaCl in the drinking water. Table 2 shows that at the end of 2-3 weeks cortex had an increase in binding, while other regions showed no significant change or a small decrease (combined HTM) ($p < 0.01$). No significant changes were seen in the DOCA-treated rats. Saturation ANG II binding and Scatchard analysis studies on the HTSM from normal, low salt diet and DOCA-treated rats also revealed no significant changes either in the B max or Kd. The mean ± SEM (n=3) were, for B max (fmol mg^{-1} protein) and Kd (nM/L) respectively, normal diet 8.1 ± 1.7 and 0.61 ± 0.6; low salt diet 7.1 ± 1.3 and 0.52 ± 0.03 and for DOCA-treated rats, 7.2 ± 1.3 and 0.59 ± 0.05.

Table 2. Changes in specific ANG II binding in regions of rat brain after treatments which increase plasma ANG II (low salt diet, pentobarbitone, renal hypertension) or decrease plasma ANG II (DOCA). Ratios of paired comparisons of specific ANG II binding (mean $\pm$ SEM). Number of rats (n)

	Low salt diet (7)	Pentobarbitone (7)	Renal hypertension (4)	DOCA (6)
Cortex	2.60 $\pm$ 0.50	-	-	1.20 $\pm$ 0.40
Septum	1.30 $\pm$ 0.30	1.50 $\pm$ 0.10	0.80 $\pm$ 0.10	0.80 $\pm$ 0.10
Superior colliculus	0.90 $\pm$ 0.10	0.90 $\pm$ 0.10	1.70 $\pm$ 0.30	1.10 $\pm$ 0.10
Remainder of midbrain	1.00 $\pm$ 0.10	1.00 $\pm$ 0.10	1.30 $\pm$ 0.10	0.90 $\pm$ 0.10
Thalamus	0.80 $\pm$ 0.10	0.90 $\pm$ 0.10	1.50 $\pm$ 0.20	1.00 $\pm$ 0.10
Hypothalamus	0.70 $\pm$ 0.10	0.90 $\pm$ 0.10	1.60 $\pm$ 0.20	1.10 $\pm$ 0.20
Posterior medulla	0.80 $\pm$ 0.10	-	-	0.90 $\pm$ 0.10
Combined HTM	0.85 $\pm$ 0.06	0.88 $\pm$ 0.04	1.50 $\pm$ 0.11	0.99 $\pm$ 0.06

In acute experiments lasting for 2-3 hours, rats were anesthetized with pentobarbitone, a treatment which elevates plasma renin and ANG II. As Table 2 shows, septum showed an increase (54%, $p < 0.01$) and for the combined HTM there was a significant reduction (12%, $p < 0.01$). These changes are in the same direction as for the low salt diet.

Renal hypertension which increases plasma ANG II was produced in rats by coarctation of the aorta with a metal clip placed between the renal arteries. Four days later ANG II binding in brain regions was determined. As Table 2 shows, changes in renal hypertensive rats were in the opposite direction to those seen after low salt or pentobarbitone treatments. Specific binding was increased in all regions except septum. Because the between-experiment variations in our assays are high, as they were for others (6), we would question the relevance of the small changes observed after the various treatments (Table 2). Clearly the changes in brain ANG II receptor binding do not follow the changes in plasma ANG II and our results suggest that central ANG II receptor binding is not regulated by changes in plasma ANG II.

ANG II IN RAT BRAIN AND ITS REGIONAL LOCALIZATION

Extracts of regions of rat brain were prepared and assayed in a radioimmunoassay with a useful range of sensitivity between 5 and 200 pg/0.5 ml incubate (18).

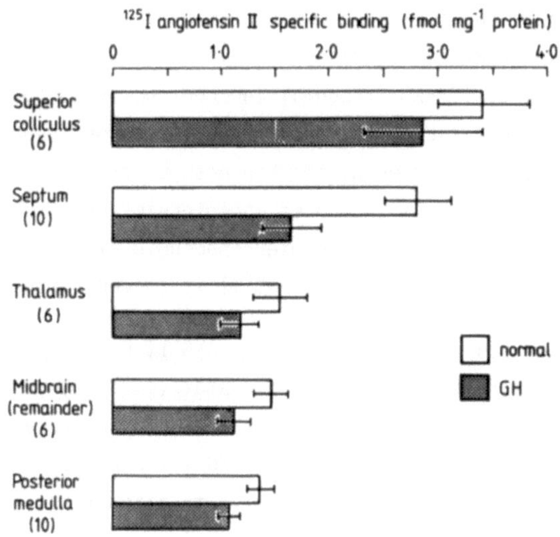

Fig. 2. Specific [^{125}I] angiotensin II binding (mean $\pm$ SEM) in brain regions from normal (open bars) and genetically hypertensive (stippled bars) rats (17). Numbers of rats (n)

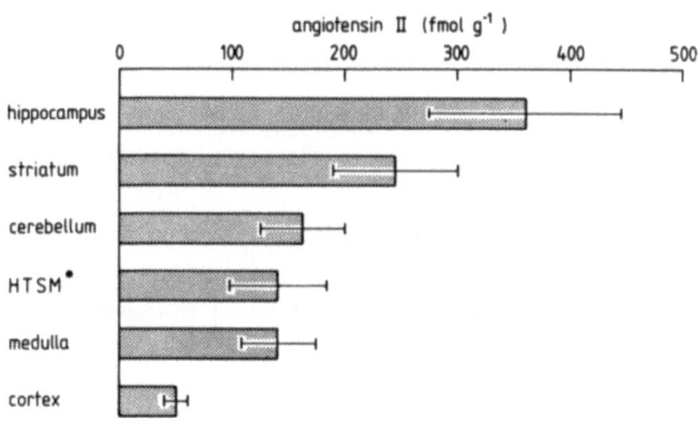

Fig. 3. Regional distribution of immunoreactive angiotensin II in rat brain (mean $\pm$ SEM), 7 experiments (18). HTSM is hypothalamus, thalamus, septum and midbrain

A regional distribution of ANG II was determined (Fig. 3). Hippocampus had the highest concentration and cortex the least. The distribution of ANG II could be explained, at least in part, if ANG II containing neurons projected to regions with receptors. For instance, receptors are concentrated in regions (septum, preoptic region, hypothalamus, midbrain) to which the hippocampus projects via the fornix (18).It is encouraging that a localization of angiotensin-converting enzyme, angiotensinases, and a capacity to generate angiotensinogen in vitro (18) are properties of hippocampus and striatum, areas in which ANG II is concentrated.

Not all of the in vitro studies of brain-specific ANG II receptor binding can be immediately related to a functioning brain renin-angiotensin system, i.e. the inconsistencies in the distribution of receptors between species and a failure to influence receptor binding by changes in plasma ANG II. Perhaps, because of the existence of the blood-brain barrier, it was naive to attempt to manipulate ANG II receptors by changing plasma ANG II levels. Nevertheless, a number of acceptable correlations have been observed, particularly in localization of receptors in ANG II-sensitive regions and the possibility of neural pathways between regions of high ANG II concentration and high ANG II receptor binding (18). The simplest explanation of these observations is that ANG II is an excitatory neurotransmitter in limbic pathways.

REFERENCES

1. AGUILERA G, SCHIRAR A, BAUKAL A, CATT KJ (1980) Angiotensin II receptors, properties and regulation in adrenal glomerulosa cells. Circ Res 46:Suppl 1 I118-I127

2. BAXTER CR, HORVATH JS, DUGGIN GG (1980) Effect of age on specific angiotensin II-binding sites in rat brain. Endocrinology 106:995-999

3. BENNETT JP, SNYDER SH (1976) Angiotensin II binding to mammalian brain membranes. J Biol Chem 251:7423-7430

4. BROOKS MJ, SIRETT NE, HUBBARD JI (1980) Area postrema units not responsive to serotonin, angiotensin II, dopamine or osmotic pressure changes. Proc Univ Otago Med Sch 58:71-72

5. BURANARUGSA P, HUBBARD JI (1979) The neuronal organization of the rat subfornical organ in vitro and a test of the osmo- and morphine-receptor hypotheses. J Physiol 291:101-116

6. COLE FE, BLAKESLEY HL, GRACI KA, FROHLICH ED and MacPHEE AA (1980) Brain angiotensin II receptor affinity and capacity in SHR and WKY rats: effects of acute dietary changes in NaCl. Brain Res 190:272-277

7. HARDING JW, STONE LP, WRIGHT JW (1981) The distribution of angiotensin II binding sites in rodent brain. Brain Res 205:265-274

8. HAYWOOD JR, FINK GD, BUGGY J, PHILLIPS MI, BRODY MJ (1980) The area postrema plays no role in the pressor action of angiotensin in the rat. Am J Physiol 239: H108-H113

9. MANN JFE, SCHIFFRIN EL, SCHILLER PW, RASCHER W, BOUCHER R, GENEST J (1980) Central actions and brain receptor binding of angiotensin II: influence of sodium intake. Hypertension 2:437-443

10. McDONALD W, WICKRE C, AUMANN S, BAN D, MOFFITT B (1980) The sustained antihypertensive effect of chronic cerebroventricular infusion of angiotensin antagonist in spontaneously hypertensive rats. Endocrinology 107:1305-1308

11. McLEAN AS, SIRETT NE, BRAY JJ, HUBBARD JI (1975) Regional distribution of angiotensin II receptors in the rat brain. Proc Univ Otago Med Sch 53:19-20

12. PHILLIPS MI (1978) Angiotensin in brain: Neuroendocrinology 25:354-377

13. RAIZADA MK, YANG JW, PHILLIPS MI, FELLOWS RE (1981) Rat brain cells in primary culture: characterization of angiotensin binding sites. Brain Res 207:343-355

14. SIRETT NE, McLEAN AS, BRAY JJ, HUBBARD JI (1977) Distribution of angiotensin II receptors in rat brain. Brain Res 122:299-312

15. SIRETT NE, HUBBARD JI (1979) Specific angiotensin II binding in the brains and adrenals of juvenile rats. Proc Aust Physiol and Pharmacol Soc 10:79P

16. SIRETT NE, THORNTON SN, HUBBARD JI (1979) Angiotensin binding and pressor activity in the rat ventricular system and midbrain. Brain Res 166:139-148

17. SIRETT NE, THORNTON SN, HUBBARD JI (1979) Brain angiotensin II binding and central [Sar^1, Ala^8] angiotensin responses in normal rats and the New Zealand strain of genetically hypertensive rats. Clin Sci 56:607-611

18. SIRETT NE, BRAY JJ, HUBBARD JI (1981) Localization of immunoreactive angiotensin II in hippocampus and striatum of rat brain. Brain Res 217:405-411

19. STAMLER JF, RAIZADA MK, FELLOWS RE, PHILLIPS MI (1980) Increased specific binding of angiotensin II in the organum vasculosum of the laminae terminalis area of the spontaneously hypertensive rat brain. Neurosci Lett 17:173-177

Brain Angiotensin II Receptors: Possible Physiological Implications

J.F.E. Mann

INTRODUCTION

Over the last two decades, evidence has been accumulated that angiotensin II exerts actions on the central nervous system (CNS) (18). These effects may be brought about by circulating angiotensin II or by angiotensin II generated locally in brain tissue (8). In any event, the central actions of angiotensin II are mediated by interaction of angiotensin II with its brain receptor site. The present paper will review the data on the in vitro characterization of brain angiotensin II receptors in mammals, their localization in the CNS and their possible role in physiological mechanisms of drinking and blood pressure (BP). In particular, a regulatory role of brain angiotensin II receptors in dietary sodium deprivation and in hypertensive disease of the spontaneously hypertensive rats (SHR) will be discussed.

RADIOLIGAND BINDING STUDIES

Using radiolabelled angiotensin II, several groups (1,5,14,19) have reported binding of the octapeptide to brain tissue of calf and rat, and to isolated brain cells in tissue culture (17). The binding characteristics are consistant with the presence of specific angiotensin II receptors in brain. The binding was shown to be time-, pH- and temperature-dependent, reversible and saturable. The apparent dissociation constant (K_D) was in the range of 0.2-1.0 nM (1,5,14,19). There appears to be one type of binding site with a maximal concentration in rats of about 12-20 fmol angiotensin II/mg tissue-protein (5,14,19), the latter being dependent on the age of the rats (5). For the calf cerebellum, higher maximal concentrations of angiotensin II receptor sites have been reported (1), however, there is no known action of angiotensin II in this particular region of the CNS. Because of breakdown of the radiolabelled angiotensin II in brain homogenates (1, and unpublished observations), various enzyme inhibitors had to be used for the in vitro characterization of the angiotensin II receptor. It should be noted, however, that these inhibitors also reduce the angiotensin II binding capacity (unpublished observations). In vitro, the

Experimental Brain Research, Suppl. 4

binding characteristics of angiotensin II to its brain receptor depend on the concentrations of various cations (1,2) in particular on the concentration of sodium (Fig. 1).

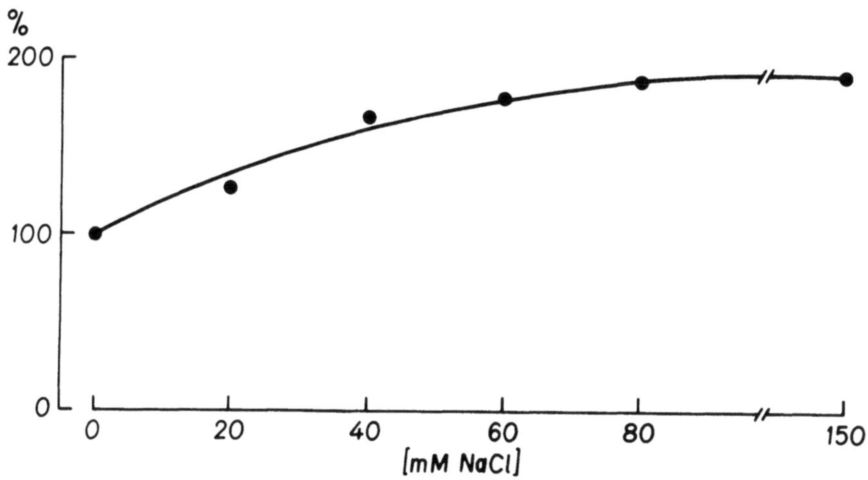

Fig. 1. Brain homogenates were incubated in a 0.1 M sodium-phosphate buffer (pH 7.0) containing 5 mM EDTA, 5 mM dithiothreitol, 5 mM phenylmethylsulfonylfluorid, 0.4% bovine serum albumin, and sodium chloride at the concentrations indicated. ^{125}I-angiotensin II was added at 0.05 nM. The y-axis gives the amount of ^{125}I-angiotensin II binding in %. The amount of angiotensin specifically bound when no sodium chloride was present is given as 100%

LOCALIZATION OF BRAIN ANGIOTENSIN II RECEPTORS

The density of angiotensin II receptors in brain tissue varies to a sizable extent between distinct brain regions. There is little binding to the cortex both in calf and rat brain (1,19). In the rat, angiotensin II receptor concentration was highest in the following regions: Thalamus, hypothalamus, midbrain, and brainstem, the latter exhibiting only half the binding capacity of thalamus and midbrain (1,19).

The distribution of the angiotensin receptors in rat brain co-incides partly with the immunohistological localization of angiotensin-like immunoreactivity. There is, however, some disparity in the reports on the localization of angiotensin II in rat brain between the different laboratories (4,7).

To define the anatomic localization of brain angiotensin receptors which interact with blood-borne angiotensin II, we designed an autoradiographic study. A threshold pressor dose of radiolabelled angiotensin II was injected intracardially and the brain was fixed within 3 min. The radioactive angiotensin was taken up by the circumventricular organs with a deficient blood brain barrier (median eminence, subfornical organ, organum vasculosum laminae terminalis, area postrema). The uptake was specific, because infusion of unlabelled [Sar^1, Ala^8]-angiotensin II reduced autoradiographic reaction by more than 90%. During this short term experiment, crossing of the blood brain barrier could not be demonstrated (22). Using a similar approach with intracerebroventricularly (i.c.v.) injected fluorescent angiotensin II, specific uptake for the i.c.v. route was mainly within the anterior part of the third brain ventricle (12).

Circulating angiotensin II has therefore access to brain receptor sites in the circumventricular organs where angiotensin is in fact acting to increase BP and water intake. For the rat, the subfornical organ appears to mediate the central effects of blood-borne angiotensin II (10), and for the dog, the bulk of data indicates a similar role for the area postrema (18). When angiotensin is injected into the cerebroventricular system (a pharmacological approach to imitate angiotensin generated in the brain), the anterior part of the third brain ventricle and the midbrain (subnucleus medialis) emerged as the most prominent sites of action (3,10), coinciding partly with the relative density of angiotensin receptors in a study on anesthetized rats (20).

POSSIBLE PHYSIOLOGICAL ROLE OF BRAIN ANGIOTENSIN RECEPTORS

It is obvious now that circulating angiotensin can combine with a specific receptor site in the brain, localized where angiotensin is known to exert its effects on BP and drinking. The question arises whether the receptor may function at physiological concentrations of circulating angiotensin II. As reviewed above, the K_D of the brain angiotensin II receptor is in the range of 0.2-1.0 nM (0.2 in our own experiments). In the blood of the rats, angiotensin II circulates at 0.05-6.0 nM (subnormal to grossly pathophysiological values). Furthermore, the threshold concentration for eliciting drinking by circulating angiotensin II is about 0.2 nM, a concentration typical for water deprivation and renal hypertension (11,15). If one accepts that angiotensin II-induced drinking is due to interaction of angiotensin II with its brain receptor, and that at least part of these receptors are characterized in the in vitro system, the potential physiological significance of the latter becomes apparent.

To further test the hypothesis that the receptors which were characterized in vitro, may be relevant for the central actions of angiotensin II in vivo, a series of studies with angiotensin II peptide analogues was designed. The analogues were analyzed in vitro for their relative potency to displace radiolabelled angiotensin II from its brain receptor site (Fig. 2). [Des-Asp^1]-angiotensin II was only slightly less potent than angiotensin II itself in these experiments (16). Angiotensin I displayed about 50 times less activity than angiotensin II (16). These binding characteristics are quite different from the ones reported for the smooth muscle of resistance vessels (9), possibly indicating, that the binding to brain homogenates was not due to uptake by vascular elements.

Angiotensin analogues were also injected into the cerebroventricular system (i.c.v.) of normal rats. Blood pressure and drinking were measured (14). As is evident from Fig. 3, the data obtained with the angiotensin II analogues in vitro and in vivo,displayed a close parallelism for the relative binding affinity and for the relative change of blood pressure and water intake. Although this is indirect evidence, it supports the view that the receptors studied in vitro may be indicative for the ones mediating the central effects of angiotensin II on pressure and fluid homeostasis.

As shown above, the sodium concentration in the incubation medium greatly affects the binding of angiotensin II to its brain receptor. In addition, there are numerous reports that angiotensin II actions on peripheral organs and peripheral receptor binding is sodium dependent. We therefore investigated effects of chronic dietary sodium deprivation on receptor binding and central activity of angiotensin II. In severely sodium-restricted rats (3 weeks of low sodium diet), we found a reduction in the binding capacity of brain angiotensin II receptors (Fig. 4). This was due to a reduction in the maximal number of binding sites, leaving binding affinity unchanged (14). In vivo, the drinking response to centrally injected angiotensin II was reduced by about 50% (Fig. 5), but the response to carbachol was unaltered (Fig. 6), indicating the specificity of the sodium induced changes for angiotensin. Considering the fact that the sodium deprived rats showed a lower daily intake of

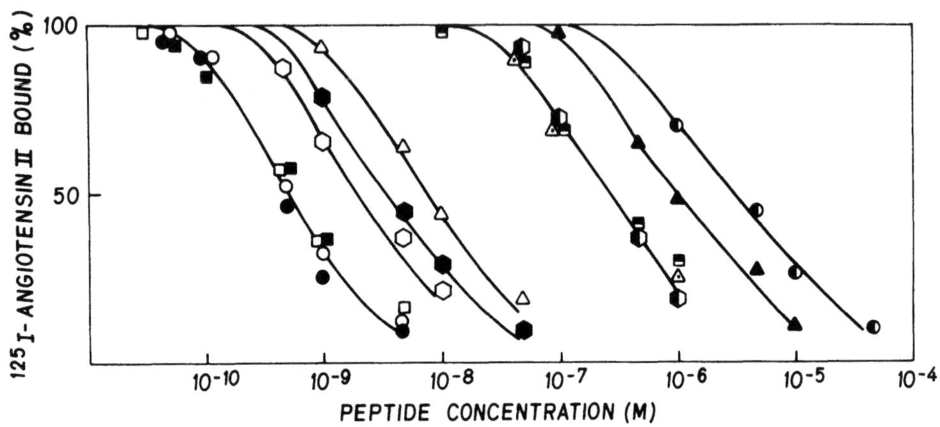

Fig. 2. Displacement of specific ^{125}I-angiotensin II binding from rat brain homogenate by various angiotensin analogues. For incubation conditions see Fig. 1.

Regression lines were calculated for the five active analogues only (y = -139.7 - 7.54 lnx, y = -66.73 - 3.52 Inx, respectively).

● = [Ile^5]-Ang II, ○ = [Val^5]-ANG II. ■ = [des-Asp^1, Ile^5]-ANG II, ⬢ = [Val^5, Trp^8]-ANG II, ⬡ = [Trp^1, Val^5]-ANG II, ▣ = [Phe^4, Val^5, Tyr^8]-ANG II, ◐ = [Trp^1, Phe^4, Tyr^8]-ANG II, △ = [des -Asp^1, Val^5, Trp^8]-ANG II, ◐ = [4-8]-ANG II-pentapeptide, ▲ = [3-8]-ANG II-hexapeptide

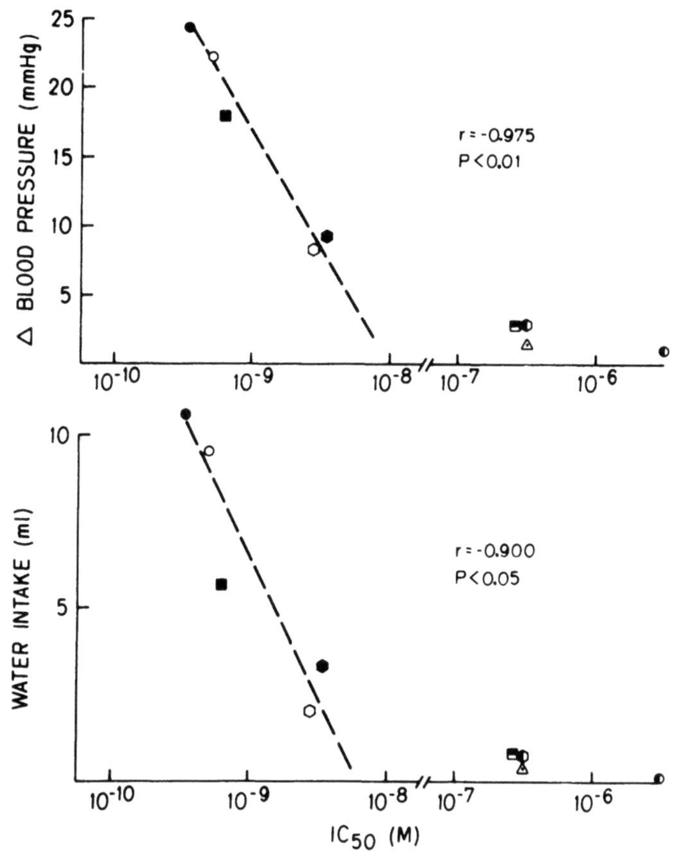

Fig. 3. Upper panel:Correlation between the central pressor effects of angiotensin analogues in vivo in normotensive rats, and their IC_{50} for specific angiotensin binding to brain receptors in vitro, as determined from Fig. 1. The IC_{50} is the concentration of the respective peptides which inhibits specific angiotensin binding by 50%, and is considered inversely proportional to the affinity of the analogues for the brain angiotensin receptor.

Lower panel: Correlation between the central dipsogenic effects of angiotensin analogues in vivo and their IC_{50} for the brain angiotensin receptor in vitro. Regression lines were calculated for the 5 active analogues only. Peptides were tested in vivo at a dose of $2x10^{-10}$ moles. For symbols, see Fig. 2

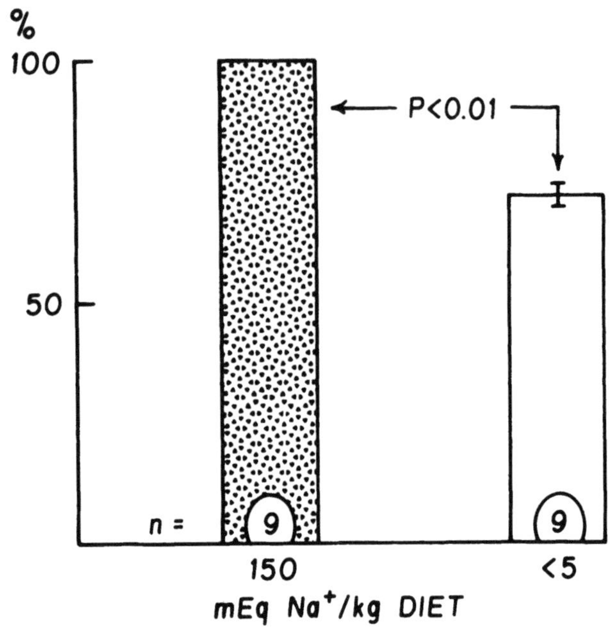

Fig. 4. Effect of low sodium intake on specific ^{125}I-angiotensin II binding to brain membranes. Normotensive rats were fed on a low sodium diet for 3 weeks (open column). Sodium content of the diet is given at the bottom. Binding of ^{125}I-angiotensin II (0.05 nM) was evaluated as described in Fig. 1. On each day, the brain of a rat on a normal sodium intake and of a rat on a low sodium intake were analyzed, and the binding in controls taken as 100%. Binding capacity during sodium deprivation was calculated accordingly (14)

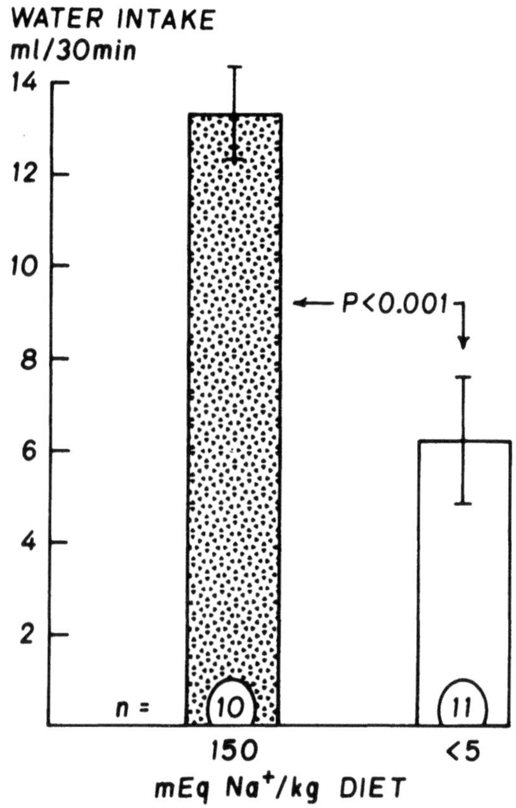

Fig. 5. Effect of a low sodium intake on the central dipsogenic response to angiotensin II in normotensive rats. Angiotensin II was i.c.v. injected at $4x10^{-10}$ moles. Rats were on a normal, or low sodium diet as described in Fig. 4 (14)

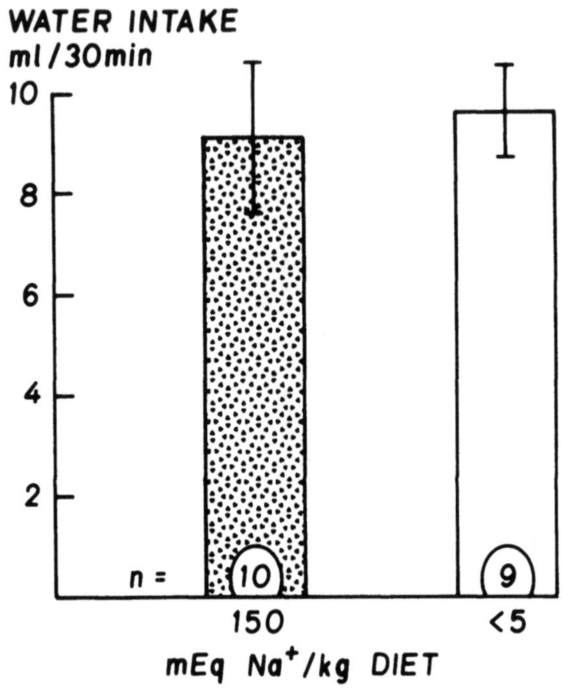

Fig. 6. Effect of a low sodium intake on the central dipsogenic response to carbachol in normotensive rats. For further details see Fig. 4 and 5 (14)

water (unpublished observations), we might advance the highly speculative hypothesis that in the sodium deprived state, the amount of water intake might partly depend on angiotensin. Accordingly, water intake might be suppressed because of central angiotensin receptor "down regulation". However, this hypothesis is completely unproven and the lower osmotic load of the low sodium diet might also be a good explanation for the reduced water intake. Interestingly, water intake of sodium deprived dogs is reported to be actually higher than in controls (7a) and the dipsogenic response to renin is enhanced (see Schölkens et al., this volume).

The BP response to centrally injected angiotensin II was reduced in normotensive rats on a low sodium diet (14), and even more so in SHR (Fig. 7). In the case of the SHR, the low sodium regimen was only maintained for one week. On a normal sodium diet, the SHR exhibit a higher central pressure response to angiotensin II as compared to their normotensive Wistar Kyoto controls (WKR). This hyperresponsiveness is blunted when the rats are sodium-deprived (Fig. 7). This raises the possibility that the hyperresponsiveness observed is a receptor-related phenomenon, directly or indirectly depending on the sodium status (21). In fact, hypernatremia and sodium retention have been reported for the SHR strain used in the present experiments (6) and higher brain angiotensin binding capacity following dietary sodium loading in SHR has been reported. Furthermore, injection of the competitive angiotensin II antagonist [Sar^1, Ala^8]-angiotensin II i.c.v. reduces blood pressure acutely (13), and may even practically normalise it on a chronic infusion schedule (8). These effects were also observed after removal of circulating renin by bilateral nephrectomy, and it was therefore hypothesized, that extrarenally formed angiotensin II is important for the maintenance of blood pressure in SHR. Biochemical, pathophysiological, pharmacological and anatomical data indicate that angiotensin is generated locally in the brain. However, because the exact local concentrations of brain angiotensin II are still unknown at the present time, we cannot state unambiguously that endogenous brain angiotensin II is the ligand for the brain angiotensin receptors which were investigated in the present study.

ACKNOWLEDGEMENT

This study was supported by a grant from the Deutsche Forschungsgemeinschaft to J.F.E. Mann (Ma 720/3). We thank Ms. Stelz for expert secretarial help.

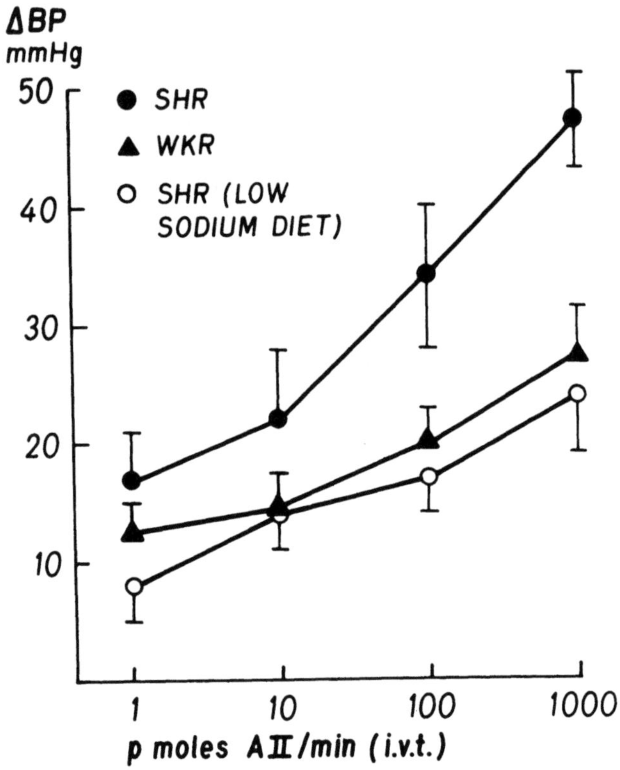

Fig. 7. Central pressor response in SHR on diets containing 100 mmol/kg (●, n=8) or less than 5 mmol/kg of sodium (○, n=9). WKR on a 100 mmol/kg sodium diet were also investigated (▲, n=6). Rats were on the diet for 1 week prior to testing. For BP-testing, SHR and WKR were anesthetized with alpha-chloralose and angiotensin II was infused for 5-10 min

REFERENCES

1. BENNETT JP Jr, SNYDER SH (1976) Angiotensin II binding to brain mammalian membranes. J Biol Chem 251:7423-7430

2. BENNETT JP Jr, SNYDER SH (1980) Regulation of receptor binding interactions of ^{125}I-angiotensin II and ^{125}I-[Sar^{1}, Leu^{8}]-angiotensin II, an angiotensin antagonist, by sodium ion. Eur J Pharmacol 67:1-10

3. BUCKLEY JP, SMOOKLER HH, SEVERS WB, DEUBEN RR (1977) A central site of action of angiotensin II and its possible role in the central regulation of the cardiovascular system. In: BUCKLEY JP, FERRARIO CM (eds) Central actions of angiotensin and related hormones. Pergamon Press, New York, p 149-156

4. CHANGARIS DG, SEVERS WB, KEIL LC (1978) Localization of angiotensin in rat brain. J Histochem Cytochem 26:593-607

5. COLE FE, FROHLICH ED, MACPHEE AA (1978) Angiotensin binding affinity and capacity in the midbrain area of spontaneously hypertensive and normotensive rats. Brain Res 154:178-181

6. DIETZ R, SCHÖMIG A, HAEBARA H, MANN JFE, RASCHER W, LÜTH JB, GRÜNHERZ N, GROSS F (1978) Studies on the pathogenesis of spontaneous hypertension of rats. Circ Res 43 (I): I-98-I-106

7a FITZSIMONS JT (1972) The physiology of drinking. Physiol Rev 52:484-559

7. FUXE K, GANTEN D, HÖKFELT T, BOLME P (1976) Immunohistochemical evidence for the existence of angiotensin II-containing nerve terminals in the brain and spinal cord of the rat. Neurosci Lett 2:229-234

8. GANTEN D, SPECK G, MANN JFE, UNGER TH (1981) Role of the brain renin-angiotensin system in central mechanisms of blood pressure control. In: LARAGH JH, BÜHLER FR, SELDIN DW (eds) Frontiers in hypertension research, Springer, New York, p 268-273

9. GUNTHER S, GIMBRONE MA, ALEXANDER RW (1980) Regulation by angiotensin II of its receptors in resistance blood vessels. Nature 287:230-232

10. JOHNSON AK, BUGGY J (1977) A critical analysis of the site of action for the dipsogenic effect of angiotensin II. In: BUCKLEY JP, FERRARIO CM (eds) Central action of angiotensin and related hormones. Pergamon Press, New York, p 357-386

11. JOHNSON AK, MANN JFE, RASCHER W, JOHNSON JK, GANTEN D (1980) Plasma angiotensin II concentrations and experimentally induced thirst. Am J Physiol 240: R229-R234

12. LANDAS S, PHILLIPS MI, STAMLER JF, RAIZADA MK (1980) Visualization of specific angiotensin II binding sites in the brain by fluorescent microscopy. Science 210:791-793

13. MANN JFE, PHILLIPS MI, DIETZ R, HAEBARA H, GANTEN D (1978) Effects of central and peripheral angiotensin blockade in hypertensive rats. Am J Physiol 234: H629-H637

14. MANN JFE, SCHIFFRIN EL, SCHILLER PW, RASCHER W, BOUCHER R, GENEST J (1980) Central actions and brain receptor binding of angiotensin II: influence of sodium intake. Hypertension 2:437-443

15. MANN JFE, JOHNSON AK, GANTEN D (1980) Plasma angiotensin II: dipsogenic levels and angiotensin generating capacity of renin. Am J Physiol 238:R372-R377

16. MANN JFE, SCHILLER PW, SCHIFFRIN EL, BOUCHER R, GENEST J (1981) Brain receptor binding and central actions of angiotensin analogs in rats. Am J Physiol 241: R124-R129

17. RAIZADA MK, YANG JW, PHILLIPS MI, FELLOWS RE (1981) Rat brain cells in primary culture: characterization of angiotensin II binding sites. Brain Res 207: 343-355

18. SEVERS WB, DANIELS-SEVERS AE (1973) Effects of angiotensin on the central nervous system. Pharmacol Rev 25:415-449

19. SIRETT NE, McLEAN AS, BRAY JJ, HUBBARD JI (1977) Distribution of angiotensin receptors in rat brain. Brain Res 122:299-312

20. SIRETT NE, THORNTON SN, HUBBARD JI (1979) Angiotensin binding and pressor activity in the rat ventricular system and midbrain. Brain Res 166:139-148

21. STAMLER JF, RAIZADA MK, PHILLIPS MI, FELLOWS RE (1980) Increased specific binding of angiotensin II in the organum vasculosum laminae terminalis area of the spontaneously hypertensive rat. Neurosci Lett 17:173-177

22. Van HOUTTEN M, SCHIFFRIN EL, MANN JFE, POSNER BI, BOUCHER R (1980) Radioautographic localization of specific binding sites for blood borne angiotensin II in rat brain. Brain Res 186:480-485

Angiotensin and Single Neurons

D. Felix, P. Schelling, and H. L. Haas

INTRODUCTION

For a peptide to be designated a neurotransmitter many criteria have to be fulfilled. First, it must be shown that it is formed intracellularly and is present in the appropriate nerve terminals. Peptides are probably synthetized only on the ribosomes of the cell body, possibly by degradation of a larger precursor molecule without local synthesis in the nerve endings (24). Neuronal angiotensin II appears to be generated from a reaction cascade comparable to that found for the classical renal renin-angiotensin system (19,33,38). Second, an important fact for a substance to be accepted as a neurotransmitter is its release from nerve terminals by incoming impulses and an action on postsynaptic membranes similar to that of the naturally released transmitter. A further characteristic of course is that exogenous substances which modify the postsynaptic action of the proposed transmitter have a similar effect at the appropriate central synapses in vivo. In the simplest case, if a given synaptic input is found to be blocked on the postsynaptic side by an agent, the same blockade should be manifested when the suspected transmitter is applied exogenously.

The use of electrophysiological techniques to study the mechanisms of neural peptides has distinct advantages. They measure directly the electrical activity of a single neuron, which is the important functional parameter of the nervous system. Furthermore, they offer a high resolution in time and space and permit the conclusion which neurons are different by virtue of their connections to the brain circuitry mediating different behavioral effects. There are, of course, dangers and difficulties of interpretation especially when substances are applied directly onto a neuron. Recording from individual cells, particularly intracellular recording, is likely to be from the region of the soma; whereas, many neurons possess elaborate dendritic trees with synapses far removed from the cell body. Nevertheless, recording from individual neurons and manipulation of their extracellular environment is a powerful tool in identifying the chemical nature of a transmitter. The present review deals exclusively with the action of angiotensin at the level of the single neuron.

Experimental Brain Research, Suppl. 4

ANGIOTENSIN II (ANG II)-SENSITIVE NEURONS: IN VIVO RECORDINGS

A convenient method to look for ANG II-sensitive neurons seems to be to follow single neuron discharges under the influence of microelectrophoretically ejected ANG II from multibarreled micropipettes. This approach is widely used and has proven successful in several structures of the brain. The first description of an ANG II related effect comes from supraoptic region (30). Almost all the responsive, antidromically-invaded supraoptic neurosecretory cells, and those exhibiting recurrent inhibition, were excited by ANG II. In contrast to the excitatory responses of supraoptic cells, cortical neurons were either unaffected or slightly depressed, suggesting that the effect seen in the supraoptic nucleus was related to a specific action of ANG II in this area. The high proportion of neurons in the medial preoptic area excited by ANG II suggests also a direct mechanism of action on neuronal elements of this region (22). The presence of sensitive units in this area is consistent with evidence identifying the anterior third ventricle region as an important receptor site for the dipsogenic action of the peptide (9).

Neurons from different brain regions in mammals differ in their reaction to ANG II. Wayner and his colleagues (45) found that angiotensin affects many different cells within the lateral hypothalamus, zona incerta, ventromedial and dorsomedial hypothalamic nuclei, dentate gyrus and thalamus. They described cells of the lateral hypothalamus and zona incerta being the most sensitive to ANG II. Lateral hypothalamic neurons lying within the medial forebrain bundle increased in discharge rate when angiotensin was applied, whereas the low, spontaneously active ventral lateral hypothalamic cells predominantly decreased. In unanesthetized rabbits most of cortical and hypothalamic neurons as well as those of the locus coeruleus were found to be sensitive to ANG II (3). The peptide also gave rise to an excitatory effect on neurons of the medulla; however, these neurons do not show a topographically or functionally characteristic distribution related to the central vasomotor control mechanisms (44). Interestingly, neurons in the area postrema were not responsive to iontophoretically applied ANG II suggesting that the peptide, after passing through the blood-brain barrier, does not act directly on the neurons in the area postrema.

The Subfornical Organ

Intracranial injection of ANG II into the anterior hypothalamic, septal or preoptic regions causes animals in a normal state of water balance to drink (9). In his review of thirst, Fitzsimons (16) suggested that the subfornical organ (SFO) might be a likely candidate as a dipsogenic receptor for ANG II. Evidence to support this hypothesis came from studies by Simpson and Routtenberg (41) in which ANG II, applied directly to the SFO in a dose as small as 0.1 ng, elicited a short-latency drinking response. Lesions of the organ abolished at least temporarly, drinking induced by injections of ANG II (5). Although this observation was concerned with central angiotensin effects, the question remained whether these behavioral studies represent a direct response of SFO neurons to the peptide action. The microiontophoretic studies on the SFO (Fig. 1A) carried out in our laboratory indicated that neurons specifically sensitive for ANG II exist in this organ (10,12). These studies were carried out on cats in which the SFO was visualized by careful removal of one hemisphere in an anesthetized preparation. Activation of cell discharges was obtained from the majority of cells (Fig. 1B) and there was a dose-response relationship. In addition, intravenous injections of ANG II excited neurons in this area. Excitation of neurons similar to those observed after ANG II was obtained also with acetylcholine. We could differentiate between those cells which are specifically responsive to ANG II and those that responded to both ANG II and acetylcholine.

Cells that were specifically responsive to ANG II not only responded to low current ejection levels but were also blocked in their response by administration of the competitive antagonist [Sar^1, Ala^8] -ANG II (Saralasin, P-113, Fig. 1C). In contrast, acetylcholine-stimulated neurons were not affected by this antagonist (31).

The experimental system in the SFO was used for structure-activity studies with angiotensin fragments (Fig. 2A). We found that the ANG [2-8] heptapeptide was more potent than the native octapeptide ANG II (Fig. 2B), while the ANG [5-8] tetrapeptide also had activity but was weaker than either the hepta- or octapeptide (Fig. 2C). Both the action of the heptapeptide and the tetrapeptide were blocked by saralasin. In contrast, the ANG [6-8] tripeptide failed to enhance the firing rate of the same neuron (14). These results indicate that the tetrapeptide C-terminal sequence of ANG II is primarily responsible for receptor recognition at central neurons.

Other Circumventricular Organs

Recently we have also recorded from neurons in the organum vasculosum laminae terminalis (OVLT). A ventral approach had to be taken to provide accurate penetration of this small structure. The results show that with very low current injection levels of ANG II there is predominantly excitation and a dose-response relationship. Although one cannot say how many molecules of ANG II are required for a response, the microiontophoretic method allows only small quantities of ions to be released (27). Since as little as 5-10 nA caused excitation, the neurons must be highly sensitive to the peptide hormone (33). Further, the response was specific as luteinizing hormone-releasing hormone (LH-RH) applied at the same location produced predominantly inhibitory responses (13), while the response to ANG II was blocked by saralsin.

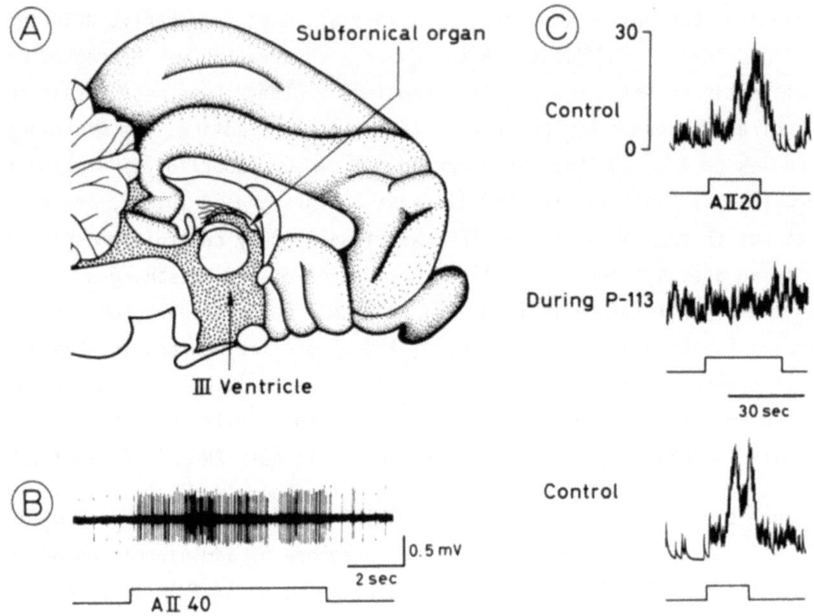

Fig. 1. A. Diagram of cat subfornical organ. B. ANG II evokes increased spike discharge. C. ANG II antagonism by saralasin (P-113) illustrated on frequency histograms (from ref. 12: by permission of Elsevier)

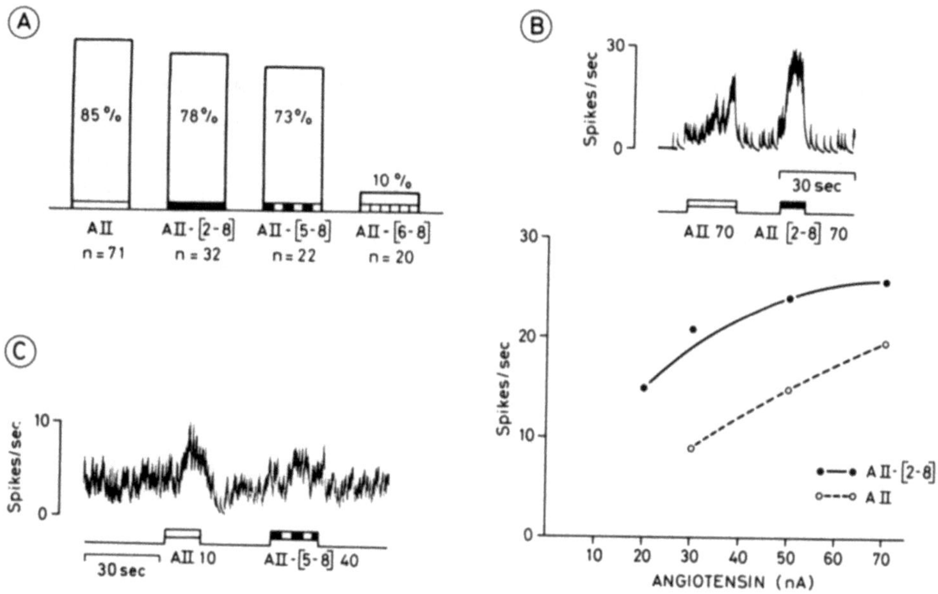

Fig. 2. A. Excitatory action of ANG II peptide fragments on SFO neurons. The percentages listed in A are based on the number of cells (n) on which the fragments were tested. B. Action of ANG II and [2-8]-ANG II ejected with equal amounts of current (70 nA) on the firing rate (insert). Lower panel shows dose-response curves of the two peptides; each point representing the firing rate produced by the peptides ejected with different currents. C. Integrated firing frequency of a neuron excited by ejection of ANG II (10 nA) and [5-8]-ANG II tetrapeptide (40 nA) (from ref. 14, by permission of Elsevier)

There appeared to be excitation of OVLT cells by sodium (28). It cannot be ruled out that there are Na^+ ions sensitive neurons in other regions of the brain (45). However, many more cells are sensitive to angiotensin than sodium; whereas, all sodium-sensitive neurons are also sensitive to angiotensin. Furthermore, it is possible that some of the sodium-induced excitation resulted from a sodium-induced increase in the sensitivity of the neurons to ANG II (29). An interaction of ANG II and hypertonic NaCl was reported on the activation of supraoptic neurosecretory cells (1) supporting the report by Andersson and coworkers (2) that ANG II might influence periventricular Na^+ transport in a manner which facilitates the excitation of primarily sodium-sensitive receptors. However, the lack of synergistic effect on thirst of combined angiotensin and hypertonic NaCl injections in the rat is an argument against a unitary neural mechanism sensitive to both ANG II and sodium (6).

SEARCH FOR RECEPTORS IN THE BRAIN

The route by which ANG II reaches the receptors in periventricular regions may be either via the blood, since the circumventricular organs have an impaired blood-brain barrier, or via the cerebrospinal fluid (CSF). Although it is considered that there is no barrier between CSF in the ventricle and adjacent brain tissue, the diffusion rate of ANG II from the ventricle to the brain tissue seems slow (26) and cannot be accounted for the relatively short delay of activating supraoptic neurosecretory cells by intraventricular application of ANG II (1). Thus, it seems more likely that intraventricular ANG II stimulates receptors located in the periventricular region and activates neurosecretory cells via a neuronal pathway (32). A similar model was suggested recently of the neuronal organization of the SFO, in which ANG II sensitive neurons lying superficially, synapse with deeper carbachol-sensitive neurons (7).

Although the emphasis has been on a ventricular site for the location of angiotensin receptors there is increasing evidence for the existence of angiotensin-receptive sites in regions lying within the blood-brain barrier. The description of a complete renin-angiotensin system in the brain raised the concept that angiotensin is locally synthesized (15,20). The brain possesses abundant angiotensin receptor binding, higher than in some other tissues of the body (4,35,42), and immunohistochemical studies have revealed neuronal systems that stain with antiserum to angiotensin II (8,17). The finding of angiotensin-sensitive neurons in regions with intact blood-brain barriers is of importance in offering further evidence for the existence of an independent brain angiotensin system. Such cells were found in areas of the brain such as the nigrostriatal system. It is of certain interest that, in this area, the general effect produced by ANG II was an inhibitory one and the inhibition was blocked by specific antagonists as well (40). Recently we have investigated the septal areas of the rat. Again, this study enabled a search for ANG II receptors in a system with an intact blood-brain barrier, which contains cells having significant angiotensin binding activity (42) and which is one of the most sensitive sites for the dipsogenic responses to ANG II (9). Our results on rats, anesthetized with urethane, indicate that the septum contains cells having a specific chemosensitivity to angiotensin. The results are strengthened by the finding that even low doses led to an immediate response and that the ANG II-analogue, saralasin, revealed a specific blockade of angiotensin-sensitive neurons. In contrast to the results obtained from the SFO, both activation and depression of the firing rate were observed. Excitation was confined to the lateral part of the septum, whereas a low percentage of neurons (regularly distributed over the whole septum) decreased their activity in reaction to the peptide (25). In contrast, an univocal decrease of septal neurons by ANG II was obtained in rats anesthetized with Ketamine (39).

One important aspect of biological function of ANG II is its potential role as a transmitter or a modulator substance. Lateral neurons respond to fimbria stimulation with activation followed by a lengthy inhibition while medial septal neurons respond with direct inhibition. Iontophoretic application of ANG II results in an enhanced effect of the synaptic activity; a second phase of excitation was often produced following the first activation. The inhibitory period, in which GABA may be involved as a transmitter, was not influenced by the peptide. Likewise intravenously and iontophoretically applied saralasin did not influence the fimbria input to the septal area. Thus, the synaptic transmitter transferring information from hippocampus to the septum is not identical to angiotensin. However, the peptide may influence the synaptic transmission in a modulatory way (25).

THE EFFECT OF SARALASIN ON SPONTANEOUS ACTIVITY

A controversial aspect exists whether ANG II antagonists would affect the spontaneous activity of neurons, i.e., whether endogenous ANG II participates in the maintenance of basal activity. So far, the only clear cut results were reported from supraoptic neurosecretory cells on which intraventricular injection of 1 μg [Sar^1, Ile^8] -ANG II inhibited the firing rate but did not affect thalamic or non-neurosecretory hypothalamic cells (1). Such a direct action, however, could not be confirmed in the neostriatum (40), whereas on neurons in the preoptic area, saralasin alone excited 3 out of 10 cells and decreased it in none (22). In the SFO (31), the OVLT (33) and the septum (25) we found either no effect or a decrease in unit firing. An excitatory effect was so far never observed. The specificity of saralasin-induced depressant effect on cultured hypothalamic neurons has been doubted as it was also seen on cultured cerebellar and hippocampal cells (18). However, one can argue about the validity of such controls since in both the cerebellum and the hippocampus ANG II might be present as well (4,23).

ANGIOTENSIN-SENSITIVITY IN SPONTANEOUSLY HYPERTENSIVE RATS

It has been shown by Ganten and coworkers (21) that blockade of central ANG II receptors or inhibition of cerebral ANG II synthesis lowers blood pressure in spontaneous hypertensive rats. Thus, there is evidence for a functional role of the brain renin-angiotensin system in central blood pressure control. Furthermore, the specific binding of ANG II was reported to be elevated in these animals (43). We intended to answer the question whether the receptor properties were changed in hypertensive animals by the microiontophoretic approach. ANG II-evoked neuronal firing occurred at much lower thresholds in hypertensive rats and showed a clearly extended postactivity as compared to age-matched normotensive Wistar Kyoto rats. In contrast, acetylcholine-induced activation remained unchanged in both strains (Fig. 3). In conclusion, the higher sensitivity of septal neurons in hypertensive rats contributes to the view of an activated brain renin-angiotensin system being involved in the genesis of hypertension in these rats (37).

INTRACELLULAR RECORDINGS FROM HIPPOCAMPAL SLICES OF THE RAT

As pointed out earlier, it is of crucial importance to demonstrate the localization of ANG II in structures with an intact blood-brain barrier. In a recent study we have shown histochemical evidence for the presence of the octapeptide in hippocampal pyramidal cells of the rat. Immunoreactive cell bodies were detectable on CA 1 and CA 3 hippocampal regions and in the stratum granulosum of the dentate area. Electrophysiological experiments were carried out on the hippocampal slice. Such a preparation offers unique advantages for such a study, because drugs can be applied in known concentrations by perfusion, the sites for stimulation and recording are under visual control and the anatomy and functional connections are well established. Furthermore, in in vitro preparations many of the limitations of the in vivo studies are absent such as influence of anesthetics, vascular effects and mechanical instability for intracellular recordings. We could show a direct depolarizing effect of angiotensin which was resistant to tetrodotoxin, and further, an enhancement of excitatory transmission depending probably on a disinhibitory mechanism. The actions of angiotensin were antagonized by saralasin. A large proportion of the fibers stimulated in our experiments originates from ipsilateral CA 3 and contralateral CA 1 pyramidal cells. A subpopulation of both these cell types displayed a strong

		SHR-sp n=20	WKR n=10
ACh	Threshold	71 ± 28 nA	64 ± 33 nA
AII	Threshold	16 ± 4 nA ***	69 ± 10 nA
	Postactivity	113 ± 19 sec **	33 ± 15 sec

means ± S.E.M. ***p < 0.001 **p < 0.01

Fig. 3. ANG II - and acetylcholine (ACh) - evoked responses of septal neurons from stroke-prone spontaneously hypertensive rats (SHR-sp) as compared to normotensive Wistar Kyoto rats (WKR). Thresholds are expressed in amount of currents (nA), post-activity is shown in seconds immunoreactivity to renin and ANG II antibodies. The transmitter released from the en passant boutons, which is responsible for the recorded excitatory postsynaptic potentials, is most probably an excitatory amino acid. It may well be that angiotensin is released together with this transmitter from the same or separate varicosities in order to modulate its efficacy. The angiotensin effect could be through an interference with inhibitory transmission (23).

NEURONS IN CULTURE

The past few years have seen an increasing interest in using tissue culture systems. Cell lines and cell cultures, which have the lowest levels of organization, offer the advantage of great accessibility for physiological studies and such preparations are employed in studies requiring detailed cellular mechanisms. So far, there are numerous reports demonstrating the ability of isolated neurons to generate postsynaptic potentials and action potentials. However, the question remains whether neurons grown under culture conditions faithfully duplicate the responses of in vivo cells. There are only few reports on the effect of angiotensin on cultured cells (36). Gähwiler and Dreifuss (18), using cultures of the hypothalamic supraoptic nucleus area, could demonstrate two actions of ANG II. First, it increased the average firing frequency of randomly discharging neurons. Second, in cells displaying a tendency for phasic activity, ANG II increased the duration of bursts. These results indicate that angiotensin can modify neuronal firing patterns in cultures; however, it has to be pointed out that the effects were usually short-lived. A similar transition from random to phasic firing was also observed by Phillips and coworkers (34) on cultured spinal neurons.

CONCLUSION

Although no single piece of evidence by itself proves or disproves the presence of an endogenous angiotensin producing system in the brain, the present electrophysiological experiments offer further evidence in favor of such a system. Angiotensin-sensitive neurons have been found in many areas of the brain and this specific chemosensitive action was confined to regions where highest angiotensin binding activity has been observed. We do not yet know under what circumstances these receptors are normally activated. However, by defining these receptors, we should eventually be able to study neural circuits that control thirst, blood pressure and other physiological mechanisms in as precise a way as the sensory system are now being studied.

So far, there is evidence for different access for angiotensin reaching the receptor sites (Fig. 4). Circumventricular organs of the anterior hypothalamus may be

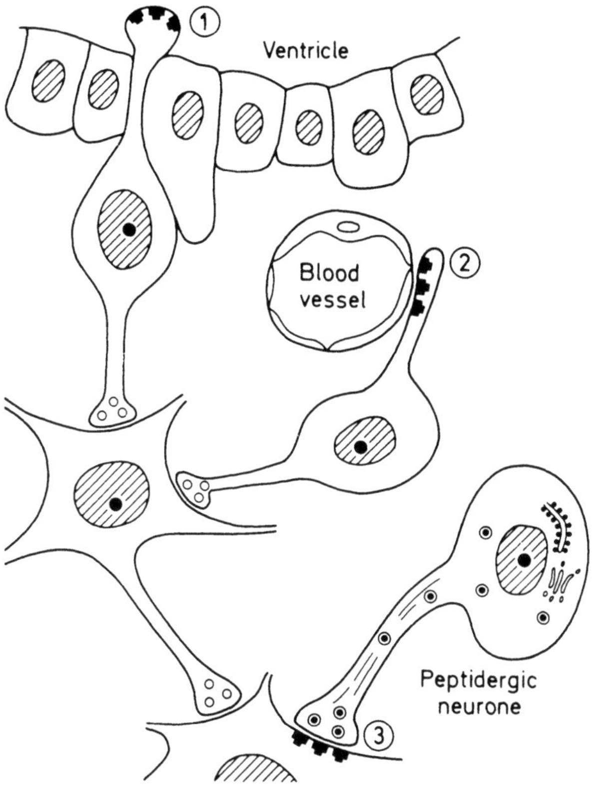

Fig. 4. Schematic diagram of possible receptor cells for angiotensin central effects (see text)

target organs for angiotensin arriving either from the periphery or from the ventricle. On the other hand, evidence also implies that angiotensin is formed intracellularly and is concentrated in nerve terminals. In this situation, the peptide hormone could act as a peptide neurotransmitter being released from peptidergic neurones.

In the next few years we should see the resolution of many questions currently unanswered about angiotensin and, given the rapid progress in research on this peptide, it may serve as a model for the action of peptides on neuronal function (19,32).

ACKNOWLEDGEMENT

These studies were supported by grants 3.948.80, (PS), 3.271.78 and 3.017.81 (DF) of the Swiss National Science Foundation.

REFERENCES

1. AKAISHI T, NEGORO H, KOBAYASHI S (1980) Responses of paraventricular and supraoptic units to angiotensin II, [Sar^1, Ile^8] -angiotensin II and hypertonic NaCl administered into the cerebral ventricle. Brain Res 188:499-511

2. ANDERSSON B, ERIKSSON L, FERNANDEZ O (1977) Reinforcement by Na^+ of centrally mediated hypertensive response to angiotensin II. Life Sci 10:633-638

3. ANISSIMOV JZ, DELEVA JI, NICOLOV NA, SUDAKOV KV, SHERSTNEV VV (1980) Neuronal sensitivity of some brain regions to angiotensin II in rabbits. Agressologie 21: 171-175

4. BENNETT JR, SNYDER SH (1976) Angiotensin II binding to mammalian brain membranes. J Biol Chem 251:7423-7430

5. BUGGY J, FISHER AE, HOFFMAN WE, JOHNSON AK, PHILLIPS MI (1975) Ventricular obstruction: effect of drinking induced by intracranial angiotensin. Science 190:72-74

6. BUGGY J, HOFFMAN WE, PHILLIPS MI, FISHER AE, JOHNSON AK (1979) Osmosensitivity of rat third ventricle and interactions with angiotensin. Am J Physiol 236: R75-82

7. BURANARUGSA P, HUBBARD JI (1979) The neuronal organization of the rat subfornical organ in vitro and a test of the osmo- and morphine-receptor hypothesis. J Physiol (Lond) 291:101-116

8. CHANGARIS DG, SEVERS WB, KEIL LC (1978) Localization of angiotensin in rat brain. J Histochem Cytochem 26:593-607

9. EPSTEIN AN, FITZSIMONS JT, ROLLS BJ (1970) Drinking induced by injection of angiotensin II into the brain of the rat. J Physiol (Lond) 210:457-474

10. FELIX D (1974) Peptide and acetylcholine action on neurones of the cat subfornical organ. Naunyn-Schmiedeberg's Arch Pharmacol 292:15-20

11. FELIX D, AKERT K (1974) The effect of angiotensin II on neurones of the cat subfornical organ. Brain Res 76:350-353

12. FELIX D, PHILLIPS MI (1978) Effects of angiotensin II on central neurones. In: RYALL RW, KELLY JS (eds) Iontophoresis and transmitter mechanisms in the mammalian central nervous system. Elsevier/North Holland, Amsterdam, p 104-106

13. FELIX D, PHILLIPS MI (1979) Inhibitory effects of luteinizing hormone releasing hormone (LH-RH) on neurons in the organum vasculosum laminae terminalis (OVLT). Brain Res 169:204-208

14. FELIX D, SCHLEGEL W (1978) Angiotensin receptive neurones in the subfornical organ. Structure-activity relations. Brain Res 149:107-116

15. FISCHER-FERRARO C, NAHMOD VE, GOLDSTEIN DJ, FINKIELMAN S (1971) Angiotensin and renin in rat and dog brain. J Exp Med 133:353-361

16. FITZSIMONS JT (1972) Thirst. Physiol Rev 52:468-561

17. FUXE K, GANTEN D, HÖKFELT T, BOLME P (1976) Immunohistochemical evidence for the existence of angiotensin II containing nerve terminals in the brain and spinal cord of the rat. Neurosci Lett 2:229-234

18. GÄHWILER BH, DREIFUSS JJ (1980) Transition from random to phasic firing induced in neurons cultured from the hypothalamic supraoptic area. Brain Res 193: 415-425

19. GANTEN D, SPECK G (1978) The brain renin-angiotensin system: a model for the synthesis of peptides in the brain. Biochem Pharmacol 27:2379-2389

20. GANTEN D, MINNICH JL, GRANGER P, HAYDUK K, BRECHT HM, BARBEAU A, BOUCHER R, GENEST J (1971) Angiotensin-forming enzyme in brain tissue. Science 173:64-65

21. GANTEN D, HUTCHINSON JS, SCHELLING P (1975) The intrinsic brain iso-renin-angiotensin system in the rat: its possible role in central mechanisms of blood pressure regulation. Clin Sci Mol Med 48:265s-268s

22. GRONAN RJ, YORK DH (1978) Effects of angiotensin II and acetylcholine on neurons in the preoptic area. Brain Res 154:172-177

23. HAAS HL, FELIX D, CELIO MR, INAGAMI T (1980) Angiotensin II in the hippocampus. A histochemical and electrophysiological study. Experientia 36:1394-1395

24. HÖKFELT T, JOHANSSON O, LJUNGDAHL A, LUNDBERG JM, SCHULTZBERG M (1980) Peptidergic neurones. Nature 284:515-521

25. HUWYLER T, FELIX D (1980) Angiotensin II-sensitive neurons in septal areas of the rat. Brain Res 195:187-195

26. JOHNSON AK, EPSTEIN AH (1975) The cerebral ventricles as the avenue for the dipsogenic action of intracranial angiotensin. Brain Res 86:399-418

27. KELLY JS (1975) Microiontophoretic application of drugs onto single neurons. In: IVERSEN L, IVERSEN S, SNYDER S (eds) Handbook of Psychopharmacology, Vol II. Plenum Press, New York, p 30-67

28. KNOWLES WD, PHILLIPS MI (1980) Angiotensin II responsive cells in the organum vasculosum laminae terminalis (OVLT) recorded in hypothalamic brain slices. Brain Res 197:256-259

29. MANN JFE, SCHIFFRIN EL, JOHNSON AK, BOUCHER R, GENEST J (1979) Modulation of the central effects of angiotensin II (ANG II) by sodium. Soc Neurosci Abstr 5:221

30. NICOLL RA, BARKER JL (1971) Excitation of supraoptic neurosecretory cells by angiotensin II. Nature New Biol 233:172-174

31. PHILLIPS MI, FELIX D (1976) Specific angiotensin II receptive neurons in the cat subfornical organ. Brain Res 109:531-540

32. PHILLIPS MI, FELIX D, HOFFMAN WE, GANTEN D (1977) Angiotensin-sensitive sites in the brain ventricular system. In: COWAN WM, FERRENDELLI JA (eds) Society for Neuroscience Symposia 2:308-339

33. PHILLIPS MI, WEYHENMEYER J, FELIX D, GANTEN D, HOFFMAN W (1979) Evidence for an endogenous brain renin-angiotensin system. Fed Proc 38:2260-2266

34. PHILLIPS MI, NELSON PG, NEAL E, QUINLAN J (1980) Angiotensin induced phasic firing of spinal cord neurons in culture. Soc Neurosci Abstr 6:619

35. RAIZADA MK, YANG JW, PHILLIPS MI, FELLOWS RE (1981) Rat brain cells in primary culture: characterization of angiotensin II binding sites. Brain Res 207:343-355

36. SAKAI KK, MARKS BH, GEORGE J, KOESTNER A (1974) Specific angiotensin II receptors in organ-cultured canine supraoptic nucleus cells. Life Sci 14: 1337-1344

37. SCHELLING P, FELIX D (to be published) Increased sensitivity of septal neurons to angiotensin II (ANG II) in stroke prone spontaneously hypertensive rats (SHR-sp) as compared to normotensive Wistar Kyoto rats (WKR). The Fourth Int Symp on rats with spontaneous Hypertension, Heidelberg, 1981

38. SCHELLING P, SPECK G, UNGER TH, GANTEN D (1980) The brain renin-angiotensin system: biochemistry, localization and functional aspects. In: PARVEZ H, PARVEZ S (eds)Advances in Experimental Medicine. Elsevier/North Holland, Amsterdam, Vol 1, p 243-288

39. SIMONNET G, BIOULAC B, RODRIGUEZ F, VINCENT JD (1980) Evidence of a direct action of angiotensin II on neurons in the septum and in the medial preoptic area. Pharmacol Biochem Behav 13:359-363

40. SIMONNET G, GIORGUIEFF-CHESSELET MF, CARAYOU A, BIOULAC B, CESSELIN F, GLOWINSKY J, VINCENT JD (1981) Angiotensin II et systême nigro-néostriatal. J Physiol (Paris) 77:71-79

41. SIMPSON JB, ROUTTENBERG A (1973) Subfornical organ: site of drinking elicitation by angiotensin II. Science 181:1172-1175

42. SIRETT NE, McLEAN AS, BRAY JJ, HUBBARD JI (1977) Distribution of angiotensin II receptors in rat brain. Brain Res 122:299-312

43. STAMLER JF, RAIZADA MK, FELLOWS RE, PHILLIPS MI (1980) Increased specific binding of angiotensin II in the organum vasculosum of the laminae terminalis area of the spontaneously hypertensive rat brain. Neurosci Lett 17:173-177

44. SUGA T, SUZUKI M, SUZUKI M (1979) Effects of angiotensin II on the medullary neurons and their sensitivity to acetylcholine and catecholamines. Jap J Pharmacol 29:541-552

45. WAYNER JM, ONO T, NOLLEY D (1973) Effects of angiotensin II on central neurons. Pharmacol Biochem Behav 1:679-691

Effects of Angiotensin and Bradykinin on Neurons of the Sensomotor Cortex

N. A. Nicolov, K. V. Sudakov, J. I. Deleva, V. I. Badikov, V. V. Sherstnev, and E. A. Yumatov

INTRODUCTION

Cardiovascular homeostasis has recently been discussed from the standpoint of central peptidergic stimulation (Ganten et al., 1979; Schölkens et al., 1980). The mechanism of action of peptides like angiotensin II (ANG II), bradykinin (BK) and others on a number of neural structures has not yet been completely elucidated, however a problem demanding further clarification for example is the interrelationship between cortex and hypothalamus during elevated plasma renin levels as is the case during stress (Kvetnansky et al., 1981). Sudakov, Sherstnev and Osipovski (1976) found in mollusca neurons an ANG II effect independent of the nervous system microvasculature. These authors observed that the microiontophoretic application of ANG II to sensomotor cortical (SMC) neurons of rabbits resulted in spike activity facilitation. By extracellular registration of neuron bioelectric activity from the supraoptic and paraventricular nucleus and upon ANG II administration into the third brain ventricle Akaishi, Negoko and Kobayasi (1980) found increased neuron firing rate.

According to Huwyler and Felix (1980) 35% of the septal neurons increased spike activity after immobilization. In our earlier investigations (Nicolov et al., 1979) we observed a complex type of interaction between microiontophoretically applied ANG II and prostaglandin E_2 effects on SMC neurons in rabbits. The studies of Anissimov et al. (1979) revealed variable sensitivity of neurons from SMC, locus coeruleus and the hypothalamus after microiontophoretic administration of ANG II. Studies with microiontophoretic application of ANG II on neurons of the nigro-neostrial region (Simonett et al., 1981) gave evidence for a prevailing inhibitory effect on spike activity. The aim of the present study was to trace changes in the spike activity of SMC neurons induced by intravenous and microiontophoretic application of ANG II before and after ES-VMH and the microiontophoretic application of BK before and after ES-VMH and VLH,proceeding from the hypothesis that SMC neurons activity is influenced by the peptides in correlation with the biologic modality of hypothalamic stimuli.

Experimental Brain Research, Suppl. 4

MATERIAL AND METHODS

Experiments were carried out in 12 unanesthetized rabbits "Chinchilla" breed of both sexes with a weight of 2.3-3.5 kg. The animals were fixed in a stereotaxic apparatus and the locus of fixation was anesthetized by subcutaneous application of 1% Novocain. Extracellular spike activity of SMC neurons was registered through glass microelectrodes, filled with 2.5 M NaCl. Reliable long-duration spike activity registration was ensured by placing the electrodes in a micromanipulator fixed to the skull of the rabbits by duracryl.

The microiontophoretic administration of ANG II and BK was carried out through five-channel glass microelectrodes 2-4 μm in diameter and 5-15 MΩ resistance. Amplified action potentials were registered by a 8-channel polygraph "San-Ei" - Japan. Bipolar insulated nichrome electrodes were used for electrical stimulation. Electrical stimulation of Ventromedial Hypothalamus (ES-VMH) and Ventrolateral Hypothalamus (VLH) was exerted by rectangular impulses at a 50 Hz frequency, amplitude - 2.5 - 10 V, pulse-width - 0.1 ms, duration - 3 s. In some of the experiments the stimulation had a duration of 10 s. Experiments with ANG II and BK were done following control investigations for testing the emotional reactions of the animals through ES-VMH and ES-VLH according to the method of selfstimulation and escape reaction.

In all experiments histological control of electrodes localization was done.

In the first series of experiments ANG II (Ciba) reconstituted in saline was injected into the femoral vein at doses 200 ng/kg and 1000 ng/kg. Control experiments were carried out with saline. In the second series of experiments with microiontophoretic administration of ANG II and BK, ANG II (Serva): 2.5×10^{-3}M, pH=5.5 and BK: 2×10^{-3}, pH=6.0 were used. Controls were done with saline. The results were presented as percentage of the control value.

RESULTS

A. Changes in SMC neurons spike activity upon intravenous administration of ANG II and ES-VMH.

Fig. 1 presents SMC neurons reactions upon 200 ng/kg ANG II and ES-VMH. The combined action of both factors results first in an activatory reaction, i.e. spike activity increase, this is followed by a spike activity inhibitory reaction after which neurons giving a phase reaction are observed. The other section of the figure presents the effects of the combined action of 1000 ng/kg ANG II and ES-VMH. In this experimental set up there is a characteristic rise in the number of phase reacting neurons compared to the number of the inhibited ones upon increasing the dose of ANG II (n=15).

Fig. 2 shows that the combined action of both pressor factors, namely 200 ng/kg ANG II i.v. and ES-VMH applied separately results in different effects: 200 ng/kg ANG II do not evoke significant alterations in the spike activity of the studied SMC neuron, ES-VMH inhibits the spike activity of the same neuron, whilst ES-VMH applied at the same time with 200 ng/kg ANG II activates the firing frequency of the neuron.

B. Changes in SMC neurons spike activity upon microiontophoretic application of ANG II and BK and upon ES-VMH and ES-VLH.

Fig. 3 shows the changes in the sensitivity of a single sensomotor cortical neuron to ANG II and BK after positive and negative hypothalamic stimulation.

Negative hypothalamic stimulation evokes a sharp increase in the firing frequency of the studied neuron following the microiontophoretic administration of ANG II and no significant reactivity to BK. Positive hypothalamic stimulation does not result in significant reactivity changes of the cells towards the introduced substances. Following three consequent positive hypothalamic stimulations a distinct response to ANG II was marked.

The effects of the microiontophoretic administration of ANG II before and after ES-VMH were equally studied.

Fig. 4 presents a record of the changes in the impulse activity following 2 consequent ANG II injections. The first injection does not induce significant alterations, while following the second dose a significant spike activity inhibition occurs. The lower section of the figure shows that ES-VMH evokes a sharp increase in the spike activity of the same SMC neuron and a certain consequent inhibition, but the microiontophoretic administration of ANG II following the stimulation results in a marked increase in the firing frequency of the studied cell.

The effects of bradykinin on the spike activity of SMC neurons after ES-VLH are illustrated in Fig. 5, which shows a SMC neuron reaction to the microiontophoretic introduction of bradykinin. ES-VLH evokes marked spike activity increase and bradykinin administration during hypothalamic stimulation results in a sharp firing frequency facilitation.

Table 1 presents the results from the reactions of 44 SMC neurons to the microiontophoretic administration of ANG II and of 30 neurons to BK.

Table 1

Substance Applied	Total of units	Character of response: Increase of unit activity	Decrease of unit activity	No response
Angiotensin II	44	14 - 31.81%	13 - 29.50%	17 - 38.69%
Bradykinin	30	11 - 36.66%	7 - 23.33%	12 - 40%

ANG II evokes spike activity increase in 31.81% of SMC neurons, spike activity decreases in 29.50% and no reaction in 38.69%. BK application results in spike activity increases in 33.66% of SMC neurons, decrease in 23.33% and no reaction in 40%.

Table 2 depicts the results of SMC neurons reactivity to the microiontophoretic administration of ANG II and BK following ES-VMH, inducing negative emotional condition and ES-VLH, inducing positive behavioural emotional condition.

Table 2

	Positive stimulation		Negative stimulation	
Unit response	Angiotensin II	Bradykinin	Angiotensin II	Bradykinin
Increase	5 - 18.51%	6 - 33.33%	11 - 40.74%	4 - 22.20%
Decrease	2 - 7.41%	-	-	-
Reversive Response	2 - 7.41%	3 - 16.66%	4 - 14.81%	2 - 11.11%
Emergence of response	3 - 11.11%	4 - 22.22%	4 - 14.81%	2 - 11.11%
No change in Reaction to Peptide	15 - 55.55%	5 - 27.77%	8 - 29.62%	10 - 55.55%
T o t a l	27 - 99.99%	18 - 99.98%	27 - 99.98%	18 - 99.99%

Effects of ANG II after ES-VLH ("positive" emotional stimulation) is shown in Table 2. ES-VLH diminishes the number of neurons, reacting with spike activity increase (Table 1) upon separate ANG II application from 31.8% to 18.51%. There is a marked increase in the number of non-reacting neurons: with separate microiontophoretic application of ANG II 38.69% did not react which compares to 55.50% during ES-VLH.

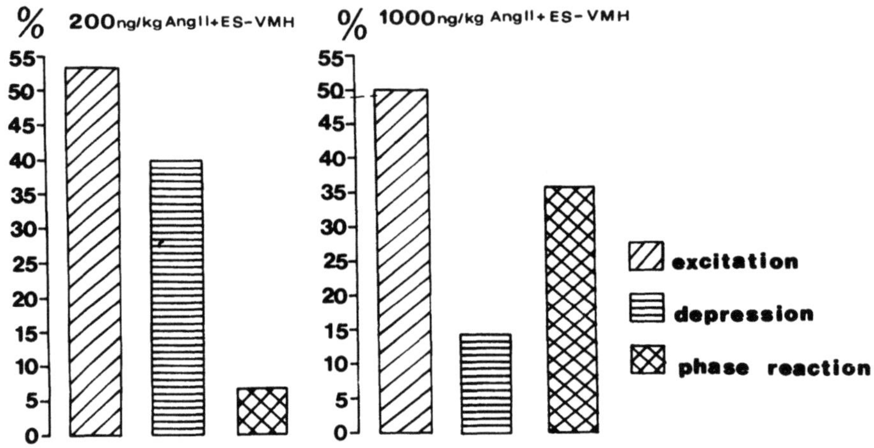

Fig. 1. Mutual action of ANG II + ES-VMH and reaction types of SMC neurons. On the ordinate is the % of reacting neurons. At 200 ng/kg + ES-VMH the number of neurons reacting by a decrease in spike activity is considerable while at 1000 ng/kg ANG II + ES-VMH the number of inhibited neurons is diminished compared to the number of phase reacting cells /+/ - or -/+/

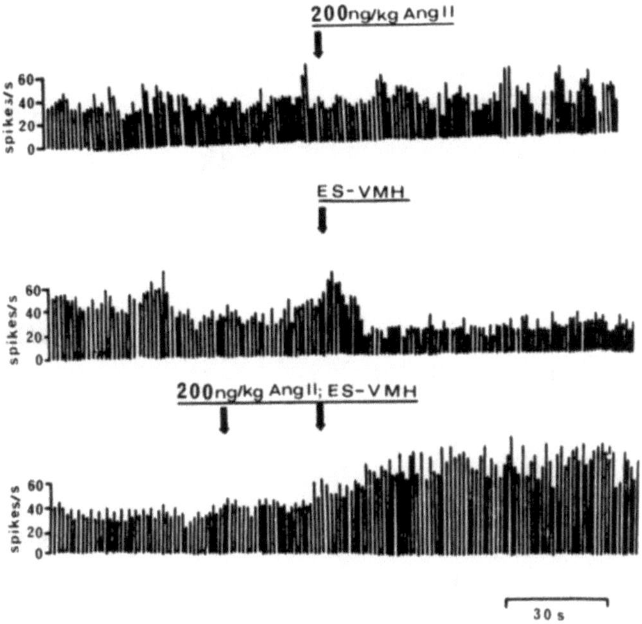

Fig. 2. Effect of 200 ng/kg ANG II, injected intravenously on spike activity. The rise in arterial pressure by 30 mm Hg is parallel to a certain spike activity facilitation in a SMC neuron, ES-VMH induces a similar arterial pressure elevation and spike activity inhibition in the neuron, while the injection of 200 ng/kg ANG II followed by ES-VMH within 30 s evokes an abrupt spike activity facilitation of the same unit and arterial pressure elevation ($p < 0.05$)

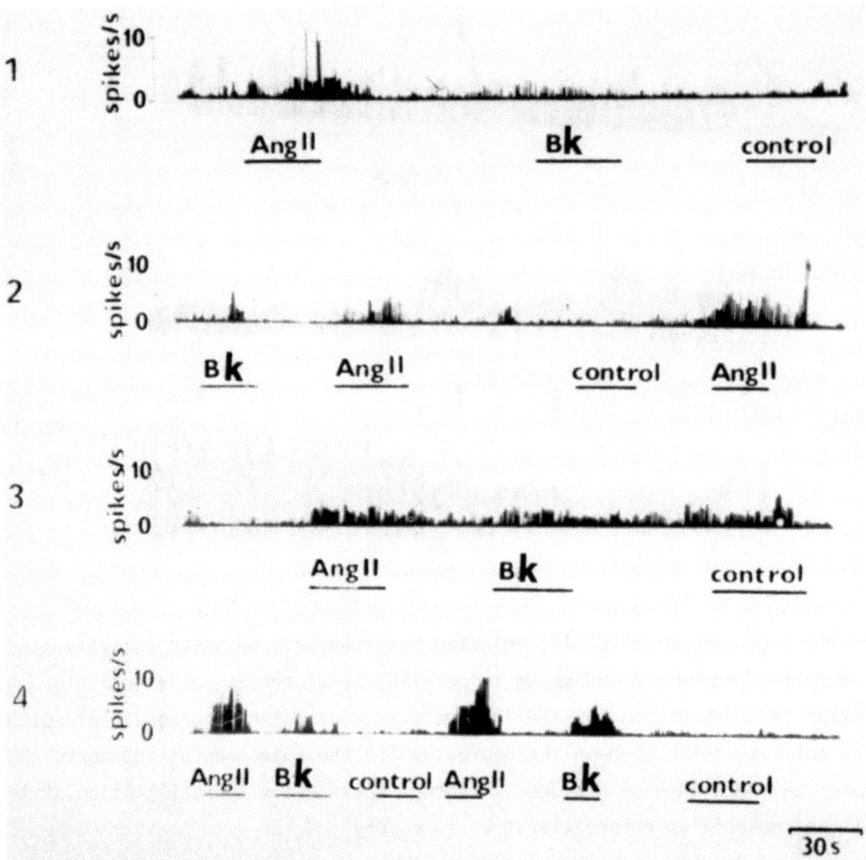

Fig. 3. Sensitivity of a single sensomotor cortical unit to ANG II and BK after positive and negative hypothalamic stimulations.

1. Unit's response to microiontophoretic application of ANG II, BK and sodium chloride control solution before hypothalamic stimulation.

2. 10 sec after negative hypothalamic stimulation (ES-VMH). Sharp increase of unit response to ANG II.

3. 10 sec after positive hypothalamic stimulation (ES-VLH). No response to ANG II and BK.

4. 10 sec after three consequent positive hypothalamic stimulations. No response to BK and distinct response to ANG II (bold line indicates the period of application of substances)

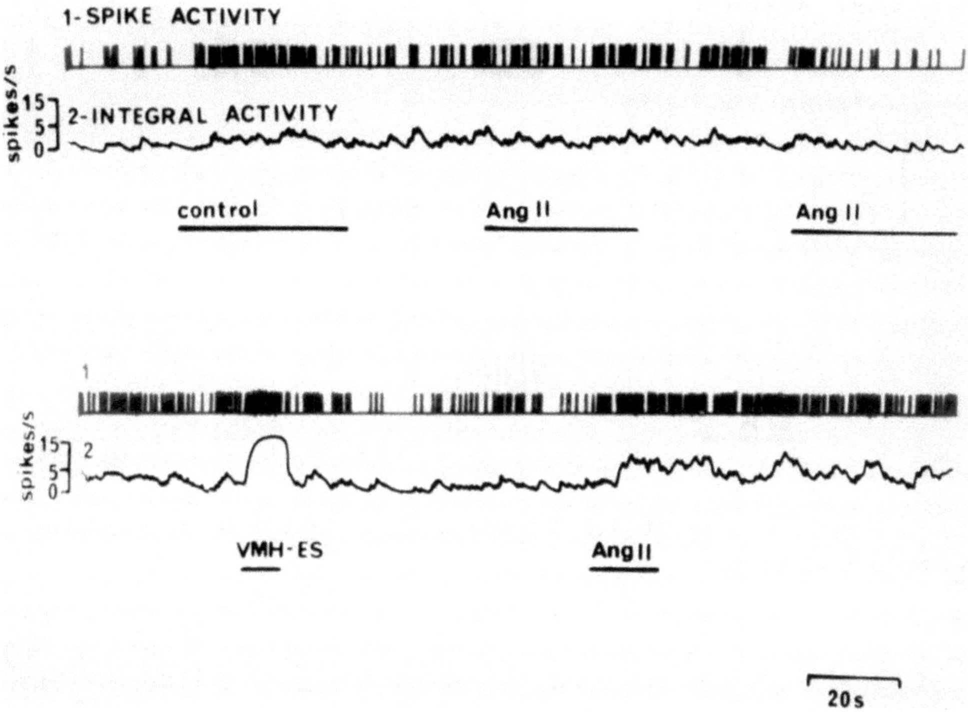

Fig. 4. Changes in the sensitivity of a sensomotor cortical neuron to ANG II after negative hypothalamic stimulation

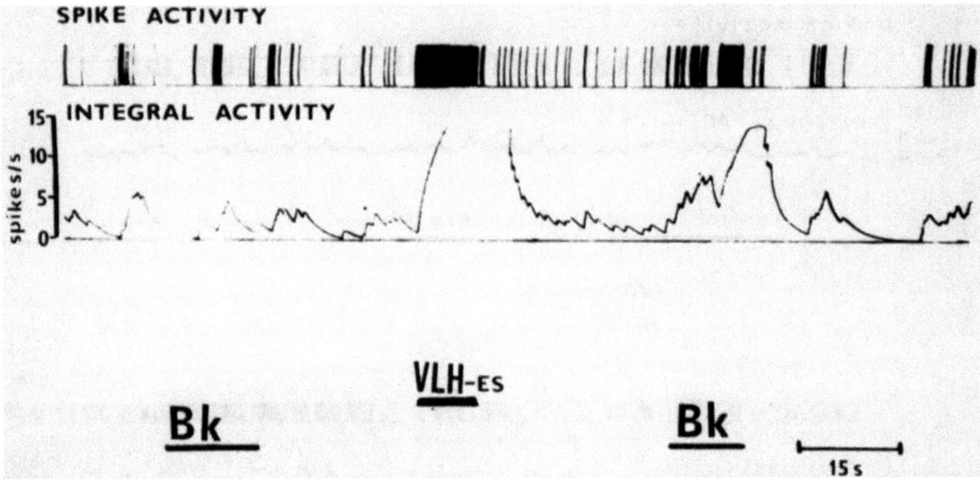

Fig. 5. Changes in the sensitivity of single sensomotor unit to BK after positive hypothalamic stimulation

Effect of ANG II after ES-VMH ("negative" emotional stimulation). The number of neurons reacting with an activated firing frequency rose from 31.81% to 40.74% upon ANG II application after ES-VMH, with a parallel decrease in the number of non-reacting SMC neurons from 38.69% to 29.62%.

These results suggest that ANG II after ES-VLH decreases the number of neurons reacting with spike activity facilitation while ES-VMH increases the number. The number of non-reacting neurons upon ANG II application rises after ES-VLH and falls after ES-VMH.

Negative emotional stimulation induced by ES-VMH increases the number of reacting neurons mainly by activating their spike frequency, while the positive emotional condition induced by ES-VLH decreases the number of neurons activated by the micro-iontophoretic application of ANG II and increases the number of cells "non-sensitive" to ANG II.

Effect of bradykinin after ES-VLH ("positive" emotional stimulation):

ES-VLH results in a diminution of the number of neurons non-reacting to BK from 40% when BK is applied alone to 27.77% after application following ES-VLH. The number of neurons reacting with activation remains within the same limits. After hypothalamic stimulation the number of units varying their responses to BK was considerably higher as compared to ANG II (72.3% and 44.4% respectively). For example, increase of unit firing to BK after positive hypothalamic stimulation (ES-VLH) was observed in 33.3% of the units, and to ANG II 18.5%. Twenty-to per cent of the units earlier exhibiting no reaction to peptides responded to BK and 11.1% to ANG II.

Effect of bradykinin after ES-VMH ("negative" emotional stimulation): Negative emotional stimulation induced by ES-VMH results in a decrease of the number of activated neurons from 36.66% following microiontophoretic application after ES-VMH; 2 units which were non-sensitive before became reactive. The number of non-reacting neurons rose from 40% following separate BK application to 55.55% after ES-VMH.

It becomes evident from these results that ES-VMH which is a model of negative emotional condition inhibits SMC neurons spike activity towards BK and increases the number of units non-reacting to BK.

Thus, negative hypothalamic stimulation more often influenced the responses to ANG II, whereas positive hypothalamic stimulation changes the reaction to BK. It is necessary to emphasize that the response to the peptides might considerably vary after several repeated stimulation of hypothalamic emotiogenic regions.

DISCUSSION

ANG II is the most potent vasoactive agent in the organism. It does not pass the blood brain barrier and exerts its vasoactive effect partly through sympathetic neuron activation, interference with noradrenaline re-uptake at the level of sympathetic nerve endings and partly through direct constrictor action on the smooth muscle cells of blood vessels. The role of angiotensin in the pathogenesis of arterial hypertension has been shown to be of importance also through its direct effects on the brain where the blood brain barrier is deficient.

A new dimension in the understanding of the pathogenesis of hypertension emerged with the discovery of an intrinsic renin-angiotensin system in the brain by Ganten et al. (1971), Ganten, Boucher and Genest (1971), Fischer-Ferraro et al. (1971). All compounds of this enzyme peptide system were found in the brain giving grounds to the assumption that stress might increase the generation of pressor substance in the brain. This opened up a new line of research namely that central peptidergic stimulation might contribute to the development of hypertension (Ganten et al., 1979). This theory includes the pressor effects of cerebral bradykinin, enkephalins, substance P and other vasopressor peptides. These studies, however, did not solve the problem about how a peptidergic mechanism may switch arterial blood pressure control centers to an activation of brain pressor systems from protective depressor mechanisms. The present study shows that the intravenous application of ANG II exerts its action on SMC neurons through a number of relay stations resulting in a depression of firing frequency. A certain part of these neurons are non-sensitive to angiotensin or give phase reactions. This suggests that neurons of the neocortex are biologically multimodal i.e. they have alternative receptors for different biologic stimulations. The latter provides the ability of neocortical neurons for participation in adaptive reactions of different biological sequence thus ensuring the homeostasis of the organism.

The ventromedial hypothalamic stimulation has been found to evoke a negative emotional reaction of the organism, i.e. this is not only a satiety center (King, 1980; Sudakov, 1976; Yumatov et al., 1978) but also a center of passive defence reactions, when escaping is not possible, and a reaction of arterial pressure elevation occurs. VMH stimulation induces arterial hypertension and tachycardia (Gauntier, Reis, Nathan, 1981) while VMH ablation prevents arterial pressure elevation even in renal hypertension (Johnson, Buggy, Fink, Brody, 1981). These findings give ground to the reevaluation of VMH functions and its perception as a center of negative emotional stress able to reinforce an unimodal pressor reactivity on the other neural formations. Our investigations with extracellular recordings of SMC neurons activity upon intravenous and microiontophoretic application of ANG II indicates that the combined action of ANG II and ES-VMH induced a maximal sensitivity of SMC neurons, reacting mainly by a spike activity increase. A pressor system arises in the organism where SMC neurons influenced by ES-VMH react at a higher level of activation to plasma ANG II through a number of relay stations or to the immediate microiontophoretic action. This brings forth the problem about an uniform pressor neuro-humoral dominant during emotional stress induced by ES-VMH.

The application of BK, the central effect of which is also pressor, evokes a more prominent response of SMC neurons after ES-VLH. ES-VLH induces a positive emotional reaction (selfstimulation), which modulates the reactivity of SMC neurons to the microiontophoretic introduction of BK.

These studies suggest that there is no isolated neuron function per se (Anokhin, 1974) but that it is a component of a determined biologically motivated reaction. The increase in the level of plasma or brain ANG II during negative emotional stress stimulation exerted by ES-VMH augments cortical neurons sensitivity to ANG II, while positive emotional stimulation by ES-VLH augments SMC neurons sensitivity to the microiontophoretically applied BK.

The following conclusions can be drawn from our results:

1. The combined action of ANG II and ES-VMH increases the number of SMC neurons reacting with spike activity facilitation to the intravenous application of ANG II.
2. ES-VMH increases the number of neurons reacting with spike activity facilitation to microiontophoretic application of ANG II to SMC neurons, while ES-VLH increases the number of neurons reacting to the microiontophoretic application of BK.

In summary, experiments were carried out in two series. First, the effects of electric stimulation of the ventromedial hypothalamus (ES-VMH) on neurons of the sensomotor cortex (SMC) and changes in the spike activity of SMC neurons under the action of ANG II and ES-VMH was studied. The combined action of these factors has been found to increase the sensitivity of SMC neurons and to induce an increase in their spike activity. SMC neurons react mostly by an activation in their discharge activity following ES-VMH during systemic ANG II infusion.

In a second series of experiments the effects of ANG II and bradykinin (BK) microiontophoretically introduced into SMC neurons was studied before and after electric stimulation of the ventromedial (VMH) and ventrolateral hypothalamus (VLH). A preliminary test indicated that ES-VMH evokes a negative emotional state and ES-VLH a positive one. It has been found that ES-VMH in anesthetized rabbits increases the number of SMC neurons reacting to ANG II application by a spike activity increase, while ES-VLH increases the number of neurons reacting by a spike activity increase to BK application. Thus, positive emotional stimulation induced by ES-VLH increases SMC neurons sensitivity to BK, while negative emotional stimulation induced by ES-VMH increases the sensitivity of the neurons to ANG II.

REFERENCES

1. AKAISHI T, NEGORO H, KOBAYASI S (1980) Responses of paraventricular and supraoptic units to Angiotensin II, Sar^1, Ala^8-Angiotensin II and Hypertonic NaCl administered into the cerebral ventricle. Brain Res 188:499-511

2. ANISSIMOV JZ, DELEVA JI, NICOLOV NA, SUDAKOV KV, SHERSTNEV VV (1980) Neuronal sensitivity of some brain regions to Angiotensin II in rabbits. Agressologie 21: 171-175

3. ANOKHIN PK (1974, in Russian) Systematic analysis of the integrative activity of the neuron. Successes in physiological sciences 5 (2):5-92

4. FISCHER-FERRARO C, NAHMOD VE, GOLDSTEIN DJ, FINKIELMAN S (1971) Angiotensin and renin in rat and dog brain. J Exp Med 133:353-361

5. GANTEN D, BOUCHER R, GENEST R (1971) Renin activity in brain tissue of puppies and adult dogs. Brain Res 33:557-559

6. GANTEN D, MARQUEZ-JULIO A, GRANGER P, HAYDUK K, KARSUNKY KP, BOUCHER R, GENEST J (1971) Renin in dog brain. Am J Physiol 221:1733-1737

7. GANTEN D, UNGER TH, ROCKHOLD R, SCHAZ K, SPECK G (1979) Central peptidergic stimulation: its possible contribution to blood pressure regulation. In: Radioimmunoassay of drugs and hormones in cardiovascular medicine. Eds. ALBERTINI A, Da PRADA M, PESKAR BA, Elsevier/North-Holland Biomedical Press, pp 33-43

8. GAUNTIER P, REIS DJ, NATHAN MC (1981) Arterial hypertension elicited by lesion or by electrical stimulations of the rostral hypothalamus in the rat. Brain Res 211:91-105

9. HUWYLER T, FELIX D (1980) Angiotensin II - sensitive neurons in septal areas of the rat. Brain Res 195:187-195

10. JOHNSON AK, BUGGY J, FINK GD, BRODY MJ (1981) Prevention of renal hypertension and of the central pressor effect of Angiotensin by ventromedial hypothalamic ablation. Brain Res 205:255-264

11. KING BM (1980) A re-examination of the ventromedial hypothalamic paradox. Neuroscience and Biobehavioural Reviews 4:151-160

12. KVETNANSKY R, BELOVA TI, IINDRA A, OPRSHALOVA Z, DUSHKIN V (1981) Changes of the blood plasma renin activity in August and Wistar rats in emotional stress. Fiziol Jh USSR 67 (6):911-915

13. NICOLOV NA, DELEVA JI, SUDAKOV KV, ANISSIMOV IZ, GROMOV AI, SHERSTNEV VV (1979) Characteristics of the interaction between the effects of prostaglandin E_2 and Angiotensin II on cortical neurons. 4th Int. Prostaglandin Conf., Washington, D.C., May 27-31, 1979, p 86

14. SCHÖLKENS BA, STEINBACH R, JUNG W, GANTEN D (1980) Induction of Angiotensin formation in brain: biological effects and pharmacological interferences. In: Enzymatic Release of Vasoactive Peptides. Eds. GROSS F, VOGEL G, Raven Press, New York, pp 365-381

15. SIMONNET G, GIORGUIEFF MF, CARAYON A, BIOULAC B, CESSELIN F, GLOWINSKI J, VINCENT JD (1981) Angiotensin II et systeme nigro-neostriatal. J Physiol, Paris, 77:71-79

16. SUDAKOV KV (1976) Emotional stress and arterial hypertension. Medizina i zdravoohranenie, Moscow (in Russian 1976), pp 1-116

17. SUDAKOV KV, SHERSTNÉV VV, OSIPOVSKY SA (1976) Direct effect of Angiotensin II on the central neurons. Bulletin Experimental Biology, Medizina, Moskow, 82/8:899-902

18. YUMATOV EA, LAVROVA EV, POLESKAJIA MM (1978) Neurochemical basis for the negative emotional reactions. Tezisi dokladov IV Vsesojuznogo seminara po rasvitiv obstei teorii functionalnich system. In: Sistemnie mechanizmi emozionalnich reakzii, Moscow, pp 25-27

The Brain Renin Angiotensin System: Some Unresolved Problems

I. A. Reid and M. S. Brownfield

INTRODUCTION

The existence of a brain renin-angiotensin system was originally proposed when it was observed that the components required for the formation of angiotensin II are present in the brain (7). Interest in this proposal was stimulated by reports that angiotensin II exerts a variety of endocrine, cardiovascular and behavioral effects when injected into the central nervous system. The proposal proved to be attractive to many investigators and today it is frequently assumed that there is a brain renin-angiotensin system and that this system participates in the regulation of blood pressure, water balance and other variables.

This concept, however, has not gone unchallenged. In 1976, it was observed that most of the renin-like activity in brain extracts is due to a non-specific action of the acid protease, cathepsin D (3). Other problems were identified and considerable debate concerning the possible existence of a brain renin-angiotensin system ensued (21,24,25). The purpose of this paper is to review some of the recent developments in this field, showing where progress has been made and identifying problem areas where deficiencies still exist.

RENIN-LIKE ACTIVITY IN BRAIN

The existence of a brain renin-angiotensin system was first seriously questioned when it was discovered that the renin-like activity measured in brain extracts at low pH in vitro was due to the lysosomal acid protease cathepsin D (3,9). There were speculations that this enzyme might function as an angiotensin-forming enzyme in vivo but, for a variety of reasons (25), this seemed unlikely. The cathepsin D problem was partially circumvented in 1978 when Hirose and associates (10) and Ganten et al. (8) demonstrated that this enzyme is not the only brain enzyme that can form angiotensin. It has now been established that the brain actually contains several acid and neutral proteases which are capable of forming angiotensin (4,11,19).

Experimental Brain Research, Suppl. 4

One of the neutral proteases closely resembles renal renin (4,11,19). It has a molecular weight and pH optimum similar to that of the renal enzyme and, like renal renin, it is a glycoprotein. The enzyme is present in an active form as well as in an inactive form which can be activated by trypsin. The activity of the enzyme can be inhibited by an antibody raised against purified renal renin. The isoelectric focusing profile of the brain enzyme differs from that of renal renin and this has been interpreted as evidence that the enzyme is synthesized within the brain.

The regional distribution of the renin-like enzyme has been determined (11). The enzyme appears to be widely distributed, the highest concentration occuring in the choroid plexus, adenohypophysis and pineal. Immunoreactive renin in these three areas has been measured by direct radioimmunoassay (11). The distribution of immunoreactive renin in the brain has also been determined using immunocytochemistry (2,6,13, 35). Some investigators report that the immunoreactivity is distributed widely throughout the central nervous system; others report a more restricted distribution. The immunocytochemical studies have provided evidence that, at least in some parts of the brain, immunoreactive renin is present within neurons. The distribution of immunoreactive renin in the brain appears to be quite similar to the distribution of immunoreactive cathepsin D (38).

Despite this progress, questions and problems still remain. There are discrepancies concerning the distribution of immunoreactive renin in the brain. Results obtained with immunocytochemistry do not always agree with those obtained with biochemical methods and, indeed, it is not known if these methods identify the same enzyme. Despite the demonstration that renin-like activity can be separated from cathepsin D, it must be emphasized that most of the renin-like activity measured in brain extracts at low pH is due to cathepsin D and that the levels of non-cathepsin D renin-like activity are low. For example, Hirose et al. (10) reported a concentration of approximately 250 pg ANG I per gram tissue per hour. This is much less than the concentration of renin normally present in plasma. It could be argued that this small amount of renin activity might be important if it were localized in discrete areas; however, as noted above, it has been reported that the enzyme is actually distributed diffusely throughout the central nervous system (35).

Finally, it still remains to be demonstrated that the renin-like enzyme functions as an angiotensin-forming enzyme in vivo. Attempts to demonstrate such activity have involved injecting angiotensinogen into the brain, but these have yielded unconvincing or negative results (12,27). Angiotensin II formation has been observed following intraventricular injections of the synthetic tetradecapeptide renin substrate (25) but this apparently resulted from cleavage of the molecule by converting enzyme rather than by renin (34). It is probably a fair statement that the only time renin activity has been convincingly demonstrated in the brain in vivo is when renin has been injected into the central nervous system (27,31). The question of whether angiotensin formation normally occurs in the brain is pursued below.

ANGIOTENSINOGEN

Perhaps the strongest piece of evidence in favor of a brain renin-angiotensin system is that angiotensinogen (renin substrate) is present in the central nervous system. Initially, it appeared that only trace amounts of the substrate were present (7). However, in 1975, a relatively high concentration of angiotensinogen (205 ng/ml) was discovered in dog cerebrospinal fluid (29). Subsequent studies revealed that angiotensinogen is present in brain tissue and cerebrospinal fluid of several species including man (14,18,20,23,37).

The biochemical properties of brain and cerebrospinal fluid angiotensinogen have been studied and compared to those of the plasma substrate (14,23,28). No significant differences in molecular weight or electrophoretic mobility have been found. Small differences have been found with isoelectric focusing and concanavalin A chromatography; these differences probably reflect differences in carbohydrate composition. Recently it was reported that in man, cerebrospinal fluid angiotensinogen is immunologically different from plasma angiotensinogen (14).

The regional distribution of angiotensinogen in brain has been studied.It is widely distributed throughout the brain but regional differences in concentration have been reported (18). Cell fractionation studies have shown that in the midbrain and forebrain of the dog, angiotensinogen is located principally in the soluble fraction (20). Calculations based on the concentration of angiotensinogen in cerebrospinal fluid suggest that all of the angiotensinogen in the brain could be in the extracellular space. In order to localize angiotensinogen in the brain more precisely, we have recently commenced immunocytochemical studies using an antibody prepared against purified rat plasma angiotensinogen by Dr. Pierre Corvol. Our preliminary results indicate the presence of angiotensinogen in glial cells, possibly astrocytes, in the thalamus, hypothalamus and other areas. Further studies are in progress.

Factors affecting angiotensinogen concentration in the central nervous system have also been investigated. Angiotensinogen concentration in cerebrospinal fluid in the dog did not change when plasma angiotensinogen concentration was increased by nephrectomy or dexamethasone treatment (25). In the rat, decreases in the angiotensinogen content of some brain regions occured following bilateral adrenalectomy; these could be reversed by treatment with adrenal corticosteroids (37).

Despite this progress, the origin of the angiotensinogen in the central nervous system is still not known. Possibilities that have been considered include entry from the blood during the process of cerebrospinal fluid formation and local synthesis within the brain. Evidence in favor of the former possibility includes the fact that plasma and brain angiotensinogen are biochemically quite similar. As noted above, small differences between central and peripheral angiotensinogens have been described but it is conceivable that such differences arise during transport of angiotensinogen into the cerebrospinal fluid. Failure to find angiotensinogen in granule fractions of brain (20) is also consistent with a peripheral origin of the protein.

Evidence in favor of synthesis of angiotensinogen within the brain includes the observation that cerebrospinal fluid angiotensinogen concentration does not change when plasma angiotensinogen concentration changes (25) and that radioiodinated angiotensinogen injected intravenously does not appear in cerebrospinal fluid (20). The differences in the properties of plasma and brain angiotensinogens mentioned above, as well as the results of our immunocytochemical studies, are also consistent with the local synthesis hypothesis.

In order to determine more directly if the brain has the ability to synthesize angiotensinogen, we recently incubated homogenates of rat brain in vitro (33). We found that the concentration of angiotensinogen in the incubation medium increased in a linear fashion during a 3 hour incubation at 37^{0}C. When the incubation was performed at 4^{0}C, no increase occured. However, the increase in angiotensinogen concentration was not blocked by actinomycin D, cycloheximide or puromycin, suggesting that it did not result from secretion of newly synthesized protein. Thus the question of whether angiotensinogen is synthesized within the brain remains open for further study.

Another important question that remains to be answered is whether or not brain angiotensinogen is normally utilized for the formation of angiotensin. The angiotensinogen can certainly be cleaved by exogenously administered renin (27,31) but it is not clear if any of the brain enzymes with renin-like activity ever release angiotensin in vivo. This question is discussed later in this review.

CONVERTING ENZYME

Converting enzyme is a ubiquitous enzyme, present in every tissue that has been studied. It is therefore not surprising that it is also present in the central nervous system. In most studies of converting enzyme in the brain, enzyme activity has been assessed using artificial substrates rather than angiotensin I. It is important to note that the results of converting enzyme assays based on artificial substrates frequently differ from those which utilize the natural substrate (1). Results obtained with artificial substrates should therefore be interpreted with caution.

Recently we have localized converting enzyme in the brain with immunocytochemistry, using an antibody raised against rabbit pulmonary converting enzyme by Dr. Richard Soffer. Immunoreactive converting enzyme was found on the ventricular surface of epithelial cells of the choroid plexus, in the subfornical organ and in capillary endothelial cells throughout the brain. No converting enzyme was detected in brain tissue other than the subfornical organ. Similar results have recently been reported by Rix et al. (30).

The localization of converting enzyme in choroid plexus raises further questions concerning the hypothesis that angiotensin II is formed within the brain. As discussed earlier, immunoreactive renin appears to be present in neurons in several brain regions. Any angiotensin I formation that does occur in the brain would presumably take place in or near these neurons. How then is angiotensin II formed? Angiotensin I might somehow be transported to the choroid plexus and converted to angiotensin II there, but this seems most unlikely, especially since the concentration of angiotensin I and angiotensin II in cerebrospinal fluid is verly low (see below). Angiotensin II might be formed by a pathway that does not require converting enzyme but there is no evidence for this at the present time. It is possible that there is a form of converting enzyme in brain which does not cross react with the antibody that was used in the localization studies. Again, this seems unlikely since it is known that antisera raised against pulmonary converting enzyme effectively inhibit brain converting enzyme activity (22). Finally, it is worth noting that converting enzyme has a wide substrate specificity and it is possible that the brain enzyme functions in the metabolism of peptides such as the enkephalins (36), bradykinin (15) and substance P (17) and plays no role in the putative formation of angiotensin II.

ANGIOTENSIN

The most direct test of the hypothesis that angiotensin II is formed in the brain is to determine if the peptide is present in the central nervous system. Several attempts have been made to detect and localize angiotensin in brain and cerebrospinal fluid but its presence has still not been conclusively demonstrated.

Cerebrospinal Fluid

Several investigators have used radioimmunoassays in attempts to measure angiotensin II concentration in cerebrospinal fluid but variable and conflicting results have been obtained (25). The material being assayed has usually not been characterized beyond the fact that it cross reacts with angiotensin II antibodies. Recently, Semple and associates have investigated the nature of the immunoreactive angiotensin II in human cerebrospinal fluid (32). Low levels of immunoreactive angiotensin II were detected with two antibodies but the concentration was higher with one antibody than the other. Paper chromatography revealed that the immunoreactivity measured with one of the antisera was not angiotensin I, angiotensin II or the 2-8, 3-8, or 4-8 fragments of angiotensin II. It was therefore concluded that the immunoreactive angiotensin II in human cerebrospinal fluid is an immunoassay artifact.

Brain

Bioassay and radioimmunoassay techniques have been used in attempts to detect and measure angiotensin II in extracts of brain. Some investigators have been unable to find angiotensin; others have reported high levels of angiotensin-like activity. As discussed previously (25,26),the methods that have been used are subject to a variety of problems including non-specificity of bioassay methods and destruction of tracer by angiotensinases in radioimmunoassay procedures. When care is taken to avoid these problems, angiotensin-like activity in brain extracts is generally reduced to low undetectable levels (25,37).

Recently Meyer and associates (D. K. Meyer, personal communication) measured angiotensin II-like immunoreactivity in rat brain using five extraction techniques and three angiotensin II antibodies, two of which have given positive results in immunocytochemical studies (see below). They found low levels of immunoreactivity ($<$80 pg/g tissue) in extracts of whole brain. However, characterization with high pressure liquid chromatography revealed that the immunoreactivity was neither angiotensin II nor angiotensin III. It was therefore concluded that these peptides are not synthesized in the brain.

Another approach that has been used is immunocytochemistry. Using this technique, Fuxe and associates localized immunoreactive angiotensin II in rat brain (5). We have also used this procedure in attempts to localize angiotensin but our initial attempts yielded negative results. However, when we utilized the antibodies used by Fuxe et al. we did observe staining of nerve fibers and cell bodies similar to that reported by Fuxe et al. Generally comparable results have been obtained with another antibody by Kilcoyne and associates (16).

These are important findings and support the concept of a brain renin-angiotensin system. However, problems exist. Firstly, although we obtained positive results with the antibodies of Fuxe et al., we have consistently obtained negative results with another twelve antibodies which were raised in three different laboratories. These antibodies all have high titers and work well in radioimmunoassays. There is a precedent for this discrepancy since other investigators have observed that antibodies useful in radioimmunoassays cannot always be used for immunocytochemistry. Nevertheless, it is worrying that of the more than fifteen angiotensin II antibodies that have been tested, only three have given positive results.

Secondly, immunocytochemical methods can never prove that a substance exists in a tissue. Extraction and biochemical characterization is required and, as discussed above, this has not been achieved. This is particularly important in immunocytochemical studies of angiotensin II, since the antibodies that have been utilized to date also crossreact with angiotensin I, angiotensin III and the synthetic tetradecapeptide renin substrate. More definitive evidence is required before it can be concluded that angiotensin II is present in the brain.

CONCLUSIONS

The enzymes and substrates required for the formation of angiotensin II - renin, angiotensinogen and converting enzyme - are present in the central nervous system. During recent years, considerable progress has been made concerning the nature and localization of these substances. Despite this progress, however, problems remain and it is still not clear if these components interact in vivo to form angiotensin II. Indeed, it has not been proven that angiotensin II is actually present in the central nervous system. These problems are not presented as evidence against the hypothesis that angiotensin II is formed in the brain. However, until they are satisfactorily resolved, conclusions concerning the existence of a brain renin-angiotensin system should be made with caution.

ACKNOWLEDGEMENT

Part of the research described in this review was supported by USPHS Grant AM06704.

REFERENCES

1. BAKHLE YS (1974) In: PAGE IH, BUMPUS FM (eds) Angiotensin. Springer-Verlag, New York, p 41

2. CELIO MR, CLEMENS DL, INAGAMI T (1980) Renin in anterior pituitary, pineal, and neuronal cells of mouse brain. Immunohistochemical localization. Biomed Res 1: 427-431

3. DAY RP, REID IA (1976) Renin activity in dog brain: enzymological similarity to cathepsin D. Endocrinology 99:99-105

4. DZAU VJ, BRENNER A, EMMETT N, HABER E (1980) Identification of renin and renin-like enzymes in rat brain by a renin-specific antibody. Clin Sci 59:45s-47s

5. FUXE K, GANTEN D, HOKFELT T, BOLME P (1976) Immunohistochemical evidence for the existence of angiotensin II-containing nerve terminals in the brain and spinal cord in the rat. Neurosci Letters 2:229-234

6. FUXE K, GANTEN D, HOKFELT T, LOCATELLI V, POULSEN K, STOCK G, RIX E, TAUGNER R (1980) Renin-like immunocytochemical activity in the rat and mouse brain. Neurosci Letters 18:245-250

7. GANTEN D, MARQUEZ-JULIO A, GRANGER P, HAYDUK K, KARSUNKY KP, BOUCHER R, GENEST J (1971) Renin in dog brain. Am J Physiol 221:1733-1737

8. GANTEN D, SPECK G (1980) The brain renin-angiotensin system: a model for the synthesis of peptides in the brain. Biochem Pharmacol 27:2379-2389

9. HACKENTHAL E, HACKENTHAL R, HILGENFELDT U (1978) Isorenin, pseudorenin, cathepsin D and renin: A comparative enzymatic study of angiotensin-forming enzymes. Biochim Biophys Acta 522:574-588

10. HIROSE S, YOKOSAWA H, INAGAMI T (1978) Immunochemical identification of renin in rat brain and distinction from acid proteases. Nature 274:392-393

11. HIROSE S, YOKOSAWA H, INAGAMI T, WORKMAN RJ (1980) Renin and prorenin in hog brain: Ubiquitous distribution and high concentration in the pituitary and pineal. Brain Res 191:489-499

12. HOFFMAN WE, SCHELLING P, PHILLIPS MI, GANTEN D (1976) Evidence for local angiotensin formation in brain of nephrectomized rats. Neurosci Letters 3:299-303

13. INAGAMI T, CLEMENS DL, CELIO MR, BROWN A, SANDRU L, HERSCHKOWITZ N, HOFFMAN LH, KASSELBERG AG (1980) Immunohistochemical localization of renin in mouse brain. Neurosci Letters 18:91-98

14. ITO T, EGGENA P, BARRETT JD, KATZ D, METTER J, SAMBHI MP (1980) Studies on angiotensinogen of plasma and cerebrospinal fluid in normal and hypertensive subjects. Hypertension 2:432-436

15. KARIYA K, OKAMOTO H, KAKIMOTO M, TSUDA Y, OKADA Y (1981) Inhibition of angiotensin converting enzyme of rat brain with bradykinin and its fragments. Biochem Biophys Res Commun 100:31-36

16. KILCOYNE MM, HOFFMAN DL, ZIMMERMAN EA (1980) Immunocytochemical localization of angiotensin II and vasopressin in rat hypothalamus: Evidence for production in the same neuron. Clin Sci 59:57s-60s

17. LEE CM, ARREGUI A, IVERSEN LL (1979) Substance P degradation by rat brain peptidases: Inhibition by SQ20881. Biochem Pharmacol 28:553-556

18. LEWICKI JA, FALLON JH, PRINTZ MP (1978) Regional distribution of angiotensinogen in rat brain. Brain Res 158:359-371

19. MENARD J, GALEN F-X, DEVAUX C, KOPP N, AUZAN C, CORVOL P (1980) Immunochemical differences between angiotensin I-forming enzymes in man.Clin Sci 59:41s-44s

20. MORRIS BJ, REID IA (1978) The distribution of angiotensinogen in dog brain studied by cell fractionation. Endocrinology 103:492-500

21. PHILLIPS MI, WEYHENMEYER J, FELIX D, GANTEN D, HOFFMAN WE (1979) Evidence for an endogenous brain renin-angiotensin system. Fed Proc 38:2260-2266

22. POLSKY-CYNKIN R, FANBURG BL (1979) Immunochemical comparison of angiotensin I converting enzymes from different rat organs. Int J Biochem 10:669-674

23. PRINTZ MP, PRINTZ JM, GREGORY TE (1978) Identification of angiotensinogen in animal brain homogenates. Relationships to a possible angiotensin I-generating pathway. Circ Res 43 (Suppl I):21-27

24. REID IA (1977) Is there a brain renin-angiotensin system? Circ Res 41:147-153

25. REID IA (1979) The brain renin-angiotensin system: a critical analysis. Fed Proc 38:2255-2259

26. REID IA, DAY RP, MOFFAT B, HUGHES HG (1977) Apparent angiotensin immunoreactivity in dog brain resulting from angiotensinase. J Neurochem 28:435-438

27. REID IA, MOFFAT B (1978) Angiotensin II concentration in cerebrospinal fluid following intraventricular injection of angiotensinogen or renin. Endocrinology 103:1494-1498

28. REID IA, MOFFAT B, MORRIS BJ (1978) Two forms of angiotensinogen in dog cerebrospinal fluid. IRCS Med Sci 6:383

29. REID IA, RAMSAY DJ (1975) The effects of intracerebroventricular administration of renin on drinking and blood pressure. Endocrinology 97:536-542

30. RIX E, GANTEN D, SCHÜLL B, UNGER TH, TAUGNER R (1981) Converting enzyme in the choroid plexus, brain and kidney: Immunocytochemical and biochemical studies in rats. Neurosci Letters 22:125-130

31. SCHÖLKENS BA, JUNG W, RASCHER W, SCHÖMIG A, GANTEN D (1980) Brain angiotensin II stimulates release of pituitary hormones, plasma catecholamines and increases blood pressure in dogs.Clin Sci 59:53s-56s

32. SEMPLE PF, MACRAE WA, MORTON JJ (1980) Angiotensin II in human cerebrospinal fluid may be an immunoassay artifact. Clin Sci 59:61s-64s

33. SERNIA C, REID IA (1980) Release of angiotensinogen by rat brain in vitro. Brain Res 192:217-225

34. SIMPSON JB, REID IA, RAMSAY DJ, KIPEN H (1978) Mechanism of the dipsogenic action of tetradecapeptide renin substrate. Brain Res 157:63-72

35. SLATER EE, DEFENDINI R, ZIMMERMAN EA (1980) Wide distribution of immunoreactive renin in nerve cells of human brain. Proc Natl Acad Sci USA 77:5458-5460

36. SWERTS JP, PERDRISOT R, MALFROY B, SCHWARTZ JC (1979) Is "enkephalinase" identical with angiotensin "converting-enzymes"? Eur J Pharm 53:209-210

37. WALLIS CJ, PRINTZ MP (1980) Adrenal regulation of regional brain angiotensinogen content. Endocrinology 106:337-342

38. WHITAKER JN, TERRY LC, WHETSELL WO (1981) Immunocytochemical localization of cathepsin D in rat neural tissue. Brain Res 216:109-124

Central Effects of Angiotensin

Cardiovascular and Humoral Effects of Endogenous Brain Angiotensin

B. A. Schölkens, W. Jung, W. Rascher, T. Unger, and D. Ganten

INTRODUCTION

In contrast to the state of knowledge concerning most other neuropeptides, the understanding of the biochemical pathways leading to the formation of angiotensin II (ANG II) is far advanced. Therefore, the brain renin-angiotensin system (RAS) was proposed as a model peptide system for the investigation of cardiovascular actions of neuropeptides, their biosynthesis and pharmacological interferences (9,11).

To decide whether the brain RAS has a functional significance the following questions needed to be answered:

1. Do the components of this enzyme peptide system interact in vivo and is ANG II formed through hydrolysis of local angiotensinogen by renin in the brain, resulting in biological actions?
2. Are pharmacological interferences with the formation or with the action of endogenous brain ANG II followed by biological effects, e.g., on blood pressure and drinking?

To answer these questions permanent cannulae were implanted into the brain ventricles of rats and dogs and renin was injected into the cerebrospinal fluid (CSF). Because of the high angiotensinogen concentrations in CSF, ANG I will be locally formed if the enzyme is active. ANG I is then converted to ANG II by brain converting enzyme (23,25). Results showed that indeed CSF ANG II was markedly increased following intracerebroventricular (IVT) injections of renin into the lateral brain ventricle of rats. Renin also induced dose-dependent increases of arterial blood pressure (25).

CARDIOVASCULAR EFFECTS OF ENDOGENOUS BRAIN ANGIOTENSIN

Since little was known about the cardiovascular changes of this central pressor response, studies were performed in conscious as well as in anesthetized dogs, so as to quantify the hemodynamic correlates under normal and sodium restricted diet.

Experimental Brain Research, Suppl. 4

If the effects of IVT renin on the hemodynamic profile of the pressor reaction closely resembled those of centrally administered ANG II, then it may be concluded that the effects of renin are the result of formation of ANG II.

Beagle dogs with cannulae chronically implanted into the third brain ventricle were kept on a normal diet (70 mEq sodium, 205 mEq potassium). Sodium depletion was produced in the same dogs 8-12 weeks later by a low-salt diet containing 6 mEq sodium and by daily administration of furosemide (10 mg/kg for 7 days p.o.). Arterial blood pressure was measured directly in the femoral artery via a pressure transducer (Statham P 23Db).

In conscious dogs IVT injection of 0.5 U hog kidney renin (NBC, Cleveland, Ohio) induced a significant steady increase of arterial blood pressure reaching a maximum 45 min after injection (Fig. 1). Systolic blood pressure increased by 14% (p 0.01) and diastolic blood pressure by 17% ($p < 0.02$). Similar results have been reported in anesthetized dogs (21).

After sodium restriction the pressor responses in conscious dogs were attenuated (Fig. 1) as had been observed in sodium restricted conscious rats following IVT ANG II (16). This is in harmony with the finding of a reduced binding of ANG II in rat brain during sodium restriction and a decreased number of ANG II receptors (4,16).

An attenuation of the vasoconstrictor responses to peripheral ANG II injections has also been observed in animals and in man on a low sodium diet (5,13,28,29), where ANG II-pressor reaction was correlated with the sodium balance, but inversely correlated with the level of plasma renin activity (29). The observed attenuation was thought to be in relation to the receptor occupancy by endogenous angiotensin.

In order to investigate the hemodynamic correlates of central ANG II pressor responses experiments were performed in anesthetized dogs with cannulae acutely implanted into the third brain ventricle. Beagle dogs were anesthetized with sodium pentobarbital (Nembutal^{R}), 35 mg/kg i.p., and respired with room air using a dog respirator (Harvard apparatus, model 660).

Intravenous injections were made into a saphenous vein, IVT injections through the ventricular cannula. After completion of surgery and insertion of different catheters under fluoroscopic control, several hemodynamic parameters were determined. Systolic and diastolic blood pressure (BP) was measured with a transducer (Statham P 23Db) connected to one femoral artery. A catheter tip manometer (Millar Instruments, Houston, Texas) was introduced into the left ventricular cavity to measure left ventricular pressure (LVP). Contractility of the left ventricle was judged from dp/dt max of the left ventricular pressure curve, electronically derived with

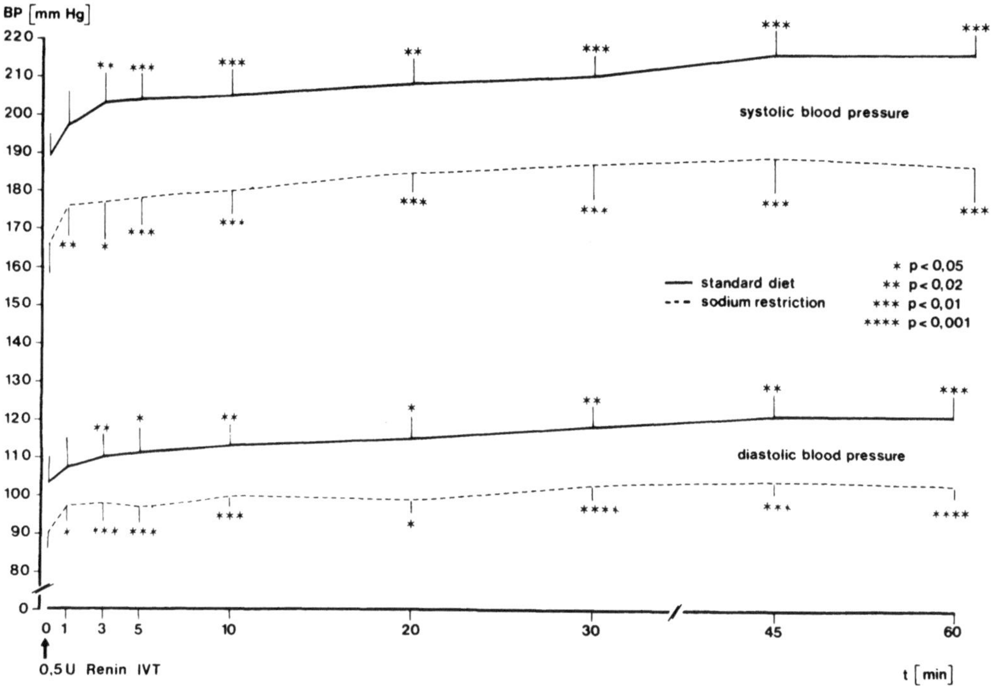

Fig. 1. Effects of intracerebroventricular (IVT) injections of renin on systemic blood pressure (BP) in conscious dogs on a normal and low-sodium diet (n=5). Hog kidney renin (0.5 U) was injected into the third brain ventricle. Illustrated are the systolic and diastolic blood pressure in dogs on a normal (straight line) and a sodium restricted (dotted line) diet. Results are given as means ± SEM

appropriate amplifiers (Hellige & Co., Freiburg, Germany). Heart rate (HR) was measured from QRS peaks using a biotachometer (Hellige & Co., Freiburg, Germany). Cardiac output (CO) was determined by the thermodilution method; saline (2.5 ml, 5-10^{0}C) was injected rapidly into the right atrium and a thermistor was placed into the aortic arch. Values for total peripheral resistance (TPR) and stroke volume (SV) were calculated from CO values using an analog computer (Hellige & Co., Freiburg, Germany). Amplifying, calibrating and recording of continuously measured data was done using a multichannel recorder (Hellige & Co., Freiburg, Germany).

Following central injection of 0.5 U renin mean arterial blood pressure, left ventricular pressure, dp/dt max and heart rate increased; these responses occurred more slowly, but were more marked and lasted much longer than after peripheral injection of renin (Table 1). IVT injection of ANG II caused similar changes in cardiovascular parameters: an increase of arterial and left ventricular pressure accompanied by an increase in dp/dt max and total peripheral resistance (2,8). The increase in blood pressure after IVT ANG II was due to increased peripheral resistance, since in the presence of increased total peripheral resistance no significant changes in cardiac output, cardiac work and stroke volume could be observed. Thus, it can be concluded that central ANG II elevates blood pressure by increasing total peripheral resistance. Central pressor effects of ANG II were accompanied by significant increases in LVP as well as dp/dt max. However, from studies in anesthetized cats it is known (3,14) that the index of contractility (dp/dt/i.p.) was essentially unchanged suggesting that the increase in ventricular pressure and dp/dt max were mainly due to an increase in after-load or total peripheral resistance rather than due to any alterations in myocardial contractility.

Table 1. Effects of intracerebroventricular injections of renin (0.5 U) and ANG II (100 ng/kg) on the hemodynamics of dogs, anesthetized with sodium pentobarbital

Parameter	Renin resting values	Renin Δ	ANG II resting values	ANG II Δ
MAP (mmHg)	146 ± 4	17 ± 4^{+}	151 ± 5	10 ± 1^{+++}
LVP (mmHg)	181 ± 3	18 ± 6	164 ± 10	10 ± 1^{+++}
dp/dt max (mmHg/sec)	4025 ± 668	875 ± 546	4600 ± 222	412 ± 360^{++}
HR (min^{-1})	138 ± 7	9 ± 8	111 ± 3	3 ± 1
CO (ml/min)	-	-	3149 ± 128	166 ± 61
SV (ml)	-	-	27 ± 3	1 ± 1
TPR (dynes x sec x cm^{-5})	-	-	6562 ± 468	585 ± 194^{+}

Results are expressed as means ± SEM (n : 5-8). Asterisk indicates significance ($^{+}p < 0.05$; $^{++}p < 0.02$; $^{+++}p < 0.001$)

Besides this similarity of the hemodynamic correlates of the central pressor reaction of renin and ANG II further evidence in support of the suggestion that the effects of IVT are mediated via the formation of ANG II, can be provided by experiments with ANG II receptor-blockers (Fig. 2).IVT pretreatment with [NSuc^1, Val^5, Phg^8] ANG II (22) in a dose of 1.0 µg/kg, a dose not modifying the blood pressure response by itself, completely abolished the cardiovascular changes of central renin injection. Saralasin had a similar effect (21). Virtually the same blocking effect could be demonstrated with respect to the hemodynamic changes of central ANG II injection. In conscious rats IVT pretreatment with saralasin (1.0 µg/kg IVT) also inhibited significantly the central pressor reaction of renin (0.01 U) (25).

DIPSOGENIC EFFECTS OF ENDOGENOUS BRAIN ANGIOTENSIN

Beside the effects of intraventricular renin on blood pressure, the dipsogenic effects of IVT renin (0.5 U) were also investigated in separate experiments in conscious dogs on a normal as well as on a low-sodium diet (Fig. 3).

Drinking was stimulated on both dietary regimes. The latency of this response was approximately 3-5 min and the mean volume of water drunk in the hour following renin was 519 ± 87 ml in normal dogs. However, under sodium restriction water consumption increased up to 935 ± 135 ml. In rats under sodium restriction the attenuated pressor reaction is accompanied by a diminished drinking response (16), in the dog on a low-sodium diet a dissociation between these two effects of brain angiotensin was observed: the diminished pressor response of IVT renin was accompanied by a marked increase in water consumption.

This could be explained by the fact that sodium restriction itself increases the daily water intake in dogs through hypovolemia (27).

Simultaneous IVT injections of renin with the converting enzyme inhibitor captopril (500 µg IVT) completely inhibited the centrally evoked drinking response in dogs on a low-sodium diet. IVT saralasin (10.0 µg/kg) given alone was without influence on water consumption, however, in combination with renin it abolished the drinking response (21). When 1.0 µg/kg ANG II was administered IVT the mean volume of water drunk within 1 hour was 366 ± 37 ml in normal dogs. This response was increased in dogs with chronic renal hypertension (two kidney, two wrapped hypertension) up to 487 ± 92 ml.

It can thus be concluded that renin induced drinking in the dog is mediated by the action of endogenously formed ANG II on central nervous system receptors. Inhibition of endogenous ANG formation abolishes the drinking response.

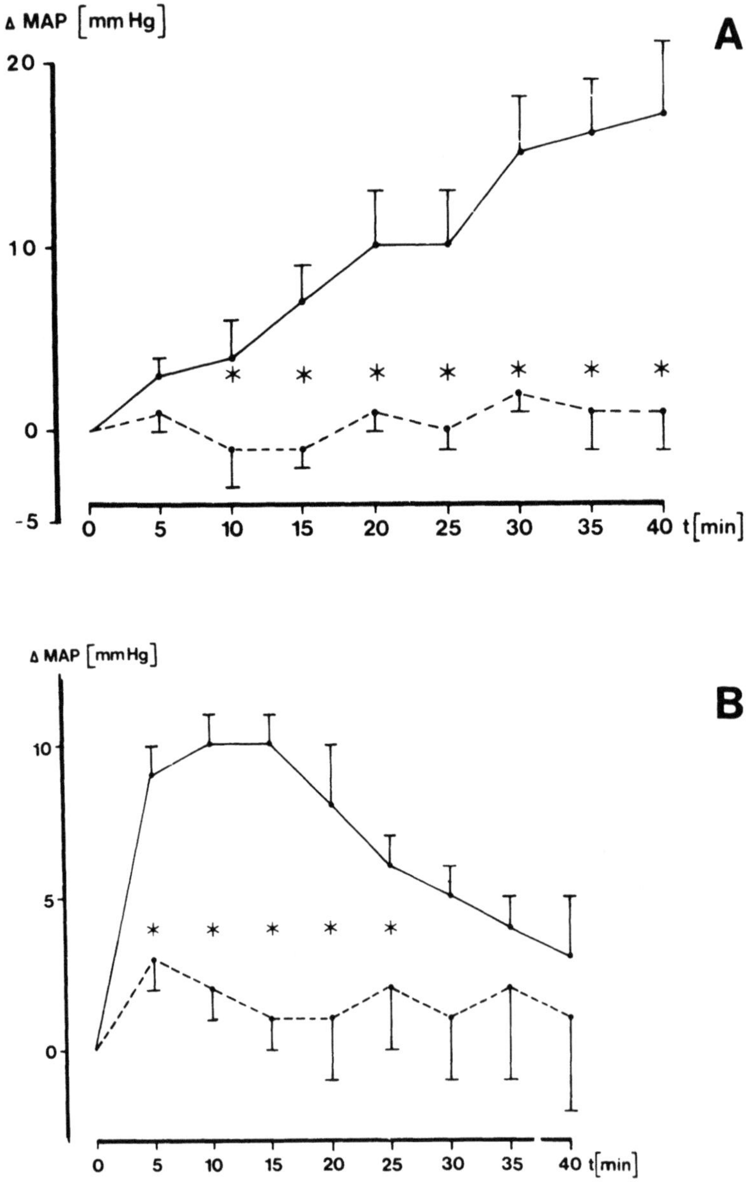

Fig. 2. Effect of intracerebroventricular [NSuc[1], Val[5], Phg[8]] ANG II (1.0 µg/kg) on the blood pressure response to intracerebroventricular renin (0.5 U) (A) and intracerebroventricular ANG II (100.0 ng/kg) (B). Straight line: Renin or ANG II; dotted line: ANG II-receptor blocker and either renin or Ang II. Results are presented as means ± SEM (n=5-8). Asterisk indicates significance ($p < 0.05$)

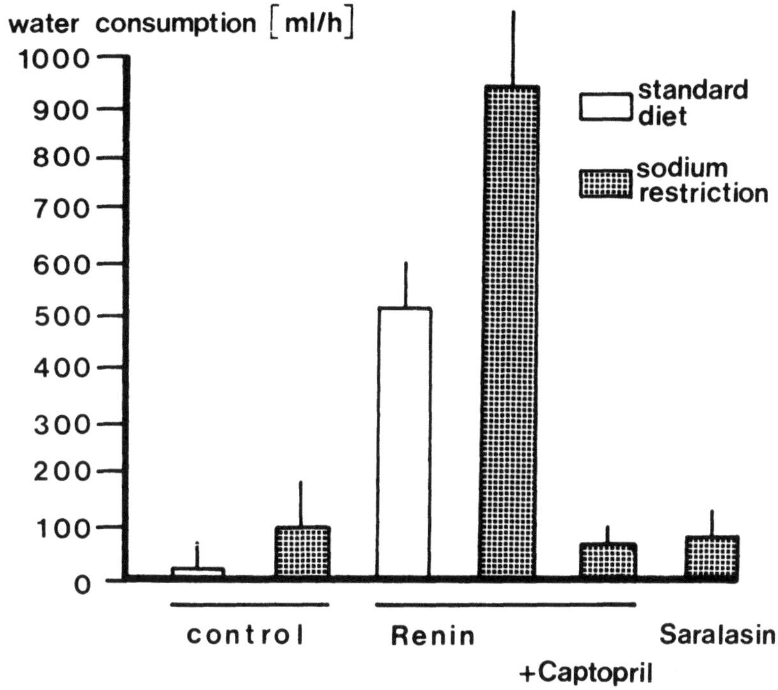

Fig. 3. Influence of intracerebroventricular injections of renin, captopril and saralasin on water consumption in dogs on a standard diet and with sodium restriction. Results are presented as means ± SEM (n=5)

HUMORAL EFFECTS OF ENDOGENOUS BRAIN ANGIOTENSIN

In addition of these blood pressure and drinking effects of brain ANG the influence of locally generated ANG II on various plasma hormones was investigated. Blood was taken in conscious dogs from the brachial vein at 60 and 20 min before (controls) and 15, 60 and 300 min after IVT renin injections. Plasma renin, ANG II (7,10,15), ADH (20), ACTH (31) and corticosterone (30) were determined by radioimmunoassay; noradrenaline and adrenaline in plasma were measured by a radioenzymatic method (6) (Fig. 4).

After IVT renin, plasma renin concentration decreased by 25% ($p < 0.05$) in dogs on a normal diet and by 58% ($p < 0.01$) in sodium depleted dogs. In the same animals plasma ANG II decreased by 55% ($p < 0.01$). IVT injection of saralasin (10.0 μg/kg) alone caused the opposite effect: an increase of plasma renin ($p < 0.01$) and of ANG II concentration ($p < 0.05$). The same changes were also seen after simultaneous injection of captopril (500.0 μg) in combination with renin.

IVT administration of renin increased ADH, noradrenaline, adrenaline and corticosterone in dogs on both dietary regimes. These central renin effects were inhibited by simultaneous IVT injections of captopril. IVT saralasin alone was without significant effect on plasma catecholamines, ADH or corticosterone (24).

The increase in plasma corticosterone is probably mediated by ACTH release, since plasma ACTH increased significantly ($p < 0.05$) in conscious dogs on a normal diet after IVT renin (Fig. 5). A similar increase in plasma ACTH with a similar latency period of 30-45 min had been observed following intravenous infusion of 10-20 ng/kg/min ANG II for 1 hour in conscious dogs (19).

Most of what is known about the central effects of ANG II has been learned from studies in rats and dogs. Only few data have been reported in primates, where neuronal fibers and cells in the brain containing ANG II like immunoreactivity with a similar distribution as in human brain have been identified (17,18). To test whether central injections of ANG II in a primate have the same effects as in non-primates, ANG II (500.0 ng/kg) was injected into the third ventricle of anesthetized rhesus monkeys (Fig. 6).

IVT injection of ANG II induced a significant increase ($p < 0.05$) of systemic blood pressure, which returned to initial values within 1 hour. Heart rate was also increased by 6%. The cardiovascular effects were accompanied by marked changes in plasma hormones.

Plasma renin concentration decreased by 56% ($p < 0.01$) and plasma ANG II by 46%. This confirms results in baboons where IVT ANG II infusions reduced PRA (1).

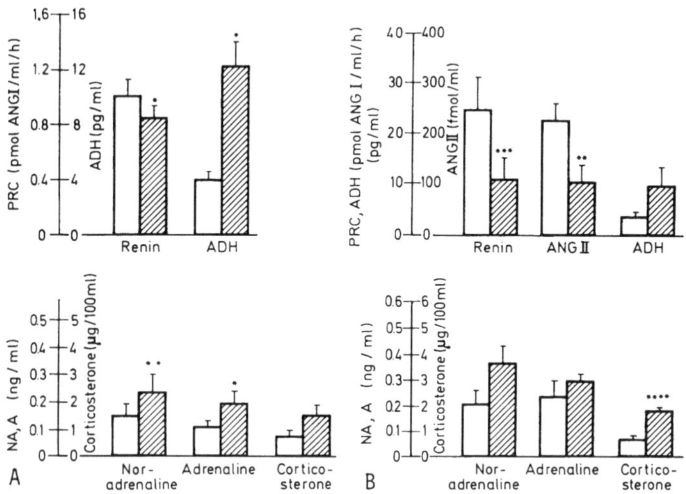

Fig. 4. Effects of central injections of renin on plasma levels of several hormones in conscious dogs on a normal diet and low-sodium diet (n=5). Hog kidney renin (0.5 U) was injected into the third brain ventricle. Illustrated are the maximal responses before (open colums) and after renin (striped columns) injection. Results are given as means ± SEM. Asterisks indicate significance (*p < 0.05;**p < 0.02; ***p < 0.01; ****p<0.001). A = normal diet; B = sodium restricted diet

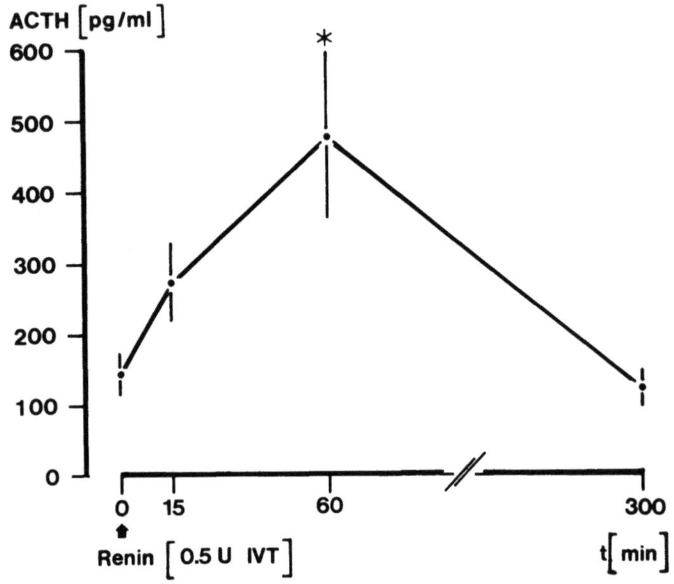

Fig. 5. Effects of intracerebroventricular injections of renin (0.5 U) on plasma concentrations of ACTH in conscious dogs (n=5). Results are given as means ± SEM. Asterisk indicates significance ($p < 0.05$)

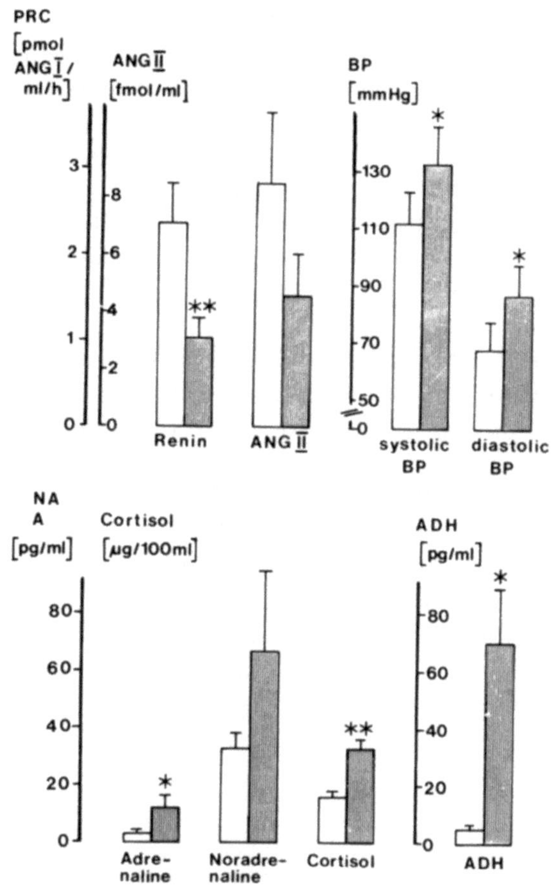

Fig. 6. Effects of intracerebroventricular ANG II on systemic blood pressure and plasma hormones in anesthetized rhesus monkeys (n=5). ANG II (500.0 ng/kg) was injected into the third brain ventricle. Illustrated are the maximal responses before (open columns) and after intracerebroventricular ANG II (striped columns) injection. Results are given as means ± SEM. Asterisks indicate significance (* $p < 0.05$; ** $p < 0.01$)

The concomitant increase of noradrenaline and adrenaline indicates that both the sympathoneuronal and the sympathoadrenal axis may be stimulated by central ANG II (11). Cortisol and ADH were also significantly ($p < 0.01$) increased.

Thus, IVT injection of ANG II in anesthetized rhesus monkeys increased systemic blood pressure and heart rate, accompanied by an increase of ADH, cortisol, adrenaline and noradrenaline. These findings demonstrate that centrally injected ANG II elicits the same cardiovascular and humoral responses in a primate species as endogenously formed ANG II in rats and dogs (26).

Central peptidergic stimulation has previously been shown to contribute to the development and maintenance of high blood pressure in experimental animals (11,12). Our results show that this possibility also exists in primates.

It is concluded that endogenous brain angiotensin participates in central mechanisms of blood pressure regulation by an activation of the sympathetic nervous system, by the release of ADH and by an elevation of plasma ACTH and subsequently increased levels of glucocorticoid hormones. This concomitant stimulation of pressor systems by central angiotensin (central peptidergic stimulation) has been shown to be involved in the maintenance of experimental hypertension in, e.g., spontaneously hypertensive rats (11).

REFERENCES

1. BROOKS VL, CHO KW, MALVIN RL, COHEN BJ (1980) Central and hormonal regulation of renin release by baboon kidneys. Proc Soc Exp Biol Med 165:147-150

2. BROSNIHAN KB, BERTI GA, FERRARIO CM (1979) Hemodynamics of central infusion of angiotensin II in normal and sodium-depleted dogs. Am J Physiol 237:H139-H145

3. BUCKLEY JP, SINGH S, STEENBERG ML, JANDHYALA BS (1977) Intraventricular infusion of angiotensin II on the hemodynamics and renal function of α-chloralose anesthetized cats. Circ Res 40 (suppl. I):I-52-I-56

4. COLE FE, BLAKESLEY HL, GRACI KA, FROHLICH ED, McPHEE AA (1980) Brain angiotensin II receptor affinity and capacity in SHR and WKY rats: effects of acute dietary changes in NaCl. Brain Res 190:272-277

5. COWLEY AW, LOHMEIER T (1978) Relationship between body fluid volume, sodium ion concentration, and sensitivity to pressor effects of angiotensin II in dogs. Circ Res 42:503-511

6. DaPRADA M, ZÜRCHER G (1976) Simultaneous radioenzymatic determination of plasma and tissue adrenaline, noradrenaline, and dopamine within the femtomol range. Life Sci 19:1161-1174

7. FAHRER A, GANTEN D (1978) Influence of additives on the renin-substrate reaction: Biochemical and practical aspects. In: KRAUSE DK, HUMMERICH W, POULSEN K (eds) Radioimmunoassay-Renin-Angiotensin. Thieme, Stuttgart, pp 82-91

8. FERRARIO CM, DICKINSON CJ, McCUBBIN JW (1970) Central vasomotor stimulation by angiotensin.Clin Sci 39:239-245

9. GANTEN D, SPECK G (1978) The brain renin-angiotensin system: a model for the synthesis of peptides in the brain. Biochem Pharmacol 27:2379-2389

10. GANTEN D, SCHELLING P, HOFFMANN WE, PHILLIPS MI, GANTEN U (1978) The measurement of extrarenal isorenins. In: KRAUSE DK, HUMMERICH W, POULSEN K (eds) Radioimmunoassay-Renin-Angiotensin. Thieme, Stuttgart, pp 144-155

11. GANTEN D, ROCKHOLD RW, UNGER TH, SPECK G, SCHÖLKENS B (1980) The central peptidergic stimulation syndrome: a role for brain peptides in central mechanisms of blood pressure regulation. In: PHILIPP TH, DISTLER A (eds) Hypertension: Mechanisms and Management. Springer-Verlag,Berlin,Heidelberg,New York,pp 75-82

12. GANTEN D, UNGER TH, SIMON W, SCHAZ K, SCHÖLKENS B, MANN JFE, SPECK G, LANG R, RASCHER W (1981) Central peptidergic stimulation: Focus on cardiovascular actions of angiotensin and opioid peptides. In: BUCKLEY JF, FERRARIO CM (eds) Central nervous system mechanisms in hypertension. Raven Press, pp 265-282

13. HEISTAD DD, ABBOUD FM, BALLARD DR (1971) Relationship between sodium concentration and vascular reactivity in man. J Clin Invest 50:2022-2032

14. JANDHYALA BS, SING S, BUCKLEY JP (1977) Further studies on centrally mediated hemodynamic effects of angiotensin II in anesthetized cats. In: BUCKLEY JP, FERRARIO CM (eds) Central actions of angiotensin and related hormones. Pergamon Press, New York, pp 205-213

15. MANN JFE, JOHNSON AK, GANTEN D (1980) Plasma angiotensin II: dipsogenic levels and the angiotensin generating capacity of renin. Am J Physiol 238:R372-R377

16. MANN JFE, SCHIFFRIN EL,SCHILLER PW, RASCHER W, BOUCHER R, GENEST J (1980) Central actions and brain receptor binding of angiotensin II: influence of sodium intake. Hypertension 2:437-443

17. PHILLIPS MI, QUINLAN JT, WEYHENMEYER J (1980) An angiotensin-like peptide in the brain. Life Sci 27:2589-2594

18. QUINLAN JT, PHILLIPS MI (1981) Immunoreactivity for an angiotensin II-like peptide in the human brain. Brain Res 205:212-218

19. RAMSAY DJ, KEIL LC, SHARPE MC, SHINSAKO J (1978) Angiotensin II infusion increases vasopressin, ACTH and 11-hydroxycorticosteroid secretion. Am J Physiol 234:R66-R71

20. RASCHER W, WEIDMANN E, GROSS F (1981) Vasopressin in the plasma of stroke-prone spontaneously hypertensive rats. Clin Sci 61:295-298

21. REID IA, RAMSAY DJ (1975) The effects of intracerebroventricular administration of renin on drinking and blood pressure. Endocrinology 97:536-542

22. SCHÖLKENS BA, WISSMANN H, LINDNER E, GEIGER R (1976) 1,8-disubstituted analogues of [Ile^5] and [Val^5] angiotensin II: difference in potency and specificity of angiotensin II-antagonistic activity. Hoppe Seylers Z Physiol Chem 357:825-838

23. SCHÖLKENS B, STEINBACH R, GANTEN D (1979) Blood pressure effects of endogenous brain angiotensin in rats are increased by inhibition of prostaglandin biosynthesis. Clin Sci 57:271s-274s

24. SCHÖLKENS BA, JUNG W, RASCHER W, SCHÖMIG A, GANTEN D (1980) Brain angiotensin II stimulates release of pituitary hormones, plasma catecholamines and increases blood pressure in dogs. Clin Sci 59:53s-56s

25. SCHÖLKENS BA, STEINBACH R, JUNG W, GANTEN D (1980) Induction of angiotensin formation in brain: biological effects and pharmacological interferences. In: GROSS F, VOGEL G (eds) Enzymatic release of vasoactive peptides. Raven Press, New York, pp 365-381

26. SCHÖLKENS BA, JUNG W, RASCHER W, SCHÖMIG A, GANTEN D (1981) Intracerebroventricular angiotensin II increases arterial blood pressure in rhesus monkeys by stimulation of pituitary hormones and the sympathetic nervous system. VIII. International Congress of Pharmacology, July 19-24, Tokyo, book of abstracts, p. 798

27. STEPHENS GA, DAVIS JO, FREEMAN RH, DeFORREST JM, SEYMOUR AA, ROWE BP, WILLIAMS GM (1980) Hemodynamic, fluid, and electrolyte changes during sodium depletion in conscious dogs. Proc Soc Exp Biol Med 163:416-420

28. STREWLER GJ, HINRICHS KJ, GUIOD LR, HOLLENBERG NK (1972) Sodium intake and vascular smooth muscle responsiveness to norepinephrine and angiotensin in the rabbit. Circ Res 31:758-766

29. THURSTON H, LARAGH JH (1975) Prior receptor occupancy as a determinant of the pressor activity of infused angiotensin II in the rat. Circ Res 36:113-117

30. VECSEI P (1974) Glucocorticoids: Cortisol, corticosterone and compound S. In: JAFFE BM, BEHRMAN HR (eds) Methods of hormone radioimmunoassay. Academic Press, New York and London, pp 393-415

31. VOIGT KH, FEHM HL, RECK R, PFEIFFER EF (1974) Spontaneous and stimulated secretion of QUSO-extractable immunoassayable ACTH in man. Klin Wschr 52:516-521

Centrally Administered Angiotensin Antagonists in Hypertensive Rats

W. J. McDonald and S. K. Aumann

INTRODUCTION

While a number of observations in the field of renin-angiotensin experimentation were important in the genesis of the studies reported in this paper, two were particularly critical. The first was that of Bickerton and Buckley (1) who noted that angiotensin could act directly on the central nervous system affecting an elevation of blood pressure. The second was the finding that replacement of the aromatic side chain on phenylalanine in position 8 of angiotensin II (Asp-Arg-Val-Tyr-Ile-His-Pro-Phe) with an aliphatic side chain invoked antagonistic rather than agonistic properties (6). This led to the development of angiotensin II (ANG II) antagonists and these in turn permitted the in vivo testing of the participation of ANG II in the pathophysiology of hypertension in both man and experimental animals. While the use of these analogues has provided insight, interpretation of ANG II antagonist studies is not without difficulties. Several limitations have become apparent (2). First, antagonists also possess agonistic activity which is both time and dose dependent. Second, these analogues have a short in vivo half-life. Third, if introduced in one physiologic compartment, e.g. the intraventricular (IVT) space, they may cross into other compartments, e.g. plasma, confounding the interpretation of results.

Despite these drawbacks, ANG II antagonist studies have helped to clarify which types of human and experimental hypertension are ANG II dependent. Some of the analogue experiments in the spontaneously hypertensive rat (8,10,11) have employed intravenous administration of ANG II antagonists. These studies lead to the conclusion that circulating ANG II plays no role in the elevation of blood pressure in this experimental model. A sizeable number of studies, on the other hand, have provided evidence that central or brain ANG II is of pathophysiologic importance in this regard (3,5,7,8,11,12,15,16).

Experimental Brain Research, Suppl. 4

The purpose of this communication is to review these data as well as additional data from our laboratory which bear on this issue. The ANG II antagonist experiments to be discussed include dose-response studies, intravenous (IV) infusion studies, as well as one and three weeks IVT infusions.

METHODS AND MATERIALS

Because the effects of angiotensin antagonists are both time and dose dependent, we decided to employ only infusions rather than bolus injections to determine the effects of the analogue.

Dose-Response Studies

These and subsequent experiments employed male SH rats obtained from Taconic Farms, N.Y. Ages ranged from 14-17 weeks and weights from 250 to 320 g. At this age, blood pressure is still increasing in this strain, not attaining near maximal levels until age 21 weeks.

Blood pressure was measured by tail plethysmography in the unanesthetized state. Measurements were taken for five days prior to cannulation of the lateral ventricles, five days during infusion of analogue, and five days following discontinuation of infusion.

Cannulation of the lateral ventricle was accomplished stereotactically under pentobarbital anesthesia. A hole was drilled in the calvarium 2 mm posterior to and 2 mm lateral to the bregma and a 21 gauge steel cannula was lowered into the lateral ventricle 4 mm deep from the dura. The cannula was secured to the skull with wood screws and methylmethacrylate cement. Test solutions were delivered to the ventricle by an osmotic minipump inserted subcutaneously between the scapulae and connected to the cannula with PE 60 tubing. Ten microcuries of triated water were introduced into the pump at the same time it was filled with experimental infusate and function of the pump in situ was monitored by measuring the daily urinary tritium excretion. Cannula placement in the lateral ventricle was confirmed after all experiments by injecting Evans Blue dye into the cannula prior to sacrifice and confirming the presence of dye in the ventricular system at autopsy.

After cannulation of the lateral ventricles, four groups of SH rats each containing seven to nine rats were infused with [Sar^1, Ile^8] ANG II of varying doses for five days.

Infusions were 1 µl/h and the groups received 100, 1000, 1200 or 3000 ng/h. At the end of the infusion period, pumps were disconnected from the cannula and removed. The cannula was sealed and blood pressures recorded for the ensuing five days.

Statistical analysis was performed using a two-way analysis of variance with repeated measures on one function.

Intravenous Antagonist Infusion

To ascertain the effect of peripherally administered ANG II antagonists, $[Sar^1, Ile^8]$ ANG II was infused into the jugular vein of seven SH rats aged 14 to 17 weeks. A dose of 1200 ng/h, the dose which resulted in the greatest decrease in blood pressure when administered centrally, was infused for five days. Blood pressure was measured for five days before and five days after infusion. Data were analyzed using analysis of variance with repeated measures.

One Week IVT Infusions in Older Animals

To compare the effects of IVT $[Sar^1, Ile^8]$ ANG II in older animals with those effects observed in younger animals during the dose response studies, nine older male SH rats were studied. Rats were 21-27 weeks old, an age when blood pressure has almost plateaued. These animals received IVT infusions of $[Sar^1, Ile^8]$ ANG II at a rate of 1200 ng/h for six days. Blood pressure measurements and cannulations were carried out in a fashion similar to that in the dose response studies. Control groups were five SH rats and six WKY rats similarly aged. The former received a six day IVT saline infusion at 1 µl/h and the latter a six day IVT infusion of $[Sar^1, Ile^8]$ ANG II (1200 ng/h).

Statistical analysis was carried out with a one-way analysis of variance.

Extended IVT Infusion Studies

Nine male SH rats ranging from 13-17 weeks were used in this study. Cannulation and blood pressure measurements were as in previous studies. Rats received 1200 ng/h of $[Sar^1, Ile^8]$ ANG II for 18 days. Infusions were accomplished by changing the minipumps every six days.

A one-way analysis of variance was employed for data analysis.

RESULTS

Dose-Response Studies

Results of this study are shown in Fig 1. Blood pressure fell significantly in the groups receiving 1000, 1200 and 3000 ng/h while the group infused at a rate of 100 ng/h showed no reduction. Animals in the three groups which showed a decrement followed similar blood pressure patterns during infusions. Pressure was reduced at 24 hours and remained at that level throughout the five days. For this reason, the pressures shown in Fig. 1 represent the mean of all intra-infusion blood pressures obtained in each group. The absolute magnitudes of the reductions at infusions of 1000, 1200 and 3000 ng/h were 12.9, 13.5 and 7.7 mmHg respectively. Based upon these data, a maximum-response infusion rate of 1200 ng/h was chosen for subsequent experiments.

Intravenous Antagonist Infusions

The effect of a 1200 ng/h intravenous infusion of [Sar^1, Ile^8] ANG II is shown in the right hand column of Table 1. In contrast to animals of similar age receiving the same IVT infusion rate, animals infused intravenously did not show a decrement in blood pressure.

One Week IVT Infusions in Older Animals

The blood pressure response of older (21-27 weeks) SH rats at IVT [Sar^1, Ile^8] ANG II is shown in Table 1. In response to 1200 ng/h these rats reduced blood pressure by an average of 32 mmHg over the six day infusion period. This compares to a 15 mmHg drop in younger animals receiving the same infusion, and no decrease was observed in older rats receiving IVT saline. WKY rats infused with antagonist via the IVT route also failed to decrease blood pressure.

Extended IVT Infusion Studies

Fig. 2 shows the reduction in blood pressure during an 18 day IVT infusion of [Sar^1, Ile^8] ANG II. These studies were technically more difficult. The decrease in number of experimental animals shown in Fig. 2 was attributable almost entirely to technical failure of the infusion system. Blood pressure during infusion differed significantly from basal ($p < 0.001$). Heart rates during infusion are shown in Fig.3. The depicted reduction in heart rate was also significant ($p < 0.001$). Heart rate and blood pressure correlated for the period of the study ($R = 0.4049$; $p < 0.01$).

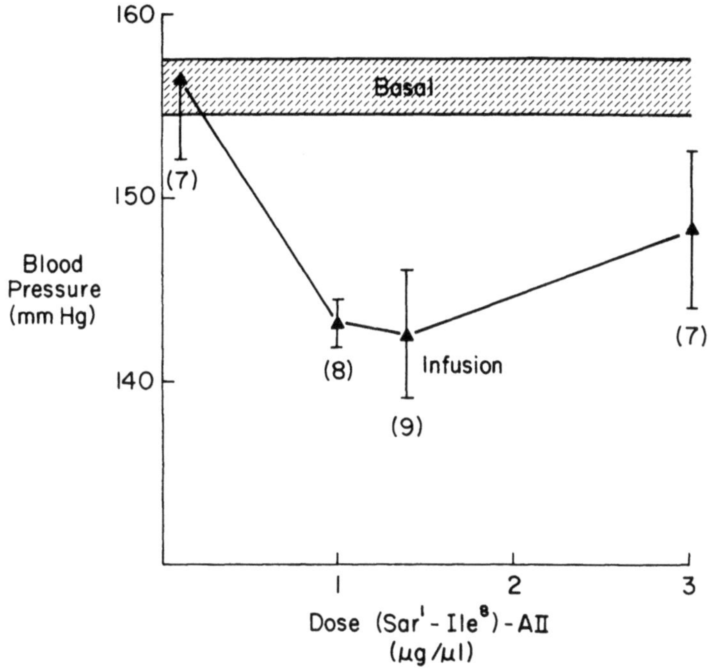

Fig. 1. Mean blood pressures during five day infusions at 100, 1000, 1200 and 3000 ng/h or 1 µg/h. N equals the number in parenthesis. Basal blood pressures shown in the stippled band, were established by pooling the basal pressures from each group, as they showed no between-group significant differences. Mean blood pressure reductions at the three large infusion rates were significant at $p < 0.005$

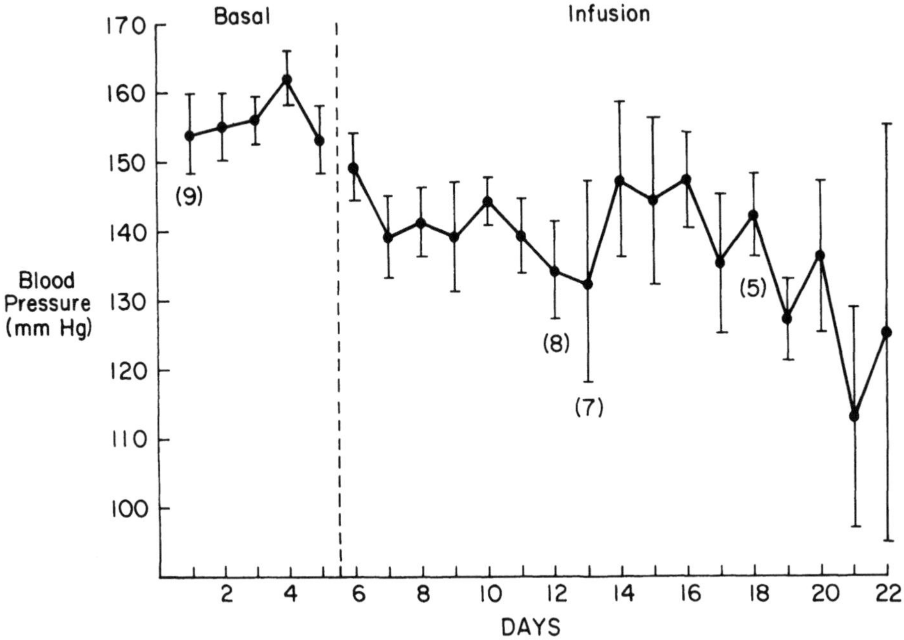

Fig. 2. Blood pressure before and during an 18 day intraventricular (IVT) infusion of [Sar[1], Ile[8]] ANG II in SH rats. Intra-infusion decrease in blood pressure was significant at $p < 0.001$. Number in parenthesis at each time point is number of surviving animals

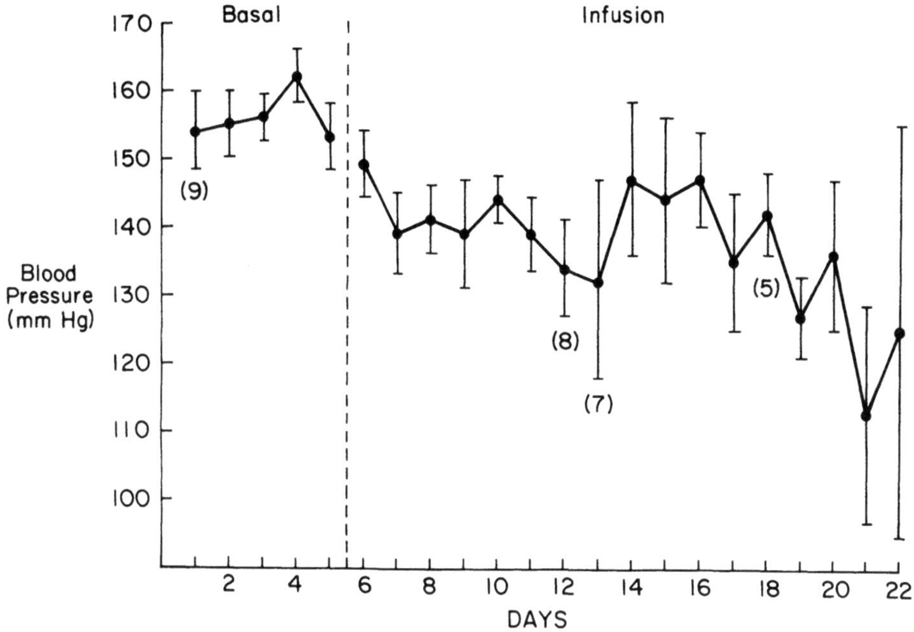

Fig. 3. Heart rate before and during 18 day intraventricular (IVT) infusion of [Sar^{1}, Ile^{8}] ANG II. The intra-infusion decrease in heart rate was significant at $p < 0.001$

Table 1. Systolic blood pressure in SH and WKY rats before, during and after intravenous (IV) or intraventricular (IVT) infusion of [Sar^1, Ile^8] ANG II or saline

Strain/Route	SH Rats/IVT			WKY Rats/IVT	SH Rats/IV
Infusate	Saline	[Sar^1,Ile^8] ANG II	[Sar^1,Ile^8] ANG II	[Sar^1,Ile^8] ANG II	[Sar^1,Ile^8] ANG II
N	5	6	9	6	7
Age (Wk)	21 - 27	21 - 27	14 - 17	21 - 27	14 - 17
Blood Pressure^{+}					
Basal (Mean)	203 ± 4	204 ± 3	156 ± 2	105 ± 2	164 ± 2
During Infusion					
Day 1	193 ± 7	161 ± 8^{++}	143 ± 6^{++}	100 ± 5	161 ± 6
2	181 ± 4	167 ± 4^{++}	139 ± 5^{++}	101 ± 2	172 ± 7
3	198 ± 10	173 ± 3^{++}	141 ± 4^{++}	91 ± 6	165 ± 5
4	198 ± 7	175 ± 7^{++}	139 ± 3^{++}	105 ± 9	172 ± 7
5	195 ± 5	174 ± 12^{++}	144 ± 7^{++}	103 ± 9	172 ± 4
6	208 ± 6	177 ± 7^{++}	139 ± 7^{++}	106 ± 7	
After Infusion					
Day 1		160 ± 5^{++}		97 ± 6	165 ± 5
2		172 ± 6^{++}		114 ± 4	173 ± 5
3				115 ± 5	170 ± 8
4		189 ± 8		96 ± 3	162 ± 4
5		190 ± 5		107 ± 7	177 ± 8
6		205 ± 9			

N = Number in each group. $^{+}$ = mmHg ± SEM. $^{++}$ = Differs from basal at $p < 0.01$

DISCUSSION

The effects of IVT ANG II antagonists have now been studied by a number of investigators. Certain conclusions may be drawn. For this purpose, as well as for comparison with the data presented in this paper, selected aspects of some of these studies are presented in Table 2.

Table 2. Selected experience with IVT angiotensin antagonist in unanesthetized genetically hypertensive rats

Investigator	SH Rat Strain	Age or Weight	ANG II Antagonist	Administration	Dose	Blood Pressure Response (mmHg)
1. Ganten et al(5)	GH	200 - 300 g	Sar^1,Ala^8	Perfusion	200 ng /min	-35
2. Phillips et al.(12)	SHR	200 - 250 g	Sar^1,Ala^8	Bolus		
3. Elghozi et al.(4)	SHR	30 wk	Sar^1,Thr^8 Sar^1,Ile^8	Bolus	1000 - 5000 ng	No change
4. Schölkens et al.(14)	SHR	10 wk 14 wk	Suc^1,Val^5-Phg^8	Bolus	300 ng /kg	- 8 (10 wk) + 8 (14 wk)
5. Phillips et al.(11) and Mann et al.(7)	SHR-SP	16 wk	Sar^1,Ala^8	Bolus	5000 ng 10000 ng 20000 ng	- 2.5 - 7.8 -12.5
6. Sirett et al.(15)	GH	12 - 24 wk	Sar^1,Ile^8		20000 - 40000 ng	None (-24.7 in anesthetized rats given 40000 ng)
7. McDonald et al.(8)	SHR	21 - 27 wk	Sar^1,Ile^8	Infusion	1200 ng /h	-31 over 6 days
8. Clough et al.(3)	SHR-SP	250 - 300 g	Sar^1,Ala^8	Infusion	10 ng/h 100 ng/h 1000 ng/h 10000 ng /h	0 0 -20 (effect lasted 3 days only) +17

Dose-Response Studies

Our data demonstrated that the SH rat responds maximally to IVT infusions of approximately 1000 ng/h while failing to decrease blood pressure at infusion rates of 100 ng/h. At rates of 3000 ng/h, blood pressure decreases significantly but to a lesser degree, perhaps related to the agonistic properties of the analogue. As noted in Table 2, other investigators have demonstrated dose-response relationships to either IVT bolus or infusions of ANG II antagonists.

Elghozi et al. (4), and Phillips et al. (12) and Sirett et al. (15) have studied the antihypertensive properties of bolus IVT injections of ANG II analogues. The former investigators (4) working with SH rats of the Okamoto strain in the unanesthetized state, failed to show a response to injections of up to 5000 ng. They employed both [Sar^1, Thr^8] ANG II and [Sar^1, Ile^8] ANG II and used 30 week old rats. In contrast, Phillips et al. (12) did find a dose-response with 0.1 μg, 5 μg and 10 μg/μl ($p < 0.05$ - 0.001). In a later study using SHR-Stroke Prone (SHR-SP) Phillips et al. (11) found significant reductions of blood pressure at 5 μg (2.5 mmHg), 10 μg (7.8 mmHg) and 20 μg (12.5 mmHg). These data pertain to younger animals (16 weeks) than those employed by Elghozi (30 weeks). Phillips, however, also found that 32 week rats responded to 20 μg injections. While it is possible that the differences observed between these studies are related to differences in the age of rats used, it is also possible that with bolus injections, the threshold dose is close to 5 μg. With slightly higher doses Elghozi et al. may have found a depressor effect. Sirett et al. (15) did not find a hypotensive effect with bolus injection of up to 40 μg of [Sar^1, Ala^8] ANG II in conscious New Zealand hypertensive (GH) rats while noting a 24.7 mmHg decrease with similar doses in anesthetized animals. The difference between this and previous studies may well be related to interstrain differences.

Because bolus injections and infusions may result in different responses, it is perhaps best to compare our data with those of Clough et al. (3). These investigators used SHR-Stroke Prone (SHR-SP) animals and IVT infusions of [Sar^1, Ala^8] ANG II. Infusions ranged from 10 ng/h to 10.000 ng/h. Rates less than 1000 ng/h did not decrease pressure. At 1000 ng/h a 20 mmHg decrease was noted. This persisted for three days before returning to baseline. At 10 μg/h, a pressor response was observed. Our data agree well with those of Clough et al. with regard to threshold infusion rate while disagreeing with regard to duration of effect.

Thus, a depressor effect has most frequently been observed with IVT ANG II competitive inhibitors in hypertensive animals, but this has not been the case with normotensive controls (4,5,7,8,14,15).

Intravenous Antagonist Infusions

Our data demonstrate that ANG II analogues when infused IV at rates sufficient to decrease blood pressure when administered centrally failed to elicit a hypotensive response. This is in close agreement with other studies which have shown either no effect (3,10,11) or a hypertensive response (11,14) to peripherally infused analogues. It would appear that only under conditions of salt depletion can one demonstrate a depressor response to ANG II analogues administered intravenously in these experimental models (3,9). Additional support for the concept of an exclusive central depressor effect of ANG II antagonists is derived from the observation that IVT ANG II analogues appear to cross into plasma only when large doses (1 mg/kg) are injected (13). Finally, centrally administered ANG II analogues are depressor in effect, even in the absence of kidneys (7). These three lines of evidence argue persuasively that ANG II antagonists exert their depressor effect only via the central nervous system under normal conditions in the genetically hypertensive rat.

Prolonged IVT ANG II Antagonist Infusions

Our studies showed that IVT infusions of five to six days lower blood pressure in both young and older animals. The decreases were of a greater magnitude in older rats but failed to attain normotensive levels. Longer infusions, while technically difficult, suggest that further reductions in pressure may occur over several weeks. Studies of Clough et al. (3) while demonstrating a blood pressure reduction for three days, did not document a sustained response. Differences between studies may involve differences in strain, antagonist or other factors such as salt balance.

CONCLUSION

Intraventricular ANG II antagonists effectively lower blood pressure in genetically hypertensive rats. This effect is in direct contrast to the effect of intravenously administered analogues which are ineffective under conditions of sodium repletion. Long term IVT infusions of the competitive ANG II inhibitors suggest that the hypotensive response may persist for the duration of the infusion. Taken together, these studies support a role for the brain renin-angiotensin system in the maintenance of hypertension in some hypertensive rat strains.

ACKNOWLEDGEMENT

Supported in part by grants from the Veterans Administration and the Oregon Heart Association.

REFERENCES

1. BICKERTON RK, BUCKLEY JP (1961) Evidence for a central mechanism in angiotensin induced hypertension. Proc Soc Exp Biol Med 106:834-836

2. BRAVO EL, KHOSLA MC, BUMPUS FM (1976) Comparative studies of the humoral and arterial pressure responses to [Sar^1, Ala^8]-, [Sar^1, Ile^8]- and [Sar^1, Thr^8]-angiotensin II in the trained unanesthetized dog. Progr Biochem Pharmacol 12: 33-40

3. CLOUGH DP, LANGMANN E, RETTIG R, ROCKHOLD R, UNGER TH, GANTEN D (1981) The effects on blood pressure of central angiotensin blockade in spontaneously hypertensive and sodium depleted rats. Proceedings of the Fourth International Symposium on the Spontaneously Hypertensive Rat. Schattauer Verlag, 1982 (in press)

4. ELGHOZI JL, ALTMAN JL, DEVYNK NA, LAIRD JF, GRUNFELD JP, MEYER P (1976) Lack of hypotensive effect on central injection of angiotensin inhibitors in spontaneously hypertensive (SH) and normotensive rats. Clin Sci Mol Med 51: 385s-389s

5. GANTEN D, HUTCHINSON JS, HAEBARA H, SPONER G, DIETZ R, SCHELLING P (1976) Effects of centrally administered angiotensin II antagonist on arterial blood pressure in different strains of rats. In: BUCKLEY JP, FERRARIO CM (eds) Central Actions of Angiotensin and Related Hormones. Pergamon Press, New York, Oxford, 527-535

6. KHAIRALLAH PA, TOTH A, BUMPUS FM (1970) Analogs of angiotensin II. Mechanisms of receptor interactions. J Med Chem 13:181-184

7. MANN JFE, PHILLIPS MI, DIETZ R, HAEBARA H, GANTEN D (1978) Effects of central and peripheral angiotensin blockade in hypertensive rats. Am J Physiol 234: H629-H637

8. McDONALD WJ, WICKRE C, AUMANN S, BAN D, MOFFITT B (1980) The sustained antihypertensive effect of chronic cerebroventricular infusion of angiotensin antagonist in spontaneously hypertensive rats. Endocrinology 107:1305-1308

9. MUNOZ-RAMIREZ H, KHOSLA M, BUMPUS FM, KHAIRALLAH PA (1978) Hypotensive effect of [Sar^1, Thr^8]-angiotensin II in spontaneously hypertensive sodium depleted rats. Am J Physiol 234:H447-H453

10. PALS OT, MASUCCI FD, DENNING GS, SIPOS F, FESSLER D (1971) Role of the pressor action of angiotensin II in experimental hypertension. Circ Res 29:673-681

11. PHILLIPS MI, MANN JFE, HAEBARA H, HOFFMAN WE, DIETZ R, SCHELLING P, GANTEN D (1977) Lowering of hypertension by central saralasin in the absence of plasma renin. Nature 270:445-447

12. PHILLIPS MI, PHIPPS J, HOFFMAN WE, LEAVITT M (1975) Reduction of blood pressure by intracranial injection of angiotensin blocker (P113) in spontaneously hypertensive rats (SHR). Physiologist 18:350

13. SCHELLING P, GANTEN U, SPONER G, UNGER TH, GANTEN D (1980) Components of the renin-angiotensin system in the cerebrospinal fluid of rats and dogs with special consideration of the origin and fate of angiotensin II. Neuroendocrinology 31:297-308

14. SCHÖLKENS BA, JUNG W, STEINBACH R (1976) Blood pressure response to central and peripheral injection of angiotensin II and 8-C-phenylglycine analogue of angiotensin II in rats with experimental hypertension. Clin Sci Mol Med 51: 403s-406s

15. SIRETT NE, THORNTON SN, HUBBARD JI (1979) Brain angiotensin II binding and central [Sar^1, Ala^8]-angiotensin responses in normal rats and the New Zealand strain of genetically hypertensive rats. Clin Sci 56:607-611

16. SWEET CS, COLUMBO JM, GANL SL, WEITZ D, WENGER HC (1977) Inhibitors of the renin-angiotensin system in rats with malignant and spontaneous hypertension: Comparative effects of central vs peripheral administration. In: BUCKLEY JP, FERRARIO CM (eds) Central Actions of Angiotensin and Related Hormones. Pergamon Press, New York, Oxford. 283-292

Contribution of Vasopressin and Renal Nerves to the Natriuresis Evoked by Centrally Administered Renin or Angiotensin

W. B. Severs, J. Y. Summy-Long, and L. C. Keil

INTRODUCTION

Exogenous angiotensin interacts with central nervous tissues to cause blood pressure elevation, vasopressin and ACTH release, natriuresis, and increased water intake (2,8,21,25). All components of the renin-angiotensin system have been detected in the brain (11,13,20,30); therefore, it is plausible that centrally administered angiotensin simulates physiologic and/or pathophysiologic functions of endogenous angiotensin (cf., 2,22).

Rats continuously infused with angiotensin via the cerebroventricles accelerate their turnover of body water, lose body mass, increase urinary sodium output and reduce plasma [Na^+] (9,10,31). Relatively little information is available on the mechanism of the natriuretic response after central administration of angiotensin. Natriuresis has been observed under experimental conditions of unchanged systemic and renal hemodynamics (3,12). A contribution of renal nerves has been suggested from results in anesthetized cats (16). Vasopressin may possess natriuretic properties since sodium excretion did not rise after hypertonic NaCl infusion into the cerebroventricles of conscious, homozygous Brattleboro rats (17). This report examines: a) whether intracerebroventricular (IVT) vasopressin infusions affect urinary Na^+ output; b) whether IVT vasopressin affects the actions of IVT angiotensin; and c) whether renal innervation affects the urinary response to IVT angiotensin.

MATERIALS AND METHODS

Male Sprague-Dawley rats (300-400 g) were used in all experiments. A cannula (28) was placed in the left lateral ventricle under pentobarbital anesthesia (45 mg/kg) four days before experimentation. Fast green dye (5 µl) was injected IVT at the end of all experiments to verify cannula placement. Data were used only if dye was distributed in the lateral, third and fourth cerebroventricles.

Experimental Brain Research, Suppl. 4

Expt. 1: Rats were continuously infused IVT with artificial CSF (18) either with or without arginine vasopressin (VP, Sigma, 0.5 or 50 ng/h) at a rate of 2 µl/min. Water intake and urine output were measured over the next 5 h. Urinary sodium and potassium were assayed by flame photometry.

Expt. 2: Rats received an IVT pretreatment (5 µl) of either CSF, VP 0.5 ng, or 2.5 ng followed 5 min later by an IVT injection (5 µl) of either CSF or angiotensin ([Ile^{5}]-angiotensin II, sigma, 100 ng). After the second IVT injection the rats were placed in a cage with water available. The onset (in seconds) and the frequency of drinking behavior were recorded. All rats were decapitated 90 sec after the second IVT injection. Trunk blood was collected into heparinized plastic tubes kept in ice water. The plasma was separated in a refrigerated centrifuge and radioimmunoassayed for vasopressin content (14).

Expt. 3: Rats were kept in metabolism cages with no food or water available. They were pretreated IVT (5 µl) with either VP (2.5 ng) or CSF. Five min later they were given a second IVT injection (5 µl) of either CSF or angiotensin (100 ng) along with 20 ml/kg tap water by stomach tube. Urine was collected hourly for 3 h and assayed for Na^{+} and K^{+} by flame photometry.

Expt. 4: During the surgery for ventricular cannulation the abdominal cavity of each rat was opened along the midline. The kidneys were denervated by the method of Colindres and Gottschalk (6) except that 5% phenol in 70% ethanol was used to paint the renal vasculature. The kidneys of control rats were handled in a "sham" manner. The abdominal wounds were closed with silk sutures. At the end of these experiments each individual kidney was assayed for norepinephrine (29); bilateral denervation ($>$85% amine depletion) was confirmed for rats included in the results.

Four days after surgery the rats were given an IVT injection (5 µl) of saline (0.85%) with or without 5 mGU renin (U.S. Biochemical Corp.), along with 20 ml/kg tap water by stomach tube. No food or water was permitted. Rats were kept in metabolism cages and urine was collected at 3 and 6 h for sodium and potassium assays.

RESULTS

Expt. 1: The effects of a 5 h IVT VP infusion are shown in Table 1. Neither dose of the peptide altered water intake, urine volume or absolute excretion of sodium and potassium ($p < 0.05$, 1-way analysis of variance).

Table 1. Effect of a five hour cerebroventricular vasopressin infusion on water intake and urine output[+]

	CSF (13)	VP, 0.5 ng/h (10)	VP, 50 ng/h (5)
U_V, ml	5.6 ± 1.0[++]	5.8 ± 0.9	5.1 ± 1.0
$U_{Na}.V$, μEq	664.0 ± 157.0	645.0 ± 108.0	598.0 ± 119.0
$U_k.U$ μEq	751.0 ± 95.0	714.0 ± 107.0	567.0 ± 159.0
Drinking, ml	1.4 ± 0.6	0.8 ± 0.4	1.8 ± 1.3

[+]Jerome, M.L. and Severs, W.B. unpublished observations
[++]Variability in table shown as ± S.E.; no means were different (1-way ANOVA)

Expt. 2: Fig. 1 shows the effects of IVT VP pretreatment on angiotensin-stimulated plasma VP concentration. A 2 x 3 analysis of variance of the data showed significant F ratios ($p < 0.05$) for pretreatment and interaction ($F = 11.5$ and 5.2, $df = 2,34$). The Newman Keuls t-test indicated that the 2.5 ng VP (VP2) pretreatment differed ($p < 0.05$) from CSF and 0.5 ng VP (VP1). Comparison of individual means revealed that the CSF + CSF and VP1 + CSF groups differed from all others ($p < 0.05$). Angiotensin elevated plasma VP ($p < 0.05$) except in the VP2 pretreatment group ($p > 0.05$).

Frequency (percent) of drinking within 90 sec in the CSF + CSF, CSF + A, VP1 + CSF, VP1 + A, VP2 + CSF and VP2 + A groups were, respectively: 0, 80, 0, 100, 0, and 75. The three angiotensin groups (80, 100 and 75) were equal ($p > 0.05$) and all differed ($p < 0.05$) from groups not receiving angiotensin (Chi Square, Fishers Exact test). Drinking latencies (sec ± S.E.) of the CSF + A, VP1 + A and VP2 + A groups were, respectively: 29.5 ± 4.4, 49 ± 5.3 and 29.5 ± 3.2. The VP1 + A group differed from the other two ($p < 0.05$, 1-way analysis of variance, $F = 7.3$, df 2,13).

Expt. 3: The effects of IVT VP pretreatment on urine production after IVT angiotensin are illustrated in Fig. 2. The data were compared by 2 x 2 analysis of variance at each time. F ratios for the pretreatments (CSF or VP) were not significant ($p > 0.05$) for urine volume or absolute sodium and potassium excretion at any time (maximum $F = 1.9$, df 1,20). No F ratio for interaction was greater than 0.9 ($p > 0.05$). Angiotensin, however, reduced urine volume at 1 and 2 h ($p < 0.05$, F's > 13). Sodium excretion was elevated by angiotensin at all times ($p < 0.05$, F's > 7.5) and potassium excretion was increased at hrs 2 and 3 ($p < 0.05$, F's > 7).

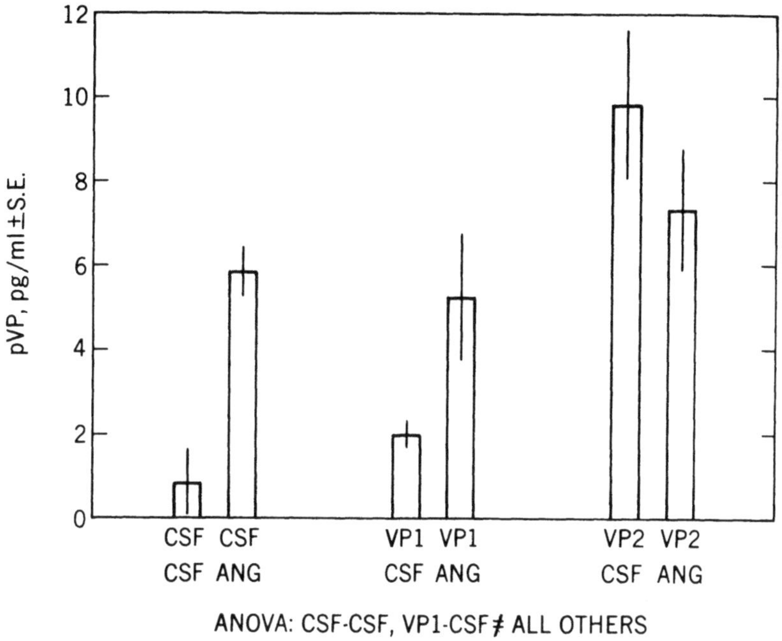

Fig. 1. Effects of IVT vasopressin pretreatment on the angiotensin stimulated increase in plasma vasopressin.

Rats received an IVT pretreatment of CSF, 0.5 ng VP (VP1) or 2.5 ng VP (VP2), followed 5 min later by a second IVT injection of CSF or angiotensin II (ANG II, 100 ng). Blood was collected 90 sec later for vasopressin radioimmunoassay. N = 5 to 8 per group. The CSF + CSF and VP1 + CSF groups differed from all others ($p < 0.05$)

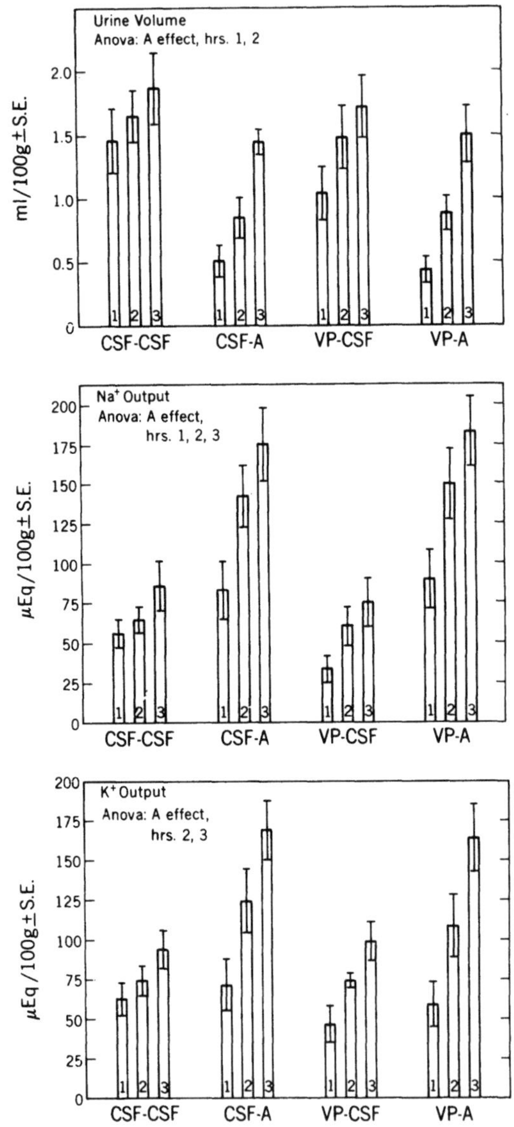

Fig. 2. Effects of IVT vasopressin pretreatment on the urinary response to IVT angiotensin.

Rats (n = 6/group) were pretreated IVT with CSF or vasopressin (VP, 2.5 ng), followed 5 min later by a second IVT injection of CSF or angiotensin (ANG, 100 ng) and 20 ml/kg p.o. water. Urine volume, sodium and potassium outputs are shown at hrs 1, 2 and 3

Expt. 4: Fig. 3 summarizes the results of the experiments with renal denervated rats. The data were compared by 2 x 2 analysis of variance at each time. At 3 h there was a tendency towards increased urine volume in denervated rats ($F = 3.7$, df 1,21 $p = 8\%$). This tendency became statistically significant at 6 h ($F = 7.9$, $p < 0.05$). Renin did not affect urine volume at either time nor was there interaction between variables (all F's < 2).

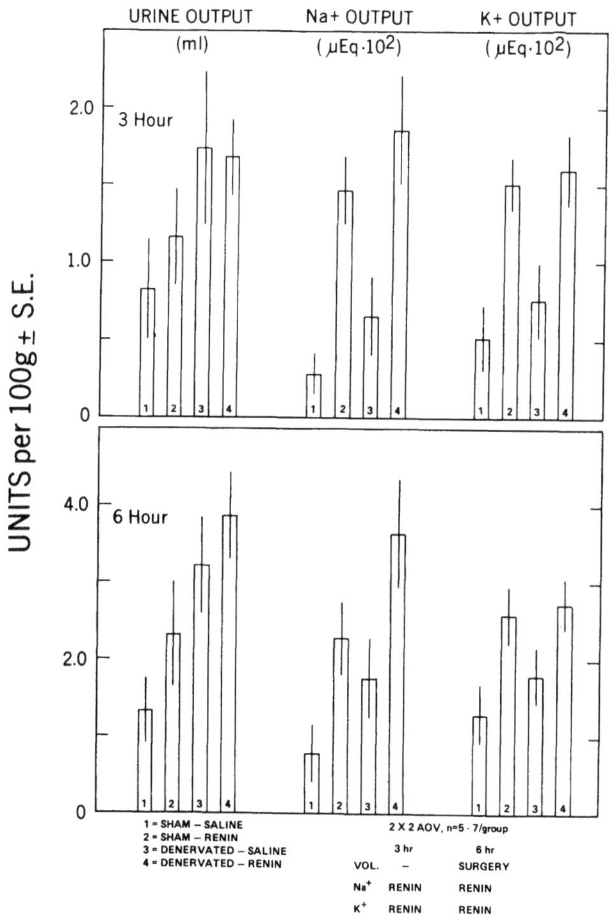

Fig. 3. Effect of renal denervation on urine production after IVT renin.

Conscious rats, sham operated or bilaterally renal denervated, were injected IVT (5 μl) with saline or renin (5 mG.U.) and given 20 ml/kg p.o. water. Urine was collected at 3 and 6 h. Significant differences ($p < 0.05$) from analysis of variance of the data are shown

Sodium output was markedly increased by IVT renin at 3 and 6 h ($p < 0.05$, F's = 19,9). There was a strong tendency (p = 6%) for an increased sodium output associated with renal denervation at 6 h. Potassium excretion was also elevated by renin at 3 and 6 h ($p < 0.05$, F's = 17, 9.4). There were no surgery or interaction effects (all F's < 1) associated with potassium excretion.

DISCUSSION

One objective of this study was to determine whether centrally administered vasopressin exerted natriuretic effects by itself or if it altered natriuresis evoked by IVT angiotensin. Mouw et al. (17) reported that homozygous Brattleboro rats have a blunted natriuretic response to IVT hypertonic NaCl. This observation raises the possibility that vasopressin has natriuretic properties. Vasopressin is present in CSF (24) and CNS structures participating in the control of blood pressure and salt/water balance (4,5). Two investigations indicated that IVT vasopressin affects water movement within the brain (19,23). One study suggested that IVT administration of the peptide functions as a negative feedback control for plasma vasopressin (1).

While these studies suggest that IVT vasopressin should affect systemic salt/water homeostasis, single injections of the hormone to conscious rats do not affect urine and electrolyte excretion (27). The experiments with single injections are limited because there is no information available defining the sites and time course of vasopressin distribution, the concentrations attained, or the duration of vasopressin's presence at potential sites of action. The experiments described herein were intended to overcome this limitation by continuously infusing the hormone IVT and thus minimize potential pharmacokinetic reasons for prior negative data. However, even under these conditions, vasopressin failed to alter the output of urine, sodium and potassium in normally hydrated conscious rats. Therefore, the data provide no evidence for a role of the hormone in CSF in the control of urine/electrolyte excretion.

Data from experiments designed to determine if IVT VP affects the response to IVT angiotensin raised a paradox. Pretreatment of moderately hydrated rats with 2.5 ng IVT VP did not affect urine/electrolyte excretion nor did it alter the antidiuresis, natriuresis and kaliuresis evoked by IVT angiotensin (Fig. 2). On the other hand data in Fig. 1 show that 2.5 ng IVT VP was associated with leakage of the hormone into plasma. Thus, plasma VP in the VP + CSF treated rats was similar ($p > 0.05$) to plasma levels in rats given angiotensin alone, or VP + angiotensin. The question can then be raised: why was IVT VP (2.5 ng) ineffective in altering urine/electrolyte output when the plasma VP levels were equivalent to those in rats receiving angiotensin alone or in combination with IVT VP? Both angiotensin-treated groups were associated with marked renal effects.

Two general explanations for this paradox emerge. Firstly, VP might have an unkown central effect offsetting the antidiuresis expected with such a high plasma titer of the hormone. Secondly, it is conceptually possible (albeit unlikely) that the renal response to angiotensin is independent of the plasma vasopressin titer. Either explanation would provide a rationale for why the VP, VP + angiotensin, and the angiotensin groups had equivalent plasma VP levels, but only the angiotensin groups exerted renal effects. It should be noted that angiotensin did not further increase plasma VP in rats pretreated with IVT VP (2.5 ng).

The results with the lower IVT dose of VP (0.5 ng) showed no leakage of the peptide into plasma, no renal response and no impairment of angiotensin's ability to release VP and alter urine/electrolyte excretion. IVT VP (0.5 ng) did prolong the drinking latency to IVT angiotensin (Expt. 2), but the significance of this effect is unclear. Rats in this experiment were placed in a novel environment for a fixed time (90 sec) before killing; competing exploratory behavior may be related to this effect and has not been ruled out.

Collectively the results suggest that in acute experiments IVT VP does not affect the urinary response to IVT angiotensin. High (2.5 ng) doses of IVT VP appear to leak into plasma, and angiotensin does not further increase plasma vasopressin concentration under these conditions. These experiments demonstrated a clear kaliuretic effect of IVT angiotensin. This effect, observed before (12,26), has been underemphasized compared to the natriuretic effect of IVT angiotensin. Kaliuresis may account for the large reduction in body mass which occurs after prolonged IVT angiotensin infusions in conscious rats (9,10).

Renal nerves participate in sodium excretion, but anesthesia is a complicating factor in results obtained by various investigators (7,15). Little information is available about the participation of renal nerves in the natriuretic action of IVT angiotensin. In pentobarbital anesthetized cats Lokhandwala et al.(16) observed that acute renal denervation partly reduced the rise in urinary sodium concentration and excretion rate evoked by IVT angiotensin. In our experiments with moderately hydrated conscious rats, IVT renin produced a marked natriuresis independent of renal nerves. Likewise, as noted in experiments described above, a marked kaliuresis was observed. The renal denervation procedure, in our hands, was not benign. Sham operated rats excreted less volume at 6 h compared to denervated rats and there was a tendency for this effect at 3 h ($p = 8\%$). Also, sodium excretion at 6 h was almost reduced ($p = 6\%$) in sham-operated, compared to renal denervated, rats. Thus, our results do not support the concept that renal innervation is an important determinant of the natriuretic response to IVT angiotensin. Further studies will be required to determine the efferent mechanism responsible for angiotensin natriuresis.

SUMMARY

The experiments reported herein addressed two hypotheses for the natriuresis evoked by cerebroventricular (IVT) angiotensin: a possible central effect of vasopressin and a change in renal nerve activity. Both explanations were excluded. Marked kaliuresis also occured after IVT renin or angiotensin. IVT vasopressin appears to have no effect on the urinary response to subsequent IVT angiotensin, but high VP doses can leak into plasma.

ACKNOWLEDGEMENTS

Supported by NASA-Ames Co-operative Agreement NCC-2-127.

We appreciate the assistance of Jack Wurpul, Lilian Rosella and Lidia Hernandez. Ms. Ann McCann provided excellent secretarial assistance.

REFERENCES

1. BHARGAVA KP, KULSHRESTHA VK and SRIVASTAVA YP (1977) Central mechanism of vasopressin induced changes in antidiuretic hormone release. Brit J Pharmacol 60: 77-82

2. BROOKS VL and MALVIN RL (1979) An intracerebral, physiological role for angiotensin: effects of central blockade. Fed Proc 38:2272-2275

3. BROOKS VL and MALVIN RL (in press) Intracerebroventricular infusions of angiotensin II increases sodium excretion. Proc Soc Exp Biol Med

4. BUIJS RM, SWAAB DF, DOGTEROM J and van LEEUWEN FW (1978) Intra- and extrahypothalamic vasopressin and oxytocin pathways in the rat. Cell Tiss Res 186:423-433

5. BUIJS RM (1980) Immunocytochemical demonstration of vasopressin and oxytocin in the rat brain by light and electron microscopy. J Histochem Cytochem 28: 357-360

6. COLINDRES RE and GOTTSHALK CS (1978) Neural control of renal tubular sodium resorption in the rat: single nephron analysis. Fed Proc 37:1218-1221

7. DiBONA GF (1977) Neurogenic regulation of renal tubular sodium reabsorption. Am J Physiol F233:73-81

8. GANTEN D, HUTCHINSON JS, SCHELLING P, GANTEN U, FISCHER H (1976) The isorenin angiotensin systems in extrarenal tissue. Clin Exp Pharmacol Physiol 3: 103-126

9. GRONAN RJ and YORK DH (1979) Effect of chronic intraventricular administration of angiotensin II on drinking behavior and blood pressure. Pharmacol Biochem Behav 10:121-126

10. HALPERIN ES, SUMMY-LONG JY, KEIL LC and SEVERS WB (1981) Aspects of salt/water balance after cerebroventricular infusion of angiotensin II. Brain Res 205: 219-221

11. HIROSE S, YOKOSAWA H and INAGAMI T (1978) Immunochemical identification of renin in rat brain and distinction from acid protease. Nature (Lond) 274: 392-393

12. HOFFMAN WE, WEET JF,PHILLIPS MI and SCHMIDT PG (1979) Central effects of angiotensin II in water and saline-loaded rats. Neuroendocrinology 28:289-296

13. INAGAMI T, CLEMENS DL, CELIO MR, BROWN A, SANDRU L, HERSCHKOWICZ N, HOFFMAN LH, KASSELBERG AG (1980) Immunohistochemical localization of renin in mouse brain. Neurosci Letters 18:91-98

14. KEIL LC and SEVERS WB (1977) Reduction in plasma vasopressin levels of dehydrated rats following acute stress. Endocrinology 100:30-38

15. KIM JK, LINAS SL and SCHRIER RW (1979) Catecholamines and sodium transport in the kidney. Pharmacol Rev 31:169-178

16. LOKHANDWALA MF, BUCKLEY JP and JANDHYALA BS (1978) Reduction of plasma renin activity by centrally administered angiotensin II in anesthetized cats. Clin Exp Hypertension 1:167-175

17. MOUW DR, VANDER AJ, LANDIS C, KUTSCHINSKI S, MATHIAS N and ZIMMERMAN D (1980) Dose-response relation of CSF sodium and renal sodium excretion, and its absence in homozygous Brattleboro rats. Neuroendocrinology 30:206-212

18. MERLIS JK (1940) The effect of changes in the calcium content of the cerebrospinal fluid on spinal reflex activity in the dog. Am J Physiol 131:67-72

19. NOTO F, NAKAJIMA T, SAJI Y and NAGAWA Y (1978) Effect of vasopressin on intracranial pressure of rabbit. Endocrinology Japan 25:591-596

20. OSMAN MY, SMEBY RR and SEN S (1979) Separation of dog brain renin-like activity from acid protease activity. Hypertension 1:53-60

21. PHILLIPS MI (1978) Angiotensin in the brain. Neuroendocrinology 25:354-377

22. PHILLIPS MI, MANN JFE, HAEBARA H, HOFFMAN WE, DIETZ R, SCHELLING P and GANTEN D (1977) Lowering of hypertension by central saralasin in the absence of plasma renin. Nature (Lond) 270:445-447

23. RAICHLE ME and GRUBB RL (1978) Regulation of brain water permeability by centrally released vasopressin. Brain Res 143:191-194

24. SEVERS WB, CHANGARIS DG, KAPSHA JM, KEIL LC, PETRO DJ, REID I, SUMMY-LONG J (1976) Presence and significance of angiotensin in cerebrospinal fluid. In: BUCKLEY JP and FERRARIO CM (eds) Central actions of angiotensin and related hormones. Pergamon Press, New York, p 216-223

25. SEVERS WB and DANIELS-SEVERS AE (1973) Effects of angiotensin on the central nervous system. Pharmacol Rev 25:415-449

26. SEVERS WB, DANIELS-SEVERS AE, SUMMY-LONG J and RADIO GJ (1971) Effects of centrally administered angiotensin II on salt and water excretion. Pharmacology 6: 241-251

27. SEVERS WB, KEIL LC and KLASE PA (1978) Consummatory behavior and urine production after cerebroventricular injection of vasopressin and vasopressin antiserum. Eur J Pharmacol 51:389-396

28. SEVERS WB, SUMMY-LONG J, TAYLOR JS and CONNOR JD (1970) A central effect of angiotensin: release of pituitary pressor material. J Pharmacol Exp Ther 174: 27-34

29. SHELLENBERGER MK, GORDON JH (1971) A rapid, simplified procedure for simultaneous assay of norepinephrine, dopamine and 5-hydroxy-tryptamine from discrete brain areas. Analyt Biochem 39:345-372

30. SLATER EE, DEFENDINI R and ZIMMERMAN EA (1980) Wide distribution of immunoreactive renin in nerve cells of human brain. Proc Natl Acad Sci USA 77:5458-5460

31. STERLING GH, CHEE O, RIGGS RV and KEIL LC (1980) Effect of chronic intracerebroventricular angiotensin II infusion on vasopressin release in rats. Neuroendocrinology 31:182-188

Possible Functions of Brain Renin

I. Haulica, G. Petresu, A. Stratone, D. Branisteanu, V. Rosca, C. Neamtu, and S. Slatineanu

INTRODUCTION

Compared with the biochemical (15,16,18,28,30,40,51), immunohistochemical (12,13, 23,24,27,37,41) and pharmacological data (4,8,19,22,49,54,60) on the renin-angiotensin system (RAS) in mammalian brain, the functional aspects are less known.

Increasing interest is being taken in both the variations of the components of the system in different physiopathological states, and the effects of ANG II on the periventricular structures involved in maintaining sympathetic tone, water ingestion and ADH secretion. By acting on these structures, the intrinsic brain RAS seems to participate in central regulation of volume (22) and blood pressure (28,56). Evidence for an interaction between the peptidergic and catecholaminergic mechanisms in central blood pressure control have been obtained (29,59).

The central pressor effects of ANG II prove to be more intense in spontaneous hypertensive rats than in normotensive ones (38). Moreover, the cerebral receptor blocking with saralasin causes the decrease of the systemic blood pressure only in the hypertensive animals (26,55). On this basis, the hypothesis of the cerebral RAS participation in producing and maintaining the spontaneous arterial hypertension has been advanced (28,56).

Besides the hormonal role in the central regulation of the volemic and circulatory homeostasis mediated by certain circumventricular structures like subfornical organ, organum vasculosum laminae terminalis, anterior hypothalamus and area postrema, the cerebral RAS seems to be also involved in other forms of central neuroglandular activities with important metabolic and functional consequences, as the following experimental data will demonstrate.

Experimental Brain Research, Suppl. 4

METHODS AND RESULTS

In our previous experiments on the distribution of isorenin in rat brain, an intense renin-like activity was observed in the pineal gland (33,34), shortly after the discovery of the isorenin in the rat and dog brain (21,25). The amount of isorenin was found about 15 times greater than in the rest of the brain (34).

The radioimmunoassay (RIA) and chromatographic (HPLC) identification of angiotensin-like peptides in bovine pineal gland recently performed by Ganten, Speck and Herman (unpublished data) points out that the main peptide resulting from brain renin activity on the specific natural substrate is ANG I.

The existence of intrinsic RAS in the pineal gland was confirmed both in vivo (12, 13,14,30,41) and in vitro (35) by pineal cell cultures. We have continued to study the pineal in order to obtain further information on physiologic role of the renin.

Release of Renin from the Pineal Gland

Since the pineal gland is a serotoninergic organ, innervated by sympathetic fibers from the superior cervical ganglion (42,66), we first tried to see if the release of pineal renin depends on sympathetic nervous stimulation.

The experiments were performed on 12 adult rabbits of both sexes anesthetized with urethane (i.v. 1.0 g/kg/ bodyweight). Collection of venous blood of pineal origin was made at the point of confluence of the longitudinal and transverse sinuses with the sagittal veins in the sinum vascular Herophily, after ligation of all the venous afferences, except for the prosencephalic medial vein.

Determination of the blood renin content was made before and after the sympathetic nervous stimulation of the pineal tissue. The stimulation was achieved through the central extremity of the sectioned cervical sympathetic nerve which is the main vegetative nerve activator of the indolic pineal hormone secretion (43). Stimulation was made with a rectangular electric current of 6-8 volts and low frequency (20 cycles/sec) applied for 2 minutes.

The electrical stimulation produced an increased renin release from the rabbit pineal gland into the effluent venous blood flow (Table 1).

In contrast with the mild increase in the content of renin in the cerebral venous blood flow, the cerebrospinal fluid (CSF) collected from cisterna magna did not contain significant renin activity, both in basal conditions and after nervous stimulation of the pineal gland.

Table 1. Plasma renin-like activity in the pineal effluent blood of the rabbit before and after electrical stimulation with rectangular impulses (BV, 20 cycles/sec, 2 minutes) of cervical sympathetic nerves

No.	PRA ng ANG I/ml/h before	after stimulation
1.	3.50	4.70
2.	3.20	3.25
3.	2.60	4.10
4.	3.70	3.90
5.	3.30	5.50
6.	3.20	4.10
7.	0.20	2.80
8.	0.60	3.00
9.	1.16	1.75
10.	1.25	1.66
Means $\pm$ SD	2.27 $\pm$ 1.32	3.50 $\pm$ 1.08; $p < 0.01$

Unlike renin, ANG II was found in high concentrations in the CSF by Ganten et al. (26), Hutchinson et al. (39) and others (20,58). This could be explained by the release of peptide into the intraventricular space, after local synthesis by the renin secretion from circumventricular structures, pineal gland included.

Like the renal juxtaglomerular apparatus, the pineal gland seems to represent a neurochemical transducer modulated by the cervical sympathetic innervation. Also there is participation of endogenous RAS as suggested by diurnal variations of pineal renin content.

Diurnal Variations in Renin Content of the Pineal Gland

Unlike the hypophysis, which has an endogenous circadian rhythm (1,32), the pineal gland has an exogenous nictemeral cycle, imposed by day light and night dark periods (68). These photopic influences control the formation of melatonin from serotonin, the main epiphyseal indolic hormone (2), via sympathetic fibers from the upper cervical ganglion (66). While the light stimulation of the retina inhibits both the noradrenaline local synthesis and the transformation of serotonin into melatonin, darkness causes the reverse phenomena of increasing the noradrenaline content in epiphysis and melatonin synthesis (67). In order to see if the pineal renin is involved in these neurohormonal reactions, we investigated whether the gland's capacity for forming angiotensin varies in the nocturnal and diurnal periods of the nictemer.

The experiments were performed on groups of adult rats of both sexes, kept under standardized food, temperature and light conditions. The light-exposion interval was of 12 hours, between 6 A.M. to 6 P.M.

Four subgroups of 30 rats each, were sacrificed every 6 hours, starting at 6 A.M. until midnight.

A fifth subgroup of 30 rats was sacrificed at 6 A.M. after a 12-hour light exposion during the night.

The renin content of the pineal gland decreased in the daytime, followed by a tendency to rise at night. This tendency no longer occurs in the animals exposed to light in the dark interval (Fig. 1).

The nictemeral variations in the pineal renin activity of low diurnal values and high nocturnal ones, follow same pattern as that of noradrenaline and melatonin in this gland (57,67). The rhythm of pineal renin is not related to the storage variations of the renal renin, as it persists in rats which were bilaterally nephrectomized 24 hours before they were killed.

Both reserpine 1 mg/kg bodyweight i.p. for 3 days before sacrifice, and bilateral superior cervical gangliectomy, suppressed the low diurnal level in the pineal renin (Fig. 2).

Since depletion of monoamines produced by reserpine is not specific for serotonin (9) or catecholamines (11), another four groups of 15 rats were given para-chlorophenylalanine (400 mg/kg b.w.) intraperitoneally injected for 3 days. This is a specific inhibitor of serotonin synthesis by a tryptophan-hydroxylase blocking mechanism (44).

The inhibition of serotonin synthesis with para-chlorophenylalanine also abolished or even reversed the mid-day epiphyseal values (Table 2).

Angiotensin Effect on the Pineal Serotonin

Since the mammalian pineal is rich in serotonin (31) and renin (30,34) we have studied the effects of synthetic ANG II on serotonin content of dog pineal gland.

The experiments were performed on 22 dogs of both sexes (weighing 7-10 kg), anesthetized with alpha-chloralose (0.11 g/kg bodyweight).

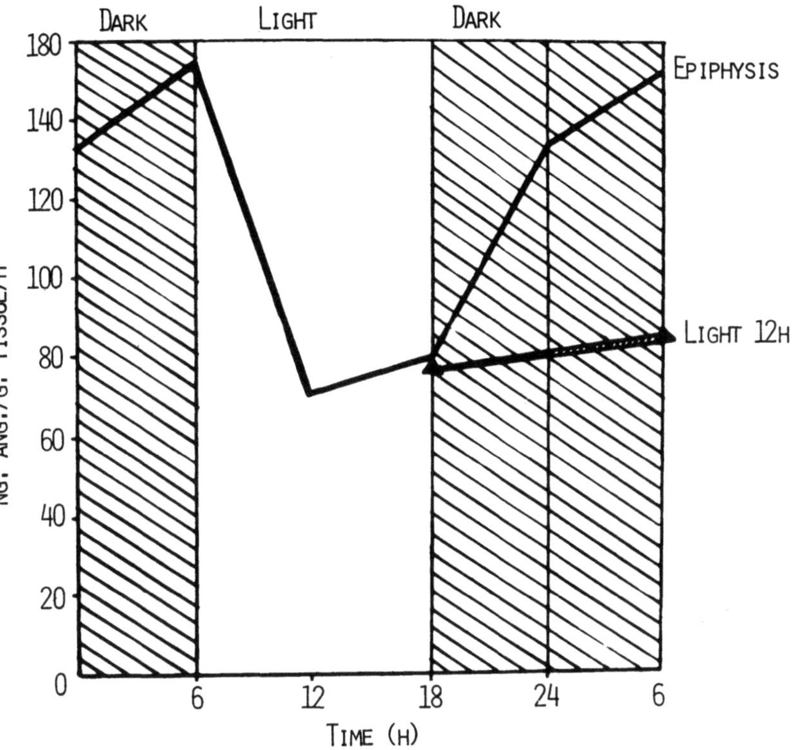

Fig. 1. Diurnal variations of the renin-like activity in the pineal gland and cerebral cortex of normal rats

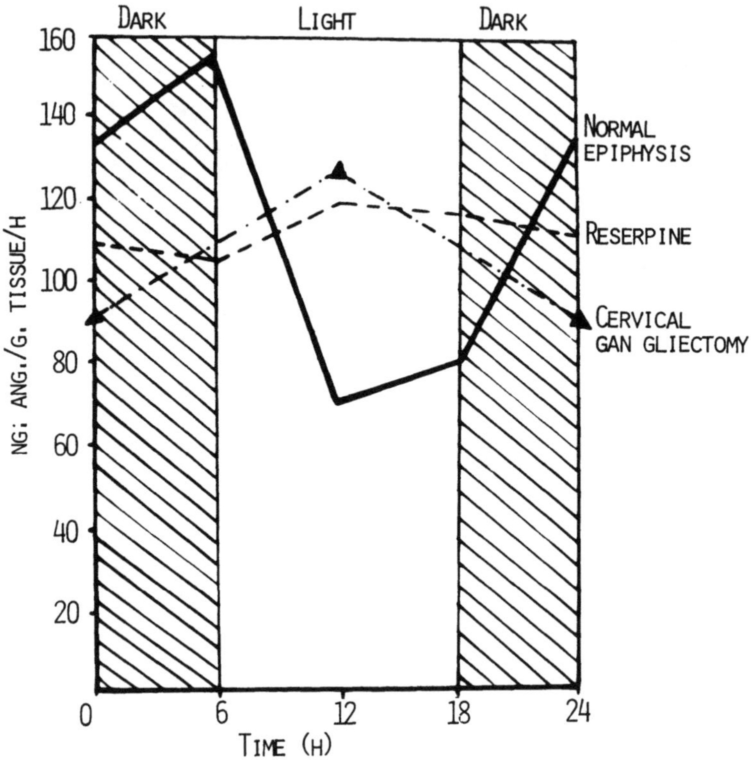

Fig. 2. Influence of reserpine and bilateral superior cervical gangliectomy on the diurnal variations of the renin-like activity in the pineal gland

Table 2. Renin-like activity of the pineal gland from adult rats along a day-night cycle

Experiment (Nr. of animals)	Pineal isorenin (ANG II/g/tissue/h)[a] 6 - 6.30[b]	12 - 12.30[b]	18 - 18.30[b]	0 - 0.30[b]
Control (120)	154 ± 36	70 ± 15	84 ± 17	116 ± 23
Bilateral Nephrectomy (40)	168 ± 38	116 ± 23	121 ± 25	137 ± 26
Para-chloro-phenylalanine 400 mg/kg b.w. (15)	139 ± 25	168 ± 32	150 ± 29	122 ± 26

[a] mean ± SEM; [b] time: clock hours

Pineal serotonin was spectrophotometrically assayed by the method of Weissbach et al. (65), after two hours of intravenous perfusion with progressive doses of synthetic ANG II (Hypertensin Ciba) in concentrations of 5, 25, 50 and 75 ng/kg body wt/min. The technical details and preliminary results were reported elsewhere (36).

Angiotensin II in all doses, except 5 ng/ml, caused a significant increase of serotonin, not only in the hypothalamus and brain stem, as was previously shown (50), but also in the pineal gland (Table 3).

Table 3. Influence of angiotensin II on serotonin content of pineal gland, hypothalamus and brain stem in adult rat (ng/g wet tissue)

	Control	Angiotensin II Perfusion 5 ng/ml	25 ng/ml	50 ng/ml	75 ng/ml
Pineal gland	1.04 ± 0.33[x] (6)[xx]	0.92 ± 0.43[x] (4)[xx]	1.97 ± 0.41[x] (4)[xx]	3.2 ± 1.1[x] (4)[xx]	2.49 ± 0.89[x] (4)[xx]
Hypo-thalamus	0.44 ± 0.28 (6)	0.47 ± 0.28 (4)	0.82 ± 0.33 (4)	1.00 ± 0.41 (4)	0.71 ± 0.37 (4)
Brain stem	0.29 ± 0.16 (6)	0.31 ± 0.32 (4)	0.64 ± 0.33 (4)	0.92 ± 0.37 (4)	0.61 ± 0.37 (4)

x = Mean ± SEM; xx = Number of animals

Taking into account the high renin activity of the pineal gland and its ability of locally producing high amounts of ANG I and ANG II, these concentrations come within the range of quasiphysiologic variations. In agreement with earlier experimental data (50), smaller increases of serotonin content were found at the hypothalamic and brain stem levels. Recently, Ganten and coworkers (unpublished data) have found a significant increase of serotonin content in rat pineal glands after intraventricular administration of renin and ANG II. Circulating ANG II may reach this circumventricular organ, since the blood-brain barrier is absent at the pineal level (63).

The mechanism by which ANG II causes serotonin content to rise in the pineal gland is yet unclear.

Whether ANG II activates the pineal serotonin synthesis by stimulating tryptophan hydroxylase (50) or inhibits the release of 5-hydroxytryptamine as in rat brain (17), remains to be elucidated.

Our experimental data suggest the existence of a local renin-angiotensin-serotonin axis in the pineal gland modulated by the cervical sympathetic innervation. A feedback mechanism may operate in the intrapineal regulation of serotonin metabolism, with the participation of endogenous RAS (Fig. 3).

While ANG II activates both serotonin and melatonin synthesis, by means of cyclic-AMP and N-acetyltransferase (7), the resulting serotonin or even melatonin seems to act as an inhibitor of renin synthesis, since the specific blockade of tryptophan-hydroxylase by para-chlorophenylalanine induced a sharp increase of the pineal renin activity.

Besides its local modulatory function in the synthesis of serotonin and indolic hormones at the level of pinealocytes, the pineal RAS seems to be a major contributor to the significant ventricular pool of ANG I and II, which may trigger the activity of the periventricular structures involved in dipsogenesis, hydroelectrolytic and blood pressure regulation.

Recently, Camacho and Phillips (10) demonstrated that alpha-adrenergic blockade with phentolamine restricted to the brain, blocked the pressor response to ANG II in a dose-related manner, while drinking remained unaffected. Reduction of central serotonin stores by para-chlorophenylalanine, had no effects on the pressor or dipsogenic actions of ANG II. Therefore the drinking response does not appear to be mediated by the monoamines.

The pressor response may be modulated by opiates, Barnes and Ferrario (3) have shown a blunting of the pressor effects of intracerebral ANG II after naloxone. A possible interaction between the endogenous brain opiates and cerebral RAS is

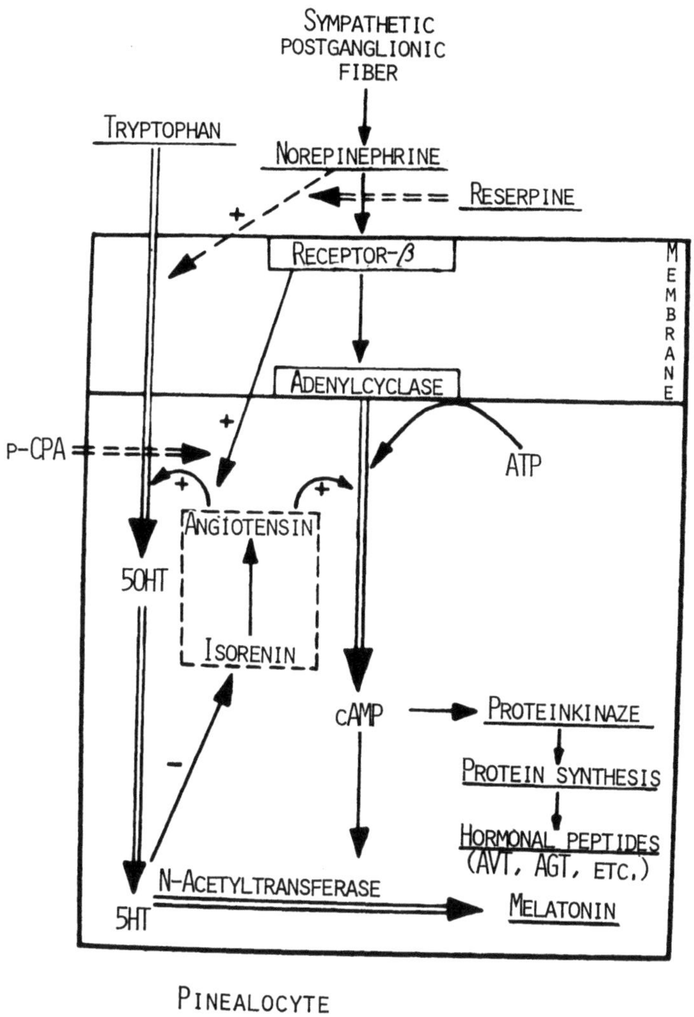

Fig. 3. Possible relationships between the renin-angiotensin system and serotonin metabolism in the rat pineal gland

also suggested by the analgesic effects of captopril recently demonstrated by Stine et al. (62). In agreement with these results, we obtained an analgesic effect after intracerebroventricular (i.c.v.) infusion of ANG II.

Evidence for an Analgesic Effect of Angiotensin i.c.v. Administration

The similarity of enkephalinase and ANG I-converting enzyme (6), both inhibited by captopril (64), suggested to us the possible participation of the endogenous cerebral ANG II, as well as enkephalin, in the analgesic effects of captopril. To check this hypothesis, we studied the analgesic effects of ANG II after i.c.v. administration in conscious rats.

Analgesia testing was carried out with the tail-flick method using a radiant heat source in 14 adult rats. For intracerebroventricular injection of angiotensin, a chronic cannula was stereotaxically implanted in the third ventricle. After 3-4 days, increasing doses of synthetic ANG II (Hypertensin Ciba) were injected i.c.v. in 10 μl.ANG II was effective in inducing a rapid-onset, dose-related and short-lasting analgesia (Fig. 4).

To discriminate whether angiotensin by itself induces analgesic effects or with mediation of enkephalin, the animals have been treated with naloxone, 10 mg/kg i.p., 30 min before the tail-flick test. In these animals angiotensin-induced analgesia was shown to be prevented by prior administration of naloxone (Fig. 5).

DISCUSSION

In this brief review of our results we have found new experimental evidence for local and modulating properties of the cerebral RAS. These investigations on its metabolic and functional implications point out some possible interrelationships between the RAS, serotonin and enkephalin in mammalian brain.

The presence of angiotensin-forming enzyme in the pineal gland, besides melatonin (2) and arginine-vasotocin (52) with their antigonadotropic (5,52,67), anticorticotropic (48), antithyreotropic (61), neurosedative (46) and hypnogenic (53) properties, leads us to propose that the RAS also participates in these effects. By its indolic and peptide hormones, the pineal induces neural and hypophyseal inhibitory effects, functioning as an antistress organ (47). Its activity is predominant under sympathetic nervous system control,and responds to changes of the adrenal medulla gland, and neural input in response to various stimuli. Activating serotonin and indolic hormone synthesis, the RAS activates the anabolic effects of pineal

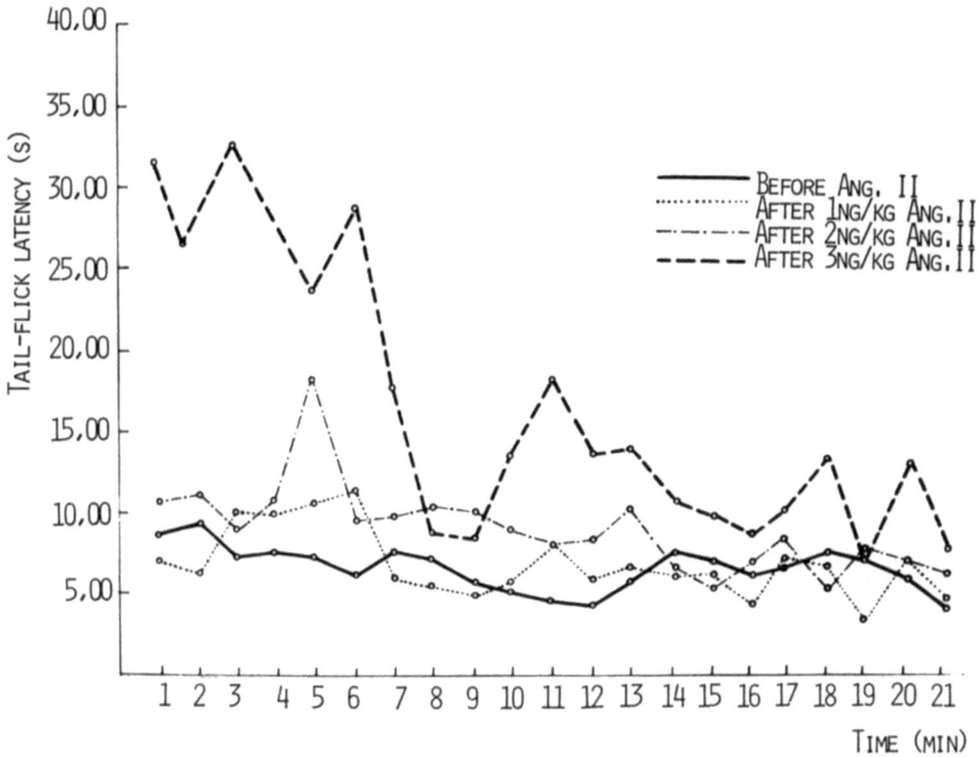

Fig. 4. Analgesic effects at different doses of angiotensin II intracerebroventricular (i.c.v.) injected in adult rats

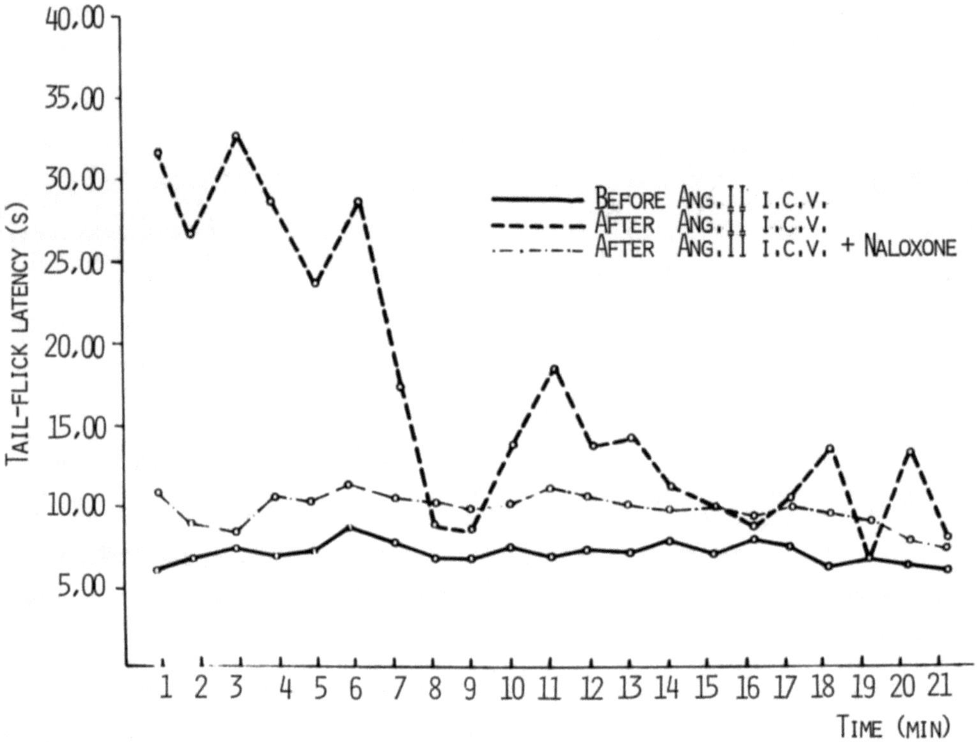

Fig. 5. Influence of naloxone on the analgetic effects of intracerebroventricular (i.c.v.) angiotensin II administration in adult rats

hormones. By its central pressor, dipsogenic and ADH releasing actions, angiotensin could also participate in repairing deficits in the extracellular fluid compartment created by catabolic reactions.

The same explanation could apply to the increase of cerebral enkephalin, expressed by naloxone inhibition of angiotensin analgesic effects. Margules has suggested (45) that the opioid peptides increase the inflows and decrease the outflows of bodily resources such as water, Na^{++}, fuels, heat, etc. Although far from complete, our preliminary results suggest some metabolic and functional relationships between brain renin, serotonin and opioid peptides, and open new ways in elucidating the physiopathological role of cerebral RAS.

SUMMARY

The renin content of pineal gland depends on the circadian rhythm and stimulation of the cervical sympathetic postganglionic fibres. Pineal renin activity is lowest during daylight and maximal during night dark. This variation follows the same pattern as the noradrenaline and melatonin in the rat pineal gland. Both chemical sympathectomy and bilateral superior cervical gangliectomy suppressed the daytime decrease of the pineal renin. The inhibiton of serotonin synthesis with para-chlorophenylalanine also abolished or even reversed the minimal mid-day values.

Intravenous infusion with angiotensin II in quasi-physiologic doses induced a significant increase of serotonin in dog pineal gland, brain stem and hypothalamus. Based on these findings, the existence of a local renin-angiotensin-serotonin axis is described in the pineal gland which is modulated by the cervical sympathetic innervation.

Evidence for an analgesic effect of angiotensin II after intracerebroventricular administration was obtained in adult rats. The angiotensin induced analgesia was prevented by prior administration of naloxone, suggesting that the analgesic effects of angiotensin II were mediated by opioid peptides. Thus, besides the hormonal role in the central regulation of the volemic and circulatory homeostasis, the renin-angiotensin system seems to modulate the cerebral serotonin and enkephalin metabolism.

REFERENCES

1. ASCHOFF J, CERESA F, IVERSEN L (1978) Angiotensin-converting enzyme: presence of high activity in choroid plexus of mammalian brain. Eur J Pharmacol 52: 147-150

2. AXELROD J, WEISSBACH H (1960) Enzymatic o-methylation of N-acetylserotonin to melatonin. Science 131:1312-1314

3. BARNES KL, FERRARIO CM (1980) Angiotensin and CNS regulation of the blood pressure. Clin exp hypertension 2(344):465-477

4. BENETATO G, HAULICA I, ULUITU M, BUBUIANU E, MOCODEAN J, STEFANESCU P, SUHACIU G (1964) L'effet central de l'angiotensine sur la pression artêrielle et sur la sêcrêtion de l'aldostêrone et de l'êquilibre êlectrolitique. Int J Neuropharmacol 2:565-570

5. BENSON B, ORTS RJ (1972) Regulation of ovarian growth by the pineal gland. In: Regulation of organ and tissue growth (GROSS RJ, ed). Academic Press, New York, p 315

6. BENUCK M, MARKS N (1979) Co-identity of brain angiotensin-converting enzyme with a membrane bound dipeptidyl carboxipeptidase inactivating met-enkephalin. Biochem Biophys Res Commun 88:215-221

7. BERG GR, KLEIN DC (1971) Pineal gland in organ culture. II-Role of 3',5'-monophosphate in the regulation of radiolabeled melatonin production. Endocrinology 89:453-459

8. BICKERTON RK, BUCKLEY JP (1961) Evidence for a central mechanism in angiotensin induced hypertension. Proc Soc Exp Biol Med 106:834-836

9. BRODIE BB, OLIN JS, KUNTZMAN RG, SHAVE PA (1957) Possible interrelationship between release of brain norepinephrine and serotonin by reserpine. Science 125: 1293-1296

10. CAMACHO A, PHILLIPS MI (1981) Separation of drinking and pressor response to central angiotensin by monoamines.Am J Physiol 240:R106-R113

11. CARLSSON A, ROSENGREN E, BERTLER A, NILSSON J (1957) Effect of reserpine on the metabolism of catecholamines. In: Psychotropic drugs (GARATTINI S, GHETTI V,eds) Elsevier Publ. Comp., Amsterdam, p 363

12. CELIO RM, CLEMENS DL, INAGAMI T (1980) Renin in anterior pituitary, pineal and neuronal cells of mouse brain, immunohistochemical localization. Biomed Res 1: 427-431

13. CHANGARIS DG, KEIL LC, SEVERS WB (1978) Angiotensin II immunohistochemistry of the rat brain. Neuroendocrinology 25:257-274

14. COCULESCU M, OPRESCU M, ZAGREAN L (1977) Angiotensin-like immunoreactive substance in pineal gland of normal and Brattleboro rats with hereditary diabetes insipidus. Rev Roum Méd Morphol Physiol 14:47-51

15. CORVOL P, DEVAUX C, MENARD J (1973) Pepstatin, an inhibitor for renin purification by affinity chromatography. FEBS Lett 34:189-192

16. DAUL CB, HEATH RG, GAREY RE (1975) Angiotensin forming enzyme in human brain. Neuropharmacol 14:75-80

17. DAVILA D, DAVILA T (1975) Effect of ANG II and its analogs on uptake and release of [^{14}C] 5-hydroxytryptamine by rat brain. Europ J Pharmacol 34: 329-335

18. DAY RP, REID IA (1976) Renin activity in dog brain: enzymological similarity to cathepsin D. Endocrinology 99:93-100

19. EPSTEIN AN, FITZSIMONS JT, ROLLS BJ (1969) Drinking caused by the intracranial injection of angiotensin into the rat. J Physiol (Lond) 200:98-100p

20. FINKIELMAN S, FISCHER-FERRARO C, DIAZ A, GOLDSTEIN DJ, NAHMOD VE (1972) A pressor substance in the cerebrospinal fluid of normotensive and hypertensive patients. Proc nat Acad Sci USA 69:3341-3344

21. FISCHER-FERRARO C, NAHMOD VE, GOLDSTEIN DJ, FINKIELMAN S (1971) Angiotensin and renin in rat and dog brain. J Exp Med 133:353-361

22. FITZSIMONS JT (1980) Angiotensin stimulation of the central nervous system. Rev Physiol Biochem Pharmacol 87:117-167

23. FUXE K, GANTEN D, HÖKFELT T, BOLME P (1976) Immunohistochemical evidence for the existence of angiotensin II-containing nerve terminals in the brain and spinal cord in the rat. Neurosci Lett 2:229-234

24. FUXE K, GANTEN D, HÖKFELT T, LOCATELLI V, POULSEN K, STOCK G, RIX E, TAUGNER R (1980) Renin-like immunocytochemical activity in the rat and mouse brain. Neurosci Lett 18:245-250

25. GANTEN D, MINNICH JL, GRANGER P, HAYDUK K, BRECHT HM, BARBEAU A, BOUCHER R, GENEST J (1971) Angiotensin forming enzyme in brain tissue. Science 173:64-65

26. GANTEN D, HUTCHINSON JS, SCHELLING P (1975) The intrinsic brain isorenin-angiotensin system: Its possible role in central mechanisms of blood pressure regulation. Clin Sci Mol Med 48:265-268s

27. GANTEN D, FUXE K, GANTEN U, HÖKFELT T, BOLME P (1977) The brain isorenin-angiotensin system: localization and biological function. In: Hypertension and brain mechanisms. Progress in Brain Res., vol. 47, pp 155-159. Elsevier, North-Holland, Biomed.Press, Amsterdam

28. GANTEN D, FUXE K, PHILLIPS MI, MANN JFE, GANTEN U (1978) The brain isorenin-angiotensin system: Biochemistry, localization and possible role in drinking and blood pressure regulation. In: Frontiers in Neuroendocrinology 5:61-99 (GANONG WF, MARTINI L, eds), Raven Press, New York

29. GANTEN D, UNGER TH, ROCKHOLD R, SHAZ K, SPECK G (1979) Central peptidergic stimulation: Its possible contribution to blood pressure regulation. In: Radioimmunoassay of drugs and hormones in cardiovascular medicine (ALBERTINI A, da PRADA M, PESKAR BA, eds),Elsevier, Amsterdam, pp 33-43

30. GANTEN D, SPECK G, MEYER D, LOOS HE, SCHELLING P, RETTIG R, UNGER TH (1980) Biochemical and functional aspects of brain renin. In: Enzymatic release of vasoactive peptides (GROSS F, VOGEL HG, eds), Raven Press, New York, pp 31-49

31. GIARMAN NJ, DAY M (1959) Presence of biogenic amines in the bovine pineal body. Biochem Pharmacol 1:235-241

32. HALBERG F (1969) Chronobiology. Ann Rev Physiol 31:675-715

33. HAULICA I, BRANISTEANU DD, ROSCA V, STRATONE A, BERBELEU V, BALAN G, BRAZDA I (1974) On the presence of the renin-like activity in the brain, pineal gland and hypophysis. Rev Roum Méd 12, 5:399-442

34. HAULICA I, BRANISTEANU DD, ROSCA V, STRATONE A, BERBELEU V, BALAN G, IONESCU L (1975) Renin-like activity in pineal gland and hypophysis. Endocrinology 96: 530-534

35. HAULICA I, COCULESCU M, GHINEA E, STRATONE A, PETRESCU GH, ROSCA V, OPRESCU M (1980) Biosynthesis of isorenin in cell culture from pineal glands of adult and new born rats. Life Sci 27:809-813

36. HAULICA I, PETRESCU GH, ULUITU M, ROSCA V, SLATINEANU S (1980) Influence of angiotensin II on dog pineal serotonin. Neurosci Lett 18:329-332

37. HIROSE S, YOKOSAWA H, INAGAMI T (1978) Immunochemical identification of renin in rat brain and distinction from acid proteases. Nature (Lond) 274:392-393

38. HOFFMAN WE, PHILLIPS MI, SCHMID PG (1977) Central angiotensin II induced responses in spontaneously hypertensive rats. Am J Physiol 232:H426-H433

39. HUTCHINSON JS, CSICSMAN J, KORNER PI, JOHNSTON CI (1978) Characterization of immunoreactive angiotensin in canine cerebrospinal fluid as des-asp-angiotensin II. Clin Sci Mol Med 54:147-151

40. INAGAMI T, YOKOSAWA H, HIROSE S (1978) Definitive evidence for renin in rat brain by affinity chromatographic separation from acid protease. Clin Sci Mol Med 52:121-123s

41. INAGAMI T, CELIO MR, CLEMENS DL, LAU D, TAKII Y, KASSELBERG AG, HIROSE S (1980) Renin in rat and mouse brain: immunohistochemical identification and localization. Clin Sci 59:49_s-51_s

42. KAPPERS AI, SCHADE PI (1965) Structure and function of the epiphysis cerebri. Elsevier Publishing Comp., Amsterdam, p 134

43. KAPPERS AJ, PEVET P (1979) Progress in brain research, vol. 52. The pineal gland of vertebrates including man. Elsevier Publishing Comp., Amsterdam

44. KOE KB, WEISSMAN A (1966) P-chlorophenylalanine - a specific depletor of brain serotonin. J Pharmacol Exp Ther 154:499-503

45. MARGULES DL (1979) Beta-endorphin and endoloxane: hormones of the autonomic nervous system for the conservation or expediture of bodily resources and energy in anticipation of famine or feast. Neurosci Behav Rev 3:155-162

46. MARCZYNSKI TJ, YAMAGUCHI N, LING GM, GRODZINSKA L (1964) Sleep induced by the administration of melatonin to the hypothalamus in unrestrained cats. Experientia 20:435-438

47. MILINE R (1980) The role of the pineal gland in stress. J Neurol Transm 47: 191-220

48. MOTTA M, SCHIAFFINI O, PIVA F, MARTINI L (1971) Pineal principles and the control of adrenocorticopin secretion. In: The pineal gland (WOLSTENHOLME GEW, KNIGHT J, eds). Churchill Livingstone, Edinburgh, p 279

49. MOUW D, BONJOUR JP, MALVIN RL, VANDER A (1971) Central action of angiotensin in stimulating ADH release. Am J Physiol 220:239-242

50. NAHMOD VE, FINKIELMAN S, BENARROCH E, PIROLA CJ (1978) Angiotensin regulates release and synthesis of serotonin in brain. Science 202:1091-1093

51. OSMAN YM, SMEBY RR, SEN S (1979) Separation of dog brain renin-like activity from acid protease activity. Hypertension 1:53-60

52. PAVEL S, PETRESCU S (1966) Inhibition of gonadotropin by a highly purified pineal peptide and by synthetic arginine vasotocin. Nature (Lond) 212:1054-1055

53. PAVEL S, PSATA P, GOLDSTEIN R (1977) Slow-waves sleep induced in cats by extremly small amount of synthetic and pineal vasotocin injected into third ventricle of the brain. Brain Res Bull 2:251-254

54. PHILLIPS MI, FELIX D (1976) Specific angiotensin II receptive neurons in the cat subfornical organ. Brain Res 109:531-540

55. PHILLIPS MI, MANN JFE, HAEBARA H, HOFFMAN WE, DIETZ R, SCHELLING P, GANTEN D (1977) Lowering of hypertension by central saralasin in the absence of plasma renin. Nature 270:445-447

56. PHILLIPS MI, WEYHENMEYER J, FELIX D, GANTEN D, HOFFMAN WE (1979) Evidence for an endogenous brain renin-angiotensin system. Fed Proc 38:2260-2266

57. QUAY WB (1974) Pineal chemistry. Ch. C. THOMAS publisher, Springfield, Illinois, p 154

58. SCHELLING P, SPECK G, UNGER TH, GANTEN D (1980) The brain renin-angiotensin system: Biochemistry, localization and functional aspects. In: Advances in experimental medicine (PARVEZ H, PARVEZ S, eds). Elsevier, North-Holland, Biomedical Press, Amsterdam, pp 243-288

59. SCHÖLKENS BA, STEINBACH R, JUNG W, GANTEN D (1980) Induction of angiotensin formation in brain: Biological effects and pharmacological interferences. In: Enzymatic release of vasoactive peptides (GROSS F, VOGEL HG, eds). Raven Press, New York, pp 365-381

60. SEVERS WB, SUMMY-LONG J, TAYLOR JS, CONNOR JD (1970) A central effect of angiotensin: Release of pituitary pressor material. J Pharmacol Exp Ther 174:27-34

61. SINGH DV, NARANGL GD, TURNER CW (1969) Effect of melatonin and its withdrawal on thyroid hormone secretion rate of female rats. J Endocrinol 49:489-492

62. STINE SM, YANG HYT, COSTA E (1980) Inhibition of in situ metabolism of (^{3}H) (met^5)-enkephalin and potentiation of (met^5)-enkephalin analgesia by captopril. Brain Res 188:295-299

63. SUMMY-LONG JY, SEVERS BW (1979) Macromolecular changes in the subfornical organ area after dehydration and renin. Am J Physiol 237 (1):R26-38

64. SWERTS JP, PERDRISOT R, MALFROY B, SCHWARTZ JC (1979) Is enkephalinase identical with angiotensin converting enzyme? Europ J Pharmacol 53:209-210

65. WEISSBACH H, WALKES TP, UDENFRIEND S (1958) A simplified method for measuring serotonin in tissues: simultaneous assay of both serotonin and histamine. J Biol Chem 230:865-871

66. WURTMAN RJ, AXELROD J, CHU EW, FISCHER JE (1964) Mediation of some effects of illumination on the rat oestrus cycle by the sympathetic nervous system. Endocrinology 75:266-271

67. WURTMAN RJ, AXELROD J (1965) The formation, metabolism and physiologic effects of melatonin in mammals. In: Structure and function of the epiphysis cerebri (KAPPERS JA, SCHADE JP, eds). Elsevier Publ. Comp., Amsterdam, p 520

68. WURTMAN RJ (1967) Effects of light and visual stimuli on endocrine function. In: MARTINI and GANONG - Neuroendocrinology II, Academic Press, New York, p 19-59

Central and Peripheral Mechanisms Mediating Angiotensin-Induced Thirst

R. W. Lind and A. K. Johnson

INTRODUCTION

When an animal enters a state of negative fluid balance, complementary physiological and behavioral response systems are mobilized for the restitution of proper hydration. An important mediating link between the receptors detecting fluid loss and the executors of compensatory action is the hormone, angiotensin. In recent years much effort has been directed toward elucidating the means by which this octapeptide interacts with the central nervous system to stimulate thirst. In this review we will discuss the major research in this area, present some new findings from our laboratory, and propose a model for the role of angiotensin in thirst.

EVIDENCE FOR A CENTRAL SITE OF ANGIOTENSIN

Primarily as a result of the work of J. T. Fitzsimons (11,12,13), by the late 1960's it was apparent that angiotensin stimulates water intake. In 1970, Epstein, Fitzsimons, and Rolls (9) reported that central application of the hormone elicits drinking at doses that are several orders of magnitude below those that are required via the peripheral route. Johnson and Epstein (19) demonstrated that the hormone was maximally dipsogenic when delivered into the ventricular space as opposed to most parenchymal sites, indicating a periventricular locus for central angiotensin receptors. Autoradiographic studies conducted by Volicer and Loew (36) and Shrager, Osborne, Johnson, and Epstein (30) showed that angiotensin does not cross the blood-brain barrier and therefore suggested that the central receptors reside within one or more of the circumventricular organs (which possess fenestrated capillaries).

Simpson and Routtenberg (33) provided direct evidence that one of the brain's extra-blood-brain barrier structures, the subfornical organ (SFO), may house receptors for angiotensin with their report that injections into this structure stimulate thirst. These authors further noted that lesions of the SFO abolished drinking to

Experimental Brain Research, Suppl. 4

intracerebral injections of angiotensin and subsequently reported a similar lesion-induced blockade with peripheral administration (32). Other workers (1,29,34) have reported similar results when SFO lesions are combined with peripheral administration of angiotensin.

In contrast to the consensus on the importance of the SFO to systemically applied angiotensin, Buggy, Fisher, Hoffman, Johnson, and Phillips (4) suggested that SFO lesions block drinking responses to intracranial (and hence intraventricular) angiotensin injections not by destroying the receptors that are normally stimulated by this route of administration, but by occluding the interventricular foramen with edema and debris and thereby inhibiting CSF flow. This reinterpretation offered by Buggy et al. (4) was largely based on studies in which cold cream plugs were applied to various sites within the ventricular space while angiotensin was injected intraventricularly at other sites (3,16). In order for drinking to result from CSF-borne ANG II, the octapeptide had to reach the antero-ventral third ventricle, or AV3V region. Preparations that allowed suffusion of the surface of the SFO, while denying its access to the AV3V region, did not result in drinking when angiotensin was injected.

It was demonstrated further that SFO lesions blocked drinking when angiotensin was injected into the lateral ventricles, but not when it was injected directly into the ventral third ventricle (3) and that SFO lesions (or any lesion near the interventricular foramen) will block water intake following lateral ventricle injections of angiotensin only during the acute period following the placement of the lesion (16). By comparison, when the AV3V region is destroyed with electrolytic lesions, angiotensin chronically fails to initiate drinking when it is delivered either centrally or peripherally (17).

LESION STUDIES TO DEFINE A CENTRAL SITE

In order to characterize the critical tissue within the region of the AV3V for mediating drinking responses to subcutaneously administered angiotensin (Hypertensin, Ciba, 1.56 mg/kg), we have recently examined the lesions of 31 rats. These rats, along with 30 animals of a sham-lesion group, were given presurgery drinking tests to establish that they would drink when presented with this stimulus. Following recovery from the post-lesion adipsia (18), the experimental and control animals were tested twice with angiotensin at the pre-lesion dose.

Rats with lesions ingested 0.3 ($\pm$ 0.2) ml (mean $\pm$ SEM) of water after a control injection of isotonic saline and these animals were considered to be unresponsive to angiotensin if they drank less than 0.5 ml during both 2-h test periods. The control rats drank 3.8 ($\pm$ 0.7) ml following angiotensin injections. At the completion of testing the rats with lesions were perfused with buffered formalin and their brains were frozen, sectioned horizontally, and stained with cresyl violet.

The brains from animals that failed to drink to angiotensin during post-surgical testing were studied to determine the common area of destruction. Overlap analyses of the original horizontal sections restricted the lateral extent of the common area to within 0.4 mm of the midline. Fig. 1 identifies structures in the AV3V region and presents a sagittal view of the tissue damage that was observed in all animals that were refractory to angiotensin. These lesions included periventricular tissue from the floor of the ventricle up to the anterior commissure, with substantial damage to the median preoptic nucleus and the organum vasculosum of the lamina terminalis (OVLT). However, several of the rats with ventrally placed lesions drank normally to angiotensin and the majority of these animals also suffered complete destruction of the OVLT. Thus, the median preoptic and periventricular preoptic nuclei appear to be crucial for angiotensin-induced thirst while the OVLT is of minor importance (22). In another study of this nature (23), the median preoptic nucleus and the OVLT were destroyed in all animals manifesting drinking deficits to ANG II (by the central and peripheral routes), while removal of the OVLT alone resulted in animals that responded normally to both routes of administration.

Regarding the possible role of the medial preoptic area as a receptive site for centrally applied angiotensin (2,21,28), our lesion data indicate that this large nucleus can be nearly completely destroyed without inhibiting drinking responses to angiotensin delivered by either the central or peripheral routes. A recent study by Richardson and Mogenson (28) suggests that injections into the medial preoptic area are most effective in eliciting thirst when diffusion from the injection site is maximized, thus implicating periventricular receptors in the AV3V region.

A recent study by Shrager and Johnson (29) examined the effects of electrolytic lesions placed at different levels in the anterior wall of the third ventricle from the OVLT up to the SFO. In agreement with work described above, damage to the OVLT was not correlated with drinking deficits to angiotensin (delivered peripherally), while destruction of the median preoptic nucleus was observed to block angiotensin-induced thirst. It was discovered further that lesions of the septum immediately anterior and dorsal to the anterior commissure and lesions of the SFO reliably blocked drinking responses to peripheral injections of angiotensin.

The finding by Shrager and Johnson (29) that electrolytic lesions placed between the SFO and the AV3V region blocked drinking responses to peripheral administrations of angiotensin suggested that connections may exist between these two areas that are relevant to angiotensin-induced thirst. As a test of this hypothesis, 28 male albino rats that were given pre-surgery screening tests for drinking responses to

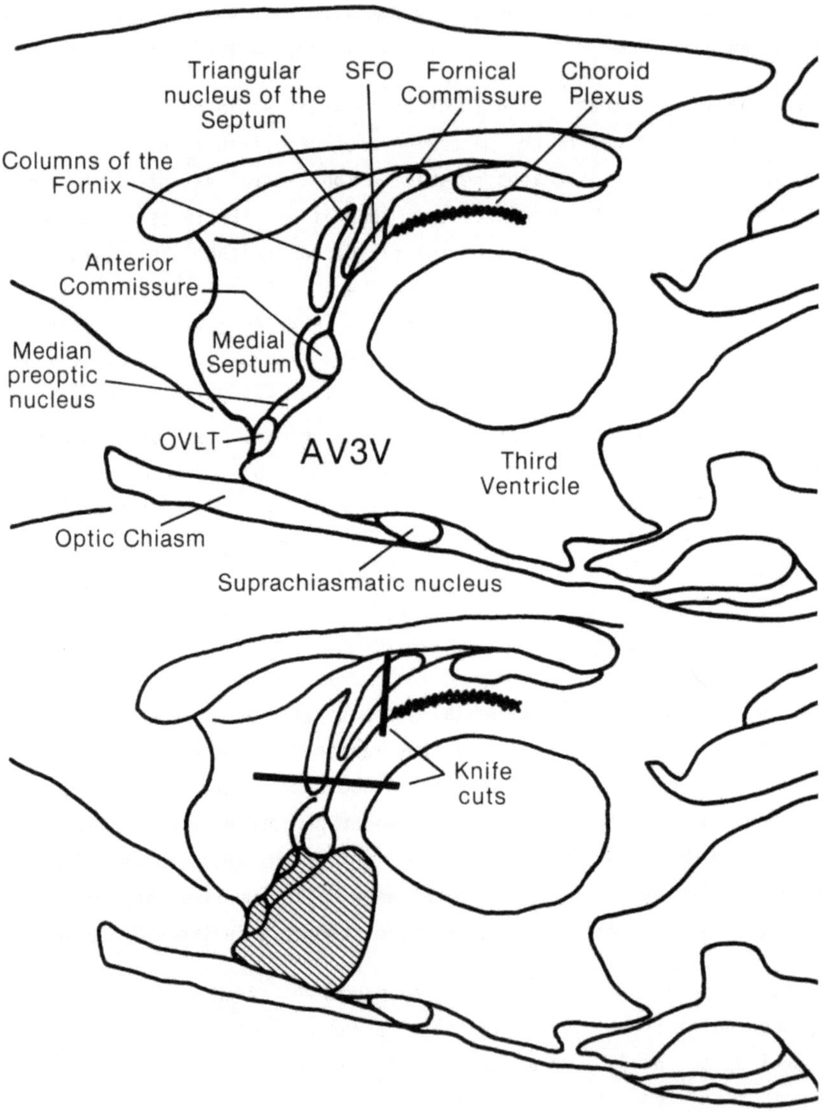

Fig. 1. Midsagittal drawings of the rat brain. Structures are identified in the upper figure (SFO, subfornical organ; OVLT, organum vasculosum of the lamina terminalis; AV3V, anterior-ventral third ventricle). The striped area in the lower figure illustrates the common area of destruction in rats that failed to drink when injected with angiotensin. The heavy bars depict the approximate size and location of knife cuts that were employed in a second experiment (see text for explanation)

angiotensin (3 mg/kg) received horizontal knife cuts between the SFO and the anterior commissure. The center of the cut was varied in the anterior-posterior direction from 2 mm anterior to bregma to 2 mm behind bregma. Two additional groups of animals consisted of six rats with vertical cuts of the dorsal stalk of the SFO (see Fig. 1) and five control animals that received trephinations but no cuts.

Post-lesion drinking tests identified six rats with horizontal cuts which were refractory to peripheral administration of two doses of angiotensin. In addition to the pre-lesion dose (3.0 mg/kg), a 1.5 mg/kg dose was also given. All of the rats with vertical cuts of the dorsal stalk of the SFO and the five control animals drank avidly when administered either dose of angiotensin (see Table 1).

Table 1. Two-hour intake (ml ± SEM) following angiotensin injections by controls and rats with vertical cuts and rats with horizontal cuts (see text and Fig. 1)

Dose of Angiotensin	Controls (n = 5)	Vertical Cut (n = 6)	Horizontal Cut (n = 6)
1.5 mg/kg	4.14 ± 0.95	4.30 ± 0.80	0
3.0 mg/kg	5.64 ± 1.53	5.91 ± 0.68	0

The horizontal knife cuts that produced drinking deficits to angiotensin were placed approximately at the dorsal-ventral level of the interventricular foramen and were centered on or near the midline. Included in effective cuts were the descending columns of the fornix, with at least partial involvement of the stria medullaris and septal areas adjacent to the fornices. Fig. 1 includes a depiction of the approximate size and location of these cuts.

NEURAL PATHWAYS FOR ANGIOTENSIN-MEDIATED DRINKING

These findings are consistent with the hypothesis that there is a functional connection for transmitting information between the SFO and the AV3V region and recent anatomical studies may have identified this substrate. Miselis, Shapiro, and Hand (26) and Miselis (25) have employed injections of tritiated amino acids into the SFO and have traced efferent connections to areas containing magnocellular neurosecretory neurons and to structures in the AV3V region (median preoptic nucleus, OVLT, periventricular preoptic nucleus). Lind, van Hoesen, and Johnson (24) have made horseradish-peroxidase injections into the SFO and find good agreement with the efferent projections reported by Miselis and his colleagues. In addition, we have also identified afferents to the SFO from both the median and medial preoptic nuclei. The former source of inputs to the SFO had previously been reported by Hernesniemi, Bruppacher, Kawana, and Sandri (15).

These anatomical studies clearly demonstrate that the SFO and the structures of the AV3V region are substantially interconnected. When considered in light of the functional experiments discussed above, these interconnections emerge as a likely conduit by which information from angiotensin receptors in the SFO is transmitted to and/or through the AV3V region for the mobilization of thirst. Considerable evidence (e.g., 6,27,35) indicates that the SFO contains angiotensin receptors, and it appears, largely on the basis of the work (some of which was discussed above) of Simpson and his colleagues (31,32,33), that these receptors mediate thirst triggered by circulating angiotensin. However, there is an ever-burgeoning literature that argues against the notion that the SFO is an exclusive target tissue for angiotensin-induced thirst. In addition to the ventricular plugging studies by Buggy and Fisher (3) and Hoffman and Phillips (16) which we have already considered, two recently reported lesion studies argue for a periventricular receptor site in addition to the SFO. The first of these (10) examined drinking responses to peripheral and central angiotensin in the opposum after complete removal of the SFO. It was found that destroying the SFO reliably eliminated drinking responses to peripherally administered angiotensin, but that substantial drinking remained with the central route of administration. Similar findings have been reported by Thunhorst, Lind, and Johnson (34) in the rat. In the latter case, however, it was noted that lesions need not destroy the entire SFO in order to block drinking responses to subcutaneous injections of angiotensin. Lesions that severed the ventral stalk which carries fibers connecting the SFO and AV3V region (24,25) were found to be equally effective.

What emerges from the foregoing discussion are the beginnings of the elucidation of a neural system for the mediation of thirst that is stimulated by circulating angiotensin. When titers of the octapeptide are sufficiently elevated to stimulate receptors in the SFO, these receptors send neural information ventrally to the AV3V region. On the other hand, angiotensin that is administered into the cerebral ventricles acts directly on receptors in the AV3V region. The extra-blood-brain barrier receptors in the SFO appear to be neither necessary nor sufficient when the hormone is injected directly into the ventricular space.

Assuming that there are at least two different sets of angiotensin receptors in the rat brain that are relevant to thirst, and that these two receptor populations are uniquely accessed by angiotensin of different sources, data reported by Johnson and Schwob (20) may be interpreted as evidence that these two systems are linked. These authors noted that intraventricular injections of saralasin (a competitive antagonist of angiotensin) significantly attenuated drinking responses to peripheral angiotensin, with no effect on drinking induced by injections of hypertonic saline. Considering the evidence indicating that circulating angiotensin acts at the SFO and that central angiotensin stimulates receptors in the AV3V region and is unable to effectively activate receptors in the SFO, it now seems unlikely that this blockade resulted from competition between the agonist (angiotensin) and antagonist (saralasin) at the same receptor site. (This conclusion rests, of course, on the assumption that centrally applied saralasin and angiotensin will reach exactly the same tissues). We would like to propose that the centrally applied blocker interrupted the efferent neural information from angiotensin receptors in the SFO at an angiotensin synapse in the AV3V region.

It has been well established by Ganten and his colleagues (14) that angiotensin, and the synthetic machinery for its manufacture, exist in the brain. Immunocytochemical studies have identified angiotensin II in nerve fibers throughout the brain (8) and localized angiotensin I within cells (7). Changaris et al. (7) detected "astrocyte-like" cells in the SFO which exhibited immunoperoxidase activity for angiotensin I. It is possible that what Changaris and his colleagues consider to be astrocytes are actually stellate neurons that transmit excitatory information from angiotensin receptors to the AV3V region.

If this speculation can be carried a bit further, we would like to suggest that these AV3V-region angiotensin-sensitive neurons reside primarily in the median preoptic nucleus (see Fig. 2). We base this suggestion, first of all, on our HRP tracing work which has identified a prominent projection from the SFO to the median preoptic nucleus. The second reason for this suggestion derives from collaborative attempts (23) to separate the tissue in the AV3V region that is crucial for mediating thirst that results from peripheral administration of angiotensin from that tissue which is important for central application. In nearly every case that we have studied, we have found that if an animal with a lesion in the AV3V region failed to drink to one route of administration, it also failed to respond by the other route. Furthermore, rats that did not drink to angiotensin had the greatest extent of common lesion damage along the anterior wall of the third ventricle from the OVLT to the anterior commissure, with the median preoptic nucleus destroyed in all cases.

An interesting observation by Buggy and Johnson (5) finds a ready interpretation with this model. In replicating and extending the early plugging studies conducted by Buggy and Fisher (3), Buggy and Johnson applied plugs to the AV3V region and tested the dipsogenic effects of peripherally administered angiotensin. This was done in order to test the hypothesis that circulating angiotensin enters the CSF and then stimulates periventricular receptors in the AV3V region. If this were

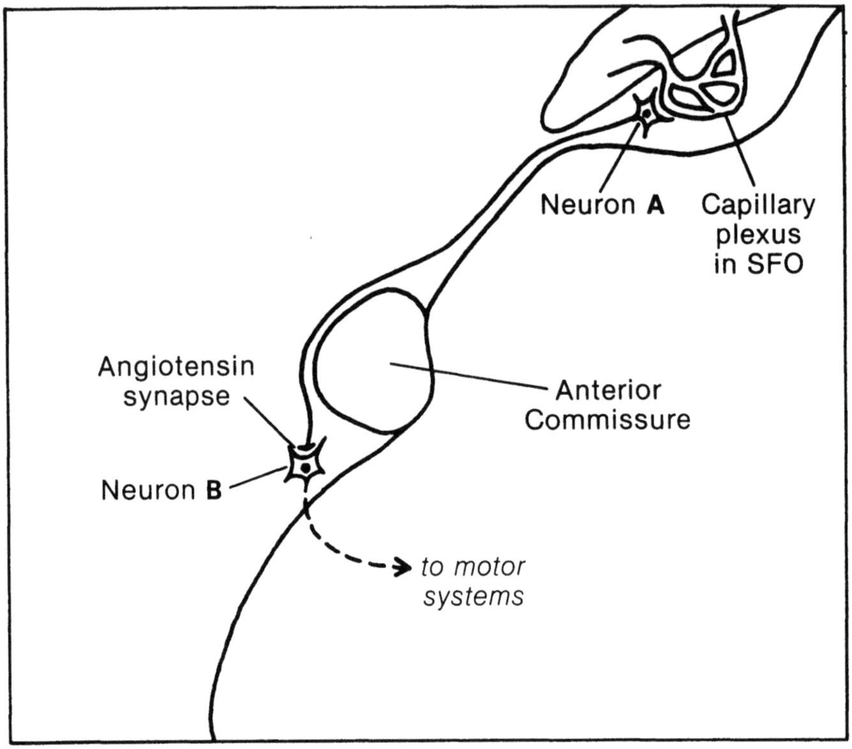

Fig. 2. A model for the dual role of angiotensin in thirst. Blood-borne angiotensin penetrates the fenestrated capillaries in the subfornical organ (SFO) and activates Neuron "A," the angiotensin receptor. Excitatory information is conducted ventrally from the SFO to the median preoptic nucleus where it is transferred to Neuron "B" by the release of angiotensin as a neurotransmitter. The information leaving Neuron "B" is channeled to motor systems for the eventual expression of the ingestive response

the route by which angiotensin reaches central receptors, an attenuation of drinking would have been expected with this preparation. Instead, these authors noted what appeared to be an enhancement of responding. According to the present model, this finding might be explained as a failure of the synaptic angiotensin to diffuse away from its receptors following release, producing prolonged stimulation and an enhanced response. More specifically, these median preoptic angiotensin synapses may lie very close to the ventricles so that advantage might be taken of the nearby CSF pool to help inactivate the transmitter by diffusion. Replacing the CSF with a plug would clearly disrupt this diluting mechanism and result in over-activation of the receptors.

An observation made by Johnson and Schwob (20) suggests that this hypothesized angiotensin synapse in the median preoptic nucleus is not relevant to thirst induced by osmotic stimulation. Whereas central application of saralasin significantly attenuated drinking response to peripheral administrations of angiotensin, drinking responses following subcutaneous injections of hypertonic saline were unaffected. If the proposed angiotensin synapse relays information from osmoreceptors for the eventual expression of the ingestive response, a similar inhibition would have been expected.

SUMMARY

To summarize our current understanding of the neural mechanisms subserving angiotensin-induced thirst, the following points can be made:

1) The SFO contains the major receptors related to thirst for circulating angiotensin.
2) Centrally applied saralasin blocks drinking to peripherally administered angiotensin, but not by competition at receptors in the SFO.
3) The AV3V contains receptors for intraventricular angiotensin.
4) The AV3V is crucial for thirst that is initiated by stimulating angiotensin receptors in the SFO.
5) The SFO sends prominent neural projections to structures of the AV3V region, especially the median preoptic nucleus.
6) Interrupting the connections between the SFO and AV3V blocks drinking responses to peripheral, but not central, angiotensin.
7) Lesions of the median preoptic nucleus block drinking responses to both central and peripheral angiotensin.

We have consolidated these conclusions and interpret them as evidence for a substrate for thirst which employs angiotensin as both a hormone and a neurotransmitter. We envisage blood-borne angiotensin signalling fluid loss to the central nervous system by activating receptors in the SFO. This information is transmitted neurally to the median preoptic nucleus where secondary neurons are excited by angiotensin as a neurotransmitter.

ACKNOWLEDGEMENTS

Original data presented in this review were collected with research funds from Public Health Service Research Grants HLP-14388 and 1 R01 HL24102. A Physiological Psychology Training Grant, 1 T32 MH 15773, provided funding for Wally Lind, and Kim Johnson is supported by an NIMH Research Scientist Development Award, 1 K02 MH 00064. We wish to thank M. J. Brody, D. K. Hartle, E. E. Shrager, and G. W. van Hoesen for their colleagueship and collaboration on the recent unpublished findings referred to in the review. We are grateful to Becky Rowe for typing the manuscript and to Lynne Ohman for her comments on a preliminary draft.

REFERENCES

1. ABDELAAL AE, ASSAF SY, KUCHARCZYK J, MOGENSON GJ (1974) Effect of ablation of the subfornical organ on water intake elicited by systemically administered angiotensin II. Can J Physiol Pharmacol 52:1217-1220

2. ASSAF SY, MOGENSON GJ (1976) Evidence that the preoptic region is a receptive site for the dipsogenic effects of angiotensin II. Pharmacol Biochem Behav 5: 697-699

3. BUGGY J, FISHER AE (1976) Anteroventral third ventricle site of action for angiotensin induced thirst. Pharmacol Biochem Behav 4:651-660

4. BUGGY J, FISHER AE, HOFFMAN WE, JOHNSON AK, PHILLIPS MI (1975) Ventricular obstruction:Effect on drinking induced by intracranial injection of angiotensin. Science 190:72-74

5. BUGGY J, JOHNSON AK (1978) Angiotensin-induced thirst: Effects of third ventricle obstruction and periventricular ablation. Brain Res 149:117-128

6. BURANARUGSA P, HUBBARD JI (1979) The neuronal organization of the rat subfornical organ in vitro and a test of the osmo- and morphine-receptor hypotheses.J Physiol (Lond) 291:101-116

7. CHANGARIS DG, DEMERS LM, KEIL LC, SEVERS WB (1977) Immunopharmacology of angiotensin I in brain. In: BUCKLEY JP, FERRARIO CM, LOKHANDWALA MF (eds) Central actions of angiotensin and related hormones. Pergamon Press, Inc., New York

8. CHANGARIS DG, KEIL LC, SEVERS WB (1978) Angiotensin II immunohistochemistry of the rat brain. Neuroendocrinology 25:257-274

9. EPSTEIN AN, FITZSIMONS JT, ROLLS BJ (1970) Drinking induced by injection of angiotensin into the brain of the rat. J Physiol (Lond) 210:457-474

10. FINDLAY ALR, ELFONT RM, EPSTEIN AN (1980) The site of the dipsogenic action of angiotensin II in the North American opposum. Brain Res 198:85-94

11. FITZSIMONS JT (1966) Hypovolaemic drinking and renin. J Physiol (Lond) 186: 130-131

12. FITZSIMONS JT (1969) The role of renal thirst factor in drinking induced by extracellular stimuli. J Physiol (Lond) 201:349-368

13. FITZSIMONS JT (1972) Thirst. Physiol Rev 52:468-561

14. GANTEN D, HUTCHINSON JS, SCHELLING P, GANTEN U, FISCHER H (1975) The iso-renin angiotensin systems in extra-renal tissue. Clin Exp Pharmacol Physiol 2: 127-151

15. HERNESNIEMI J, KAWANA E, BRUPPACHER H, SANDRI C (1972) Afferent connections of the subfornical organ and of the supraoptic crest. Acta Anat 81:321-336

16. HOFFMAN WE, PHILLIPS MI (1976) The effect of subfornical organ lesions and ventricular blockade on drinking induced by angiotensin II. Brain Res 108:59-73

17. JOHNSON AK, BUGGY J (1978) A critical analysis of the site of action for the dipsogenic effect of angiotensin II. In: BUCKLEY JP, FERRARIO CM, LOKHANDWALA MF (eds) Central actions of angiotensin and related hormones. Pergamon Press, New York

18. JOHNSON AK, BUGGY J (1978) Periventricular preoptic-hypothanlamus is vital for thirst and normal water economy. Amer J Physiol 234:R122-R129

19. JOHNSON AK, EPSTEIN AN (1975) The cerebral ventricles as the avenue for the dipsogenic action of intracranial angiotensin. Brain Res 86:399-418

20. JOHNSON AK, SCHWOB JE (1975) Cephalic angiotensin receptors mediating drinking to systemic angiotensin II. Pharmacol Biochem Behav 3:1077-1084

21. KUCHARCZYK J, ASSAF SY, MOGENSON GJ (1976) Differential effects of brain lesions on thirst induced by the administration of angiotensin II to the preoptic region, subfornical organ and anterior third ventricle. Brain Res 108:327-337

22. LIND HM, LIND RW, SHRAGER EE, BEALER SL, JOHNSON AK (1979) Critical tissues within the periventricular region of the antero-ventral third ventricle (AV3V) associated with specific thirst deficits. Neurosci Abst 5:22

23. LIND RW, HARTLE DK, BRODY MJ, JOHNSON AK (1981) Separation of the functional deficits induced by lesions of the AV3V. Fed Proc 40:390

24. LIND RW, van HOESEN GW, JOHNSON AK (to be published, 1981) Connections of the subfornical organ (SFO). Neurosci Abst 7

25. MISELIS RR (1980) New efferents and afferents of the subfornical organ: A strategic neural circuitry for the control of water balance behaviorally and physiologically. Neurosci Abst 6:32

26. MISELIS RR, SHAPIRO RE, HAND PJ (1979) Subfornical organ efferents to neural systems for control of body water. Science 7:1022-1025

27. PHILLIPS MI, FELIX D (1976) Specific angiotensin II receptive neurons in the cat subfornical organ. Brain Res 109:531-540

28. RICHARDSON DB, MOGENSON GJ (1981) Water intake elicited by injections of angiotensin II into preoptic area of rats. Am J Physiol 240:R70-R74

29. SHRAGER EE, JOHNSON AK (1980) Contributions of periventricular structures of the rostral third ventricle to the maintenance of drinking responses to humoral dipsogens and body fluid homeostasis. Neurosci Abst 6:128

30. SHRAGER EE, OSBORNE MJ, JOHNSON AK, EPSTEIN AN (1975) Entry of angiotensin into cerebral ventricles and circumventricular structures. In: DAVIES DS, REID JL (eds) Central actions of drugs in blood pressure regulation. University Park Press, Baltimore

31. SIMPSON JB, EPSTEIN AN, CAMARDO JS Jr (1978) Localization of receptors for the dipsogenic action of angiotensin II in the subfornical organ of rat. J Comp Physiol Psychol 92:581-601

32. SIMPSON JB, ROUTTENBERG A (1975) Subfornical organ lesions reduce intravenous angiotensin-induced drinking. Brain Res 88:154-161

33. SIMPSON JB, ROUTTENBERG A (1973) Subfornical organ: Site of drinking elicitation by angiotensin. Science 181:1172-1175

34. THUNHORST RL, LIND RW, JOHNSON AK (to be published, 1981) Lesions of the subfornical organ (SFO) block drinking to peripheral but not central angiotensin. Neurosci Abst 7

35. Van HOUTEN M, SCHIFFRIN EL, MANN JFE, POSNER BI, BOUCHER R (1980) Radioautographic localization of specific binding sites for blood-borne angiotensin II in the rat brain. Brain Res 186:480-485

36. VOLICER L, LOEW C (1971) Penetration of angiotensin II into the brain. Neuropharmacologia 10:631-636

Effects of Central Angiotensin on Renin Release from the Kidney

J. P. Buckley, B. S. Jandhyala, and M. F. Lokhandwala

INTRODUCTION

Ganten et al.(10) first reported the presence of renin in the dog brain in 1971, and since that time data have accumulated from various laboratories supporting the concept of a brain iso-renin-angiotensin system (9); however, this hypothesis is not universally accepted (16). If one accepts the fact that a brain renin-angiotensin system exists and that this system plays an important role in controlling various physiological systems one may ask if there is a relationship of the central renin-angiotensin system to the peripheral renin-angiotensin system. One approach would be to determine if there is a relationship between the concentration of angiotensin II in the cerebrospinal fluid and the rate of renin release from the kidney.

MATERIALS AND METHODS

Cats of either sex weighing between 2 to 3.5 kg were anesthetized with sodium pentobarbital, 30 mg/kg., i.v. A femoral artery was cannulated and a catheter inserted into the abdominal aorta to record aortic blood pressure. A femoral vein was catheterized to enable drug administration. Heart rate was monitored from the pressure pulse via a tachometer and all parameters recorded on a Narco-Bio systems physiograph. A 22-gauge unbeveled stainless steel needle was stereotoxically lowered into the left lateral ventricle and artificial cerebrospinal fluid (15) perfused as previously described (20). The perfusion rate was 0.1 ml/min and injections into the ventricular system were made through a three way stopcock placed into the ventricular cannula (11).

Two ml of blood was withdrawn from the renal vein at the time that arterial blood pressure reached peak levels following angiotensin II administration and collected in pre-chilled tubes containing 5 mg of EDTA. The tubes were immediately placed in ice and centrifuged at +4°C to recover plasma. Plasma renin activity (PRA) was determined by radioimmunoassay utilizing a Squibb angiotensin I immunotope kit.

Experimental Brain Research, Suppl. 4

The effect of intraventricular administered angiotensin II on urinary sodium excretion was studied in a separate group of cats. The animals were prepared as previously described and one of the ureters catheterized with polyvinyl tubing to collect urine. Normal saline solution was infused intravenously at the rate of 0.5 ml/min. Urinary sodium levels were estimated by means of an atomic absorption spectrophotometer. Angiotensin II, 100 ng/kg/min, was infused for approximately two to three minutes at which time pressor effects equivalent to the previous study were obtained and urine samples collected during this period. In some experiments the left kidney was acutely denervated. All connective tissue was carefully ligated, sectioned to free the kidney and nerve fibers and tissue accompanying the vasculature similarly treated so that the encapsulated kidney was connected to the body only via the renal vasculature.

Statistical significance within a group was estimated using the paired t-test and significance between two different groups determined by means of the Student's t-test.

RESULTS

Effects of Intraventricularly administered Angiotensin II, 2 µg, on Plasma Renin Activity.

Intraventricular administration of angiotensin II produced significant increases in mean blood pressure and a significant decrease in PRA (Fig. 1). The effects of angiotensin II on both blood pressure and PRA were reversible since they returned to control levels 30 to 45 minutes after administration of the octapeptide. Central administration of angiotensin II to cats in which the left kidney was acutely denervated produced significant increases in blood pressure; however, the decrease in PRA was less than that seen in cats in which the kidney was still innervated. Both the absolute as well as the percent reduction in PRA was significantly less in the denervated group as compared to innervated group. The resting values for PRA were significantly less in cats with denervated kidneys as compared with the innervated group (12.10 ng/ml/h innervated and 5.85 ng/ml/h denervated). There was no significant difference between the resting blood pressures of the two groups.

Effects of Angiotensin II, 100 ng/kg/min, on Sodium Excretion

The concentration of urinary sodium and the rate of sodium excretion in the cats were 98.08 $\pm$ 7.2 mEq/l and 0.46 $\pm$ 0.062 mEq/h respectively and following denervation they were 106.6 $\pm$ 3.6 mEq/l and 0.40 $\pm$ 0.058 mEq/h.

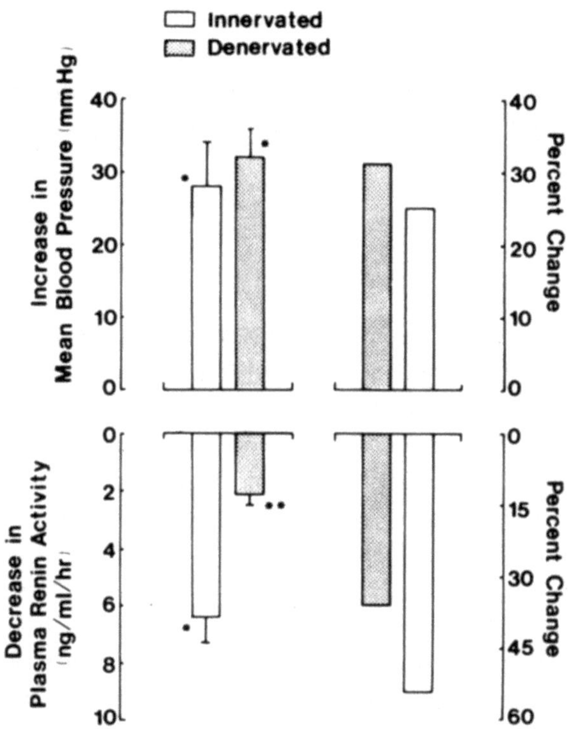

Fig. 1. Effects of intraventricular administration of angiotensin II, 2 µg, on mean arterial pressure and PRA. (*) Significantly different from pre-angiotensin levels ($p < 0.01$); (**) significantly different from response obtained prior to acute renal denervation; ($p < 0.05$) (n=5) (from Lokhandwala et al. (1978) Clin Exptl Hypertension 1:167-175)

The intraventricular infusion of angiotensin II produced significant increases in the concentration of urinary sodium as well as the rate of sodium excretion (Fig.2). Denervation of the kidney did not significantly change the concentration of urinary sodium; however, it did significantly reduce the increase in the rate of sodium excretion (Fig. 2).

DISCUSSION

The central administration of angiotensin II has been reported to increase sympathetic outflow from the central nervous system (2) whether the compound was administered via the lateral ventricles (18) or via the vertebral arteries (7). This increase in sympathetic activity would be expected to result in an activation of β-adrenergic receptors in the kidney resulting in an increase in PRA. We had previously reported that phentolamine perfusion, 50 μg/min markedly potentiated the pressor effects produced by intraventricular administration of angiotensin (19) and that saralasin, 5 μg/kg/min i.v. decreased the centrally induced pressor effects of angiotensin II following perfusion of phentolamine into the cerebroventricular system by approximately 68% (20). We concluded that angiotensin II might stimulate receptors in the central nervous system capable of activating an increase in renin production by the kidney and suggesting a positive feedback relationship between CSF renin and PRA. The pressor responses obtained in the present study to intraventricular administration of angiotensin II were accompanied by marked decreases in PRA in the renal venous blood and support the findings of Ericksson and Fyhrquist (6) who reported that the intraventricular administration of angiotensin II to conscious goats reduced PRA. Such a reduction in PRA could be secondary to and triggered in response to the elevation of systemic arterial pressure or it could be the result of a direct central action of the peptide. Either conclusion is consistent with the observation that acute renal denervation significantly attenuated the centrally administered angiotensin reductions in PRA but not in pressor activity.

The infusion of angiotensin II, 0.5 to 1 μg into the lateral ventricles in conscious goats caused a marked decrease of PRA (6) which was accompanied by an increase in arterial pressure of 25 to 30 mmHg, increased ADH release and increased renal sodium excretion. The intraventricular administration of renin, 0.1 G.U. to dogs (17) decreased PRA from 6.7 $\pm$ 2.0 to 3.9 $\pm$ 1.1 ng/ml/3 h. The same investigators administered renin, 0.1 G.U. into the third ventricle of anesthetized dogs undergoing a water diuresis and obtained a prompt increase in plasma vasopressin concentration. Malayan et al. (13) reported that the intraventricular administration of angiotensin II, 500 ng, to anesthetized dogs decreased PRA to 79.2 $\pm$ 6.1% of control values at 30 minutes and that this decrease in PRA was accompanied by

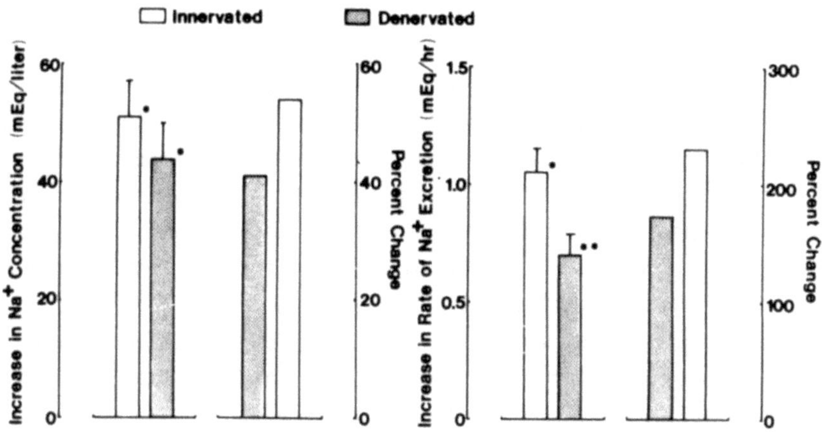

Fig. 2. Effects of intraventricular infusion of angiotensin II, 100 ng/kg/min, on the rate of sodium excretion and concentration of urinary sodium. (*) Significantly different from pre-angiotensin levels ($p < 0.01$); (**) significantly different from responses obtained prior to acute denervation ($p < 0.05$) (n=5) (from Lokhandwala et al. (1978) Clin Exptl Hypertension 1:167-175)

an increase in plasma vasopressin concentration from 9.7 $\pm$ 1.8 to 20.4 $\pm$ 3.6 pg/ml. The administration of angiotensin II centrally to hypophysectomized dogs did not significantly increase PRA or vasopressin. They concluded that the suppression of PRA by centrally administered angiotensin II could be a result of increased vasopressin secretion and/or increased blood pressure. The data obtained in the hypophysectomized dog suggest that it was the increase in the vasopressin secretion rather than the elevation in blood pressure that was responsible for the suppression of PRA and that this procedure did not significantly modify the pressor responses to centrally administered angiotensin II but completely abolished a rise in plasma vasopressin concentration and suppression of PRA. It should be noted, however, that the increase in mean arterial pressure was only 8 mmHg and that this might have been sufficient to affect renin secretion by the kidney.

The increase in urinary sodium excretion following the intraventricular administration of angiotensin II seen in the present study is similar to that reported by other investigators in goats (6) and dogs (17). We have consistently obtained both natriuresis and diuretic effects upon the administration of angiotensin II intraventricularly to cats (4). It is therefore difficult to correlate the positive effects on kidney function with increased ADH release; however, there is no question that ADH is released and that other mechanisms such as neuronal innervation may be overriding the effects of ADH on the kidney. Angiotensin II administered via the carotid inflow to the recipient's head in dog cross-circulation experiments produced marked increases in blood pressure in both the recipient and donor dogs (3). Urine production was normal in the recipient (3) but almost completely nil in the donor (unpublished data) suggesting the release of ADH from the recipient's head to the donor's body. The difference in the data obtained in these 2 experiments could be due to the fact that in the present study the angiotensin II is CSF borne and in the cross-circulation studies the peptide was blood borne or it could be a species related difference.

Investigators have reported that the administration of subpressor doses of angiotensin II, 6.25 and 12.5 ng/kg/min intraventricularly for twenty minutes to pentobarbital anesthetized rats did not decrease PRA whereas the infusion of pressor doses of angiotensin II significantly decreased plasma renin activity (14). It should be noted that in these experiments the increase in mean arterial pressure was approximately 10 mmHg. It would appear that the increase in ADH secretion plays an important role in both the pressor activity and the suppression of PRA when angiotensin II levels within the cerebrospinal fluid are increased; however, it does appear that other mechanisms are also involved and that perhaps a neuronal component contributed to the effects on PRA.

Now how can one relate the above to the hypertensive patient. There has been some limited evidence suggesting a positive correlation between CSF renin activity and the level of arterial blood pressure (8). Since increasing the levels of angiotensin II or renin in the CSF decreases PRA, this may account for the last of correlation between PRA and essential hypertension. Investigators have reported that essential hypertensive patients exhibit an abnormally high excretion of sodium and water following saline or salt load (1,5). Luft et al. (12) concluded that the degree of natriuresis is not solely a function of an elevated arterial pressure, but appears to be related to renin suppression.

The data in this present report support this view in that pressor effects to intraventricular angiotensin II were accompanied by suppression of PRA and increased sodium excretion. In addition acute renal denervation significantly attenuated the latter two effects without altering pressor activity, suggesting that sodium excretion is closely associated with renin levels. These studies suggest that the brain iso-renin-angiotensin system plays a role in the pathogenesis of essential hypertension, especially in those types associated with low PRA.

REFERENCES

1. BALDWIN DS, BIGGS AW, GOLDRING W, HULET WH, CHASSIS H (1958) Exaggerated natriuresis in essential hypertension. Am J Med 24:893-902

2. BICKERTON RK, BUCKLEY JP (1961) Evidence for a central mechanism in angiotensin-induced hypertension. Proc Soc Exp Biol Med 106:834-836

3. BUCKLEY JP (1965) Effects of synthetic angiotensin II on catecholamine levels and biological activity. Acta Physiol Scand 65:273-278

4. BUCKLEY JP, SINGH S, STEENBERG ML, JANDHYALA BS (1977) Intraventricular infusion of angiotensin II on the hemodynamics and renal function of α-chloralose anesthetized cats. Circ Res 40:152-156

5. COTTIER PT, WELLER JM, HOOBLER SW (1958) Effect of intravenous sodium chloride load on renal hemodynamics and electrolyte secretion in essential hypertension. Circulation 17:750-760

6. ERICKSSON L, FYHRQUIST F (1976) Plasma renin activity following central infusion of angiotensin II and altered CSF sodium concentration in the conscious goat. Acta Physiol Scand 98:209-216

7. FERRARIO CM, GILDENBERG PL, McCUBBIN JW (1972) Cardiovascular effects of angiotensin mediated by the central nervous system. Circ Res 30:257-262

8. FINKIELMAN S, FISCHER-FERRARIO C, DIAL A, GOLDSTEIN DJ, NAHMOD VE (1972) A pressor substance in the cerebrospinal fluid of normotensive and hypertensive patients. Proc Nat Acad Sci 69:3341-3344

9. GANTEN D, FUXE K, PHILLIPS MI, MANN JFE, GANTEN U (1978) The brain iso-renin-angiotensin system: Biochemistry, localization, and possible role in drinking and blood pressure regulation. In: GANONG WF, MARTINI L (eds) Frontiers in neuroendocrinology. Raven Press, New York, pp 61-91

10. GANTEN D, MARQUEZ-JULIO A, GRANGER P, HAYDUK K, KARSUNKY KP, BOUCHER R, GENEST J (1971) Renin in dog brain. Am J Physiol 221:1733-1737

11. LOKHANDWALA MF, BUCKLEY JP, JANDHYALA BS (1978) Reduction of plasma renin activity by centrally-administered angiotensin II in cats. Clin and Exptl Hypertension 1:167-175

12. LUFT FC, GRIM CS, WILLS LR, HIGGINS JT Jr., WEINBERGER MH (1977) Natriuretic response to saline infucion in normotensive and hypertensive man: the role of renin suppression in exaggerate natriuresis.Circulation 55:779-784

13. MALAYAN SA, KEIL LC, RAMSAY DJ, REID IA (1979) Mechanism of suppression of plasma renin activity by centrally administered angiotensin II. Endocrinology 104: 672-675

14. MATSUO H, ICHIKAWA S, SAKAMAKI T, TAJIMA Y, KOGURE M, MURATA K (1981) The effect of central iso-renin-angiotensin system on pressor responsiveness to angiotensin II. Life Sci 28:2329-2336

15. MERLIS JK (1940) The effect of changes in the calcium content of the cerebrospinal fluid on spinal reflex activity in the dog. Am J Physiol 131:67-72

16. RAMSAY DJ (1979) The brain renin-angiotensin system:A re-evaluation. Neuroscience 4:313-321

17. REID IA, DAY RP (1977) Interactions and properties of some components of the renin-angiotensin system of the brain. In: BUCKLEY JP, FERRARIO CM (eds) Central actions of angiotensin and related hormones. Pergamon Press, New York, pp 267-282

18. SMOOKLER HH, SEVERS WB, KINNARD WJ, BUCKLEY JP (1966) Centrally mediated cardiovascular effects of angiotensin II. J Pharmacol Exp Ther 153:485-497

19. VOLLMER RR, BUCKLEY JP (1977) Interaction of angiotensin II with central adrenergic receptors. In: BUCKLEY JP, FERRARIO CM (eds) Central actions of angiotensin and related hormones. Pergamon Press, New York, pp 197-204

20. VOLLMER R, BUCKLEY JP, SOLOMON TA, JANDHYALA BA (1974) Central hypertensive actions of angiotensin. Acta Physiol Lat Am 24:582-586

Nomenclature of the Renin-Angiotensin System

Report of the Nomenclature Committee of the International Society of Hypertension (ISH) on the Occasion of the 5th Meeting of the ISH in Paris, June 12 - 14, 1978 published in: Klin Wochenschr. 56 (Suppl. I), 187-190 (1978)

Biochemical Nomenclature

On the occasion of the Fourth Meeting of the International Society of Hypertension (ISH) in Sydney, 24 - 26 February 1976, a nomenclature committee was founded with the specific task to review the present nomenclature of the renin-angiotensin system (RAS) and to adopt it to the rules laid down by the Commission on Biochemical Nomenclature of the International Union of Pure and Applied Chemistry (IUPAC), and the International Union of Biochemistry (IUB).

Apart from the members of the ISH-Committee[1], the Council for High Blood Pressure Research of the American Heart Association also appointed a nomenclature committee[2] which joined the efforts of the ISH-Committee. Valuable suggestions came from scientists throughout the world in response to different memoranda which were drafted and circulated by the committee.

It was agreed upon that there may be two kinds of nomenclature, one trivial working nomenclature with usually brief and historically grown names, and a second systematic nomenclature to be introduced if the substance is well enough characterized to allow such classification.

The use of standardized nomenclature implies the danger of prematurely classifying substances that are not well characterized. This is to be avoided. On the other hand, it is exactly here where confusion and problems of terminology originate. The committee has therefore included guidelines for the naming of substances that are still hypothetical and which can only be used with the ending "-like" at the present time. Preferable is the application of rule 4 of the general comments.

Increasing knowledge and better characterization especially of the enzymes of the RAS will require revision of the present nomenclature in the future. Comments and suggestions for future revisions of these recommendations may be sent to any member of the committee.

General Comments

1. Established trivial names will be retained.
2. The names of components are based on the zymogen or precursor form of the protein.
3. The amino acid sequence of components of the kidney renin-angiotensin system from man will be used as reference, and the numbering of peptides follows that of angiotensin I or renin substrate (1-14) tetradecapeptide.
4. The source (species, tissue) and the physicochemical characteristics (e.g. iso-electric point) must be specified in the initial description of each component. In the case of measurements of inactive forms of renin (or renin-like enzymes), the method of activation (e.g. acid activation) and their molecular weight should be indicated (e.g. inactive renin, M.W. 60.000).
5. "pro" should be used whenever it is a prefix and "Pro" when it implies a proline residue.

[1] Members of the Nomenclature Committee of the International Society of Hypertension: D.R. Bangham, London; R. Boucher, Montreal; F.M. Bumpus, Cleveland; J.P. Coghlan, Melbourne; P. Corvol, Paris; D. Ganten, Heidelberg (Chairman);T. Inagami, Nashville; J.J. Morton, Glasgow; K. Poulsen, Kopenhagen

[2] Members of the Nomenclature Committee, Council fór High Blood Pressure Research, American Heart Association: D.F. Bohr, Ann Arbor; S. Oparil, Chicago; I.H. Page, Cleveland (Chairman); L. Tobian, Minneapolis

Table 1

Trivial name	Abbreviations	Systematic name	Comments
Angiotensinogen		renin substrate	Naturally occuring protein renin substrate from which angiotensin-(1-10) decapeptide is cleaved by hydrolytic action of renin. If it is found that a precursor of angiotensinogen with an additional polypeptide extension exists, this would be "pro-angiotensinogen or pro-renin substrate"
	TDP renin substrate	renin substrate-(1-14) tetradecapeptide	The amino acid sequence: 1 2 3 4 5 6 7 8 9 10 11 12 13 14 Asp-Arg-Val-Tyr-Ile-His-Pro-Phe-His-Leu-Leu-Val-Tyr-Ser serves as reference. Other synthetic renin substrates substituted analogues, or fragments follow the same rules as outlined for angiotensin.
		e.g. renin substrate-(7-13) heptapeptide	7 8 9 10 11 12 13 Pro-Phe-His-Leu-Leu-Val-Tyr
		[Pro-Pro] renin substrate-(7-13)nonapeptide	7 8 9 10 11 12 13 Pro-Pro-Pro-Phe-His-Leu-Leu-Val-Tyr If larger peptides are synthetized, the terms, e.g. renin substrate-(7-13) fragment, or renin substrate-(1-35) fragment may be used
Angiotensin I	ANG I ANG-(1-10)	angiotensin-(1-10) decapeptide	The term angiotensin (ANG) is reserved for a group of peptides with amino acid sequence and biological function similar to angiotensin-(1-8) octapeptide. The amino acid sequence of human [Ile^5]-angiotensin-(1-10) decapeptide: 1 2 3 4 5 6 7 8 9 10 Asp-Arg-Val-Tyr-Ile-His-Pro-Phe-His-Leu serves as reference for all angiotensin peptides. If not defined differently, numbers indicate the amino acid of human angiotensin I.

Table 1 (continued)

Trivial name	Abbreviations	Systematic name	Comments
	[Val^{5}] ANG I [Val^{5}] ANG-(1-10)	[Val^{5}] angiotensin-(1-10) decapeptide	for e.g. beef ANG I
Angiotensin II	ANG II ANG-(1-8)	angiotensin-(1-8) octapeptide	e.g. human, rat angiotensin II
	[Val^{5}] ANG II [Val^{5}] ANG-(1-8)	[Val^{5}] angiotensin-(1-8) octapeptide	e.g. beef, sheep angiotensin II
Angiotensin III	ANG III ANG-(2-8)	angiotensin-(2-8) heptapeptide	e.g. human, rat angiotensin III
	[Val^{5}] ANG-(2-8)	[Val^{5}] angiotensin-(2-8) heptapeptide	e.g. beef, sheep angiotensin III

Table 2 Fragments of angiotensin and analogues

Trivial name	Abbreviations	Systemic name	Comments
	ANG-(4-8)	Angiotensin-(4-8) pentapeptide	Sequence defined from human ANG I as: 4 5 6 7 8 Tyr-Ile-His-Pro-Phe
	[Phe^4] ANG-(4-8)	[Phe^4] angiotensin-(4-8) pentapeptide	4 5 6 7 8 Phe-Ile-His-Pro-Phe
Saralasin	[Sar^1, Val^5 Ala^8] ANG II [Sar^1, Val^5, Ala^8] ANG-(1-8)	[Sar^1, Val^5, Ala^8] angiotensin-(1-8) octapeptide	Angiotensin antagonist; analogous nomenclature for other angiotensin antagonists
	[Tyr^{4a}] ANG II [Tyr^{4a}] ANG-(1-8) [endo-Tyr^{4a}] ANG II	[Tyr^{4a}] angiotensin-(1-8) nonapeptide	Insertion of a residue. Amino acid sequence: 1 2 3 4 4a 5 6 7 8 Asp-Arg-Val-Tyr-Tyr-Ile-His-Pro-Phe
	[des-Pro^7] ANG II	[des-Pro^7]angiotensin-(1-8) heptapeptide	Removal of a residue, e.g. of proline in position 7: 1 2 3 4 5 6 7 8 Asp-Arg-Val-Tyr-Ile-His-□-Phe
	Pro-ANG II	[Pro]angiotensin-(1-8) nonapeptide	Extension at the N-terminus, e.g. with a proline residue: 1 2 3 4 5 6 7 8 Pro-Asp-Arg-Val-Tyr -Ile-His-Pro-Phe
	[Pro^{8a}] ANG II	[Pro^{8a}] angiotensin-(1-8) nonapeptide	Extension at the C-terminus, e.g. with a proline residue: 1 2 3 4 5 6 7 8 8a Asp-Arg-Val-Tyr-Ile-His-Pro-Phe-Pro
Renin-angiotensin system	RAS		

Table 2 (continued)

Trivial name	Abbreviations	Systemic name	Comments
Renin (E.C.3.4.23) new (E.C.3.4.99.19) old			An enzyme which cleaves angiotensin-(1-10) decapeptide from its natural renin substrate (angiotensinogen) and which has no angiotensinase activity. Renin occurs in the kidney and in extrarenal organs. The major component of the human kidney enzyme serves as renin reference. Information concerning the physicochemical characterization and source of the enzyme are to be indicated if the enzyme has not been clearly identified, and the ending "-like" (e.g. renin-like) should be used in such cases.
Iso-renin			The term iso-renin is to be reserved for forms of renin catalyzing the same reaction, but arising from genetically determined differences in primary structure. Iso-renin may be of renal or extrarenal origin
Pro-renin			Activatable precursor (zymogen) of renin which contains an additional polypeptide extension. Existence not proven for the renin-angiotensin system
Pre-pro-renin			Activatable precursor of pre-renin (pre-zymogen) which contains an additional polypeptide extension. Existence not proven for the renin-angiotensin system
Converting enzyme (E.C.3.4.15.1.)	CE	peptidyldipeptide carboxy hydrolase	Converts ANG I into ANG II by cleaving His-Leu from ANG I; to be used without the prefix ANG I, because the enzyme is not specific for ANG I
Angiotensinase activity (E.C.3.4.99.3)			A group of enzymes capable of degrading ANG I, ANG II, ANG II and other peptides; the enzymes may be amino-, carboxy-, endo- and exopeptidases of various specificities

Nomenclature of Experimental Hypertension

Recognizing the currently confused nomenclature used to describe some methods for producing experimental renovascular hypertension, we propose that it be replaced by the following terminology as approved by the Executive Committee of the Council for High Blood Pressure Research of the American Heart Association in April, 1978, and published in Hypertension (Vol. 1, No. 1, p 61) in February, 1979:

Goldblatt Hypertension

One-kidney, one clip hypertension
Two-kidney, one clip hypertension
Two-kidney, two clip hypertension

Page Hypertension

One-kidney, one wrapped hypertension
Two-kidney, one wrapped hypertension
Two-kidney, two wrapped hypertension

Grollman Hypertension

One-kidney, one figure-8 hypertension
Two-kidney, one figure-8 hypertension
Two-kidney, two figure-8 hypertension

One-kidney, one clip hypertension means one kidney has been removed and the other clipped. Two-kidney, one clip hypertension indicates that both kidneys are intact but one is clipped. Two-kidney, two clip hypertension follows logically.

REFERENCES

1. IUPAC-IUB Commission of Biochemical Nomenclature. Rules for Naming Synthetic Modifications of Natural Peptides. Tentative Rules. Biochemistry 6, 362-364 (1967)

2. Enzyme Nomenclature. Amsterdam: Elsevier 1973

3. The Nomenclature of Peptide Hormones. IUPAC-IUB Commission on Biochemical Nomenclature.Biochim. Biophys. Acta 404, 152-155 (1975)

4. The Nomenclature of Peptide Hormones. IUPAC-IUB Commission on Biochemical Nomenclature. J. Biol. Chem. 250, 3215-3216 (1975)

5. Nomenclature of Multiple Forms of Enzymes.Recommendations (1976), J. Biol. Chem. 252, 5939-5941 (1977)

Subject Index

Author Index